90 0710510 5

Cell to Cell Signals in Plant, Animal and Microbial Symbiosis

NATO ASI Series

Advanced Science Institutes Series

A series presenting the results of activities sponsored by the NATO Science Committee, which aims at the dissemination of advanced scientific and technological knowledge, with a view to strengthening links between scientific communities.

The Series is published by an international board of publishers in conjunction with the NATO Scientific Affairs Division

A Life Sciences B Physics	Plenum Publishing Corporation London and New York
C Mathematical and Physical Sciences D Behavioural and Social Sciences E Applied Sciences	Kluwer Academic Publishers Dordrecht, Boston and London
F Computer and Systems Sciences G Ecological Sciences H Cell Biology	Springer-Verlag Berlin Heidelberg New York London Paris Tokyo

Series H: Cell Biology Vol. 17

Cell to Cell Signals in Plant, Animal and Microbial Symbiosis

Edited by

Silvano Scannerini

Dipartimento di Biologia Vegetale, Università di Torino
Viale Mattioli 25, 10125 Torino, Italy

David Smith

University of Edinburgh, Old College
South Bridge, Edinbourgh EH8 9YL, United Kingdom

Paola Bonfante-Fasolo

Dipartimento di Biologia Vegetale, Università di Torino
Viale Mattioli 25, 10125 Torino, Italy

Vivienne Gianinazzi-Pearson

INRA
Station de Genetique et d'Amelioration des Plantes de Dijon
B.V 1540, 21034 Dijon, France

Springer-Verlag
Berlin Heidelberg New York London Paris Tokyo
Published in cooperation with NATO Scientific Affairs Division

Proceedings of the NATO Advanced Research Workshop on Cell to Cell Signals in Plant, Animal and Microbial Symbiosis held at Villa Gualino, Torino, Italy, May 19–22, 1987

ISBN 3-540-18555-0 Springer-Verlag Berlin Heidelberg New York
ISBN 0-387-18555-0 Springer-Verlag New York Berlin Heidelberg

Library of Congress Cataloging-in-Publication Data. Cell to cell signals in plant, animal, and microbial symbiosis / edited by Silvano Scannerini ... [et al.]. p. cm.—(NATO ASI series. Series H, Cell biology ; vol. 17) "Proceedings of the NATO advanced research workshop held at Villa Gualino, Torino, Italy, May 19–22, 1987"—T.p. verso. "Published in cooperation with NATO Scientific Affairs Division." Includes index. ISBN 0-387-18555-0 (U.S.)
1. Symbiosis—Congresses. 2. Cell interaction—Congresses. I. Scannerini, Silvano, 1940-. II. North Atlantic Treaty Organization. Scientific Affairs Division. III. Series. [DNLM: 1. Cell Communication. 2. Symbiosis—congresses. QH 548 C393 1987] QH548.C45 1988 574.5'2482—dc 19 DNLM/DLC for Library of Congress 88-6434

Printed in Germany

Printing: Druckhaus Beltz, Hemsbach; Binding: J. Schäffer GmbH & Co. KG, Grünstadt
2131/3140-543210

This book is dedicated to the memory of Prof. Mario di Lullo, secretary of the NATO Scientific Affairs Division.

PREFACE

In a mutualistic symbiosis, two (or sometimes more) organisms of very different kinds come together and establish a long-term association in which the partners show a high degree of integration with each other. Studies have been made in various types of symbiosis to understand the processes by which the partners recognise each other, but hitherto there has been no attempt to compare and correlate results from a broad range of associations to see if any common principles emerge. Furthermore, the previous lack of a comparative approach has led to inconsistences in the way in which terms are used - even such a fundamental term as 'recognition' itself.

There is frequently an assumption that molecular signals pass between the partners in the early stages of the establishment of a symbiosis, although the experimental evidence underlying this assumption requires critical evaluation. Because contact between host and symbiont becomes intimate and often morphologically complex, it may be difficult to get direct biochemical evidence for the existence of signals, and heavy reliance has to be placed on indirect evidence, especially as provided by ultrastructural, cytochemical, immunological and genetic studies.

For these various reasons, it was particularly opportune to hold a NATO Advanced Workshop on cell-to-cell signals in plant, animal and microbial symbiosis. The objective of the workshop was to identify the processes involved in contact between cells of hosts and symbionts. It involved critical reviews of the current state of knowledge of various symbiotic systems, assessment of the evidence for signals, identification of the topics worthy of further investigation, and evaluation of the most promising experimental techniques which could be used.

This book contains the papers which were presented at the Workshop, which was held at the Villa Gualino, Torino, May 19-22, 1987.

Silvano Scannerini
David Smith

ACKNOWLEDGMENTS

In sending this book to print it is our pleasure and duty to thank the institutions and people whose contribution has been essential for the achievement of the workshop and of this volume.

First of all we must thank the NATO Scientific Affairs Division which has granted us a generous contribution. In particular our thanks go, unfortunately posthumous, to prof. Mario di Lullo with whom the activity began, to Dr. Alain Jubier who has substituted him and to the NATO Scientific Representative of our workshop: Prof. J.A. Fortin.

Thanks must also be expressed to the ISI (Institute for Scientific Interchange) which has kindly hosted us at Villa Gualino; to Prof. Tullio Regge president of ISI, Prof. M. Rasetti director, and also to the cooperation of Dr. Tiziana Bertoletti and the efficiency and patience of Ms. Carmen Novella.

Thanks also to the University of Turin and to its Chancellor Prof. M.U. Dianzani whose special grant has made possible the participation of people from non-NATO countries. Among personnel and students of the Plant Biology Department a special mention goes to Ms. Marina Beorchia for secretarial work, Mr. Pantaleone Tripaldi for dealing with administration and Dr. Pietro Spanu for translations.

S. Scannerini D.C. Smith P. Bonfante V. Gianinazzi-Pearson

PARTICIPANTS

Albertano, P.

II Università degli Studi di Roma "Tor Vergata", Dipartimento di Biologia, Via O. Raimondo, 00173 (La Romanina) Roma, Italy.

Becard, G.

Université Laval, Centre de Recherche en Biologie Forestière, Faculté de Foresterie et de Géodésie, G1K 7P4 Québec, Canada.

Bermudes, D.

Boston University, Department of Biology, 2 Cummington Street, Boston, Massachusetts, 02215 U.S.A.

Berta, G.

Università di Torino, Dipartimento di Biologia Vegetale, Viale Mattioli 25, 10125 Torino, Italy.

Bertocchi, C.

Università di Trieste,Dipartimento di Biochimica, Biofisica e Chimica delle Macromolecole, P.le Europa 1, 34127 Trieste, Italy.

Bonfante, P.

Università di Torino, Dipartimento di Biologia Vegetale, Viale Mattioli 25, 10125 Torino, Italy.

Brewin, N.

John Innes Institute, AFRC Institute of Plant Science Research, Department of Genetics, Colney Lane, Norwich NR4 7UH UK.

Callow, J.A.

The University of Birmingham, Department of Plant Biology, P.O. Box 363, Birmingham B15 2TT UK.

Cheli, F.

Università degli Studi di Milano, Dipartimento di Biologia "Luigi Gorini", Sezione di Botanica Sistematica, Via Celoria 26, 20133 Milano, Italy.

Codignola, A.

Università degli Studi di Torino, Dipartimento di Biologia Vegetale, Viale Mattioli 25, 10125 Torino, Italy.

Dazzo, F.B.

Michigan State University, Department of Microbiology and Public Health, Giltner Hall, East Lansing, Michigan 48824-1101 U.S.A.

De Vecchi, L.

Università degli Studi di Milano, Dipartimento di Biologia "Luigi Gorini", Sezione di Botanica Sistematica, Via Celoria 26, 20133 Milano, Italy.

Dorritie, B.

Boston University, Department of Biology, 2 Cummington Street, Boston, Massachusetts, 02215 U.S.A.

Douglas, A.E.

John Innes Institute, AFRC Institute of Plant Science Research, Department of Cell Biology, Colney Lane, Norwich NR4 7UH U.K.

Fortin, J.A.

Université Laval, Faculté de Foresterie et de Géodesie, Centre de Recherche en Biologie Forestière, Cité Universitaire, G1K 7P4 Québec, Canada.

Galun, M.

Tel-Aviv University, Faculty of Life Sciences, Department of Botany, Tel-Aviv, Israel.

Gianinazzi, S.

Institut National de la Recherche Agronomique, Station de Génétique et d'Amélioration des Plantes de Dijon, B.V. 1540, 21034 Dijon cedex, France.

Gianinazzi-Pearson, V.

Institut National de la Recherche Agronomique, Station de Génétique et d'Amélioration des Plantes de Dijon, B.V. 1540, 21034 Dijon cedex, France.

Giovannetti, M.

Università degli Studi di Pisa, Istituto di Microbiologia Agraria e Tecnica, C.N.R., Centro di Studio per la Microbiologia del Suolo, Via del Borghetto 80, 56100 Pisa, Italy.

Green, J.R.

The University of Birmingham, Department of Plant Biology, P.O. Box 363, Birmingham B15 2TT UK.

Grilli Caiola, M.

II Università degli Studi di Roma "Tor Vergata", Dipartimento di Biologia, Via O. Raimondo, 00173 (La Romanina) Roma, Italy.

Hinde, R.

The University of Sidney, School of Biological Sciences, Macleay Building A12, NSW 2006.

Honegger, R.

Institut fur Pflanzenbiologie Cytologie, Zollikerstrasse 107, CH-8008 Zurich.

Jones, J.L.

The University of Birmingham, Department of Plant Biology, P.O. Box 363, Birmingham, B15 2TT UK.

Lefebvre, F.

Institut National des Sciences Appliquées de Lyon, Laboratoire de Biologie Appliquée, Bâtiment 406, 20 Avenue A. Einstein, 69621 Villeurbanne Cedex, France.

Maffei, M.

Department of Biological Chemistry, Washington State University, Pullman, Washington, U.S.A.

Margulis, L.

Boston University, Department of Biology, 2 Cummington Street, Boston, Massachusetts, 02215 U.S.A.

Martinotti, G.

Università di Torino, Istituto di Microbiologia, Via Santena 9, 10126 Torino, Italy.

Massicotte, H.B.

University of Guelph, College of Biological Science, Department of Botany, Guelph, Ontario, Canada N1G 2W1.

Matta, A.

Università di Torino, Istituto di Patologia Vegetale, Via P. Giuria 15, 10125 Torino, Italy.

McAuley, P.J.

University of Oxford, Department of Plant Sciences, Agricultural Science Building, Parks Road, Oxford OX1 3PF UK.

McFall-Ngai, M.J.

University of California San Diego, La Jolla, California, 92093 U.S.A.

Minganti, C.

Istituto Guido Donegani, Department of Biotechnology, Via G. Fauser 4, 28100 Novara, Italy.

Monsigny, M.

Laboratoire de Biochimie Cellulaire et Moléculaire des Glycoconjugués, Centre de Biophysique Moléculaire du Centre National de la Recherche Scientifique et U.F.R. de Sciences Fondamentales et Appliquées de l'Université d'Orléans, 1 rue Haute, 45071 Orléans Cedex 2, France.

Nardon, P.

Institut National des Sciences Appliquées de Lyon, Laboratoire de Biologie Appliquée, Batiment 406, 20 Avenue A. Einstein, 69621 Villeurbanne Cedex, France.

Nealson, K.H.

The University of Wisconsin-Milwaukee, Center for Great Lakes Studies, 600 E. Greenfield Avenue, Milwaukee, Wisconsin U.S.A.

Noris, E.

Istituto Guido Donegani, Department of Biotechnology, Via G. Fauser 4, 28100 Novara, Italy.

Nuti, M.

Università degli Studi di Padova, Dipartimento di Biotecnologie Agrarie, Via Gradenigo 6, 35131 Padova, Italy.

Okker, R.J.H.

Department of Plant Molecular Biology, Botanical Laboratory, Nonnensteeg 3 2311 VJ Leiden, The Netherlands.

Pasti, M.B.

Università degli Studi di Padova, Dipartimento di Biotecnologie Agrarie, Via Gradenico 6, 35131 Padova, Italy.

Perotto, S.

Centro di Studio sulla Micologia del Terreno del C.N.R., Viale Mattioli 25, 10125 Torino, Italy.

Piché, Y.

Université Laval, Faculté de Foresterie et de Géodesie, Cité Universitaire, G1K 7P4 Québec Canada.

Rahat, M.

The Hebrew University of Jerusalem, The Institut of Life Sciences, The Department of Zoology, 91904 Jerusalem, Israel.

Ray, T.C.

The University of Birmingham, Department of Plant Biology, P.O. Box 363, Birmingham, B15 2TT UK.

Reisser, W.

Fachbereich Biologie der Philipps-Universitat Marburg, Botanik, D-3550 Marburg-Lahnberge.

Ronhen, R.

Tel-Aviv University, Faculty of Life Sciences, Department of Botany, Tel-Aviv, Israel.

Ruby, E.

University of Southern California, Department of Biological Sciences, Los Angeles, California 90089 U.S.A.

Sacchi, L.

Università degli Studi di Pavia, Dipartimento di Biologia Animale, Piazza Botta 9, 27100 Pavia, Italy.

Savoia, D.

Università degli Studi di Torino, Istituto di Microbiologia, Via Santena 9, 10126 Torino, Italy.

Scannerini, S.

Università degli Studi di Torino, Dipartimento di Biologia Vegetale, Viale Mattioli 25, 10125 Torino, Italy.

Schubert, A.

Università degli Studi di Torino, Istituto di Coltivazioni Arboree, Cattedra di Viticoltura, Via P. Giuria 16, Torino, Italy.

Smith, D.C.

University of Edinburgh, Old College, South Bridge, Edinburgh, EH8 9YL UK.

Spanu, P.

Università degli Studi di Torino, Dipartimento di Biologia Vegetale, Viale Mattioli 25, 10125 Torino, Italy.

Squartini, A.

Università degli Studi di Padova, Dipartimento di Biotecnologie Agrarie, Via Gradenigo 6, 35131 Padova, Italy.

Testa, B.

U.S.L. 40, Servizio Fisica Sanitaria, Via Aldisio 2, 10015 Ivrea, Italy.

Torrey, J.G.

Harvard University, Harvard Forest, Petersham, Massachusetts, 01366 U.S.A.

Trench, R.K.

University of California, Santa Barbara, Department of Biological Sciences, Santa Barbara, California, 93106 U.S.A.

Trotta, A.

Università degli Studi di Torino, Dipartimento di Biologia Vegetale, Viale Mattioli 25, 10125 Torino, Italy.

VandenBosch, K.

John Innes Institute, AFRC Institute of Plant Science Research, Department of Genetics, Colney Lane, Norwich NR4 7UH UK.

Vidotto, V.

Università degli Studi di Torino, Istituto di Botanica Speciale Veterinaria, VIale Mattioli 25, 10125 Torino, Italy.

Zaat, A.J.

Department of Plant Molecular Biology, Botanical Laboratory, Nonnensteeg 3, 2311 VJ Leiden, The Netherlands.

CONTENTS

IV. RECOMMENDATIONS FOR FUTURE RESEARCH AND APPLICATIONS

CELLULAR INTERACTIONS BETWEEN HOST AND ENDOSYMBIONT IN DINITROGEN-FIXING ROOT NODULES OF WOODY PLANTS

John G. Torrey
Harvard Forest, Harvard University
Petersham, Massachusetts 01366, U.S.A.

INTRODUCTION

During the last decade increasing attention has been paid by plant scientists to temperate and tropical woody plants known to establish symbiotic relationships with soil microorganisms capable of generating and expressing the enzyme nitrogenase. In the appropriate biological context this enzyme catalyzes the reduction of molecular dinitrogen from the atmosphere into organic form assimilable by the eukaryotic partner in the association. Management of biological nitrogen fixation by tree species is viewed increasingly as an important approach to improvement of world agriculture and forestry.

Two major families of microorganisms are involved: the Rhizobiaceae, which includes the non-sporulating, rod-shaped usually flagellated, Gram-negative soil bacteria of the genera *Rhizobium* and *Bradyrhizobium*, and the Frankiaceae, which is comprised of the single genus *Frankia*, a filamentous, sporulating, Gram-variable soil bacterium of the Actinomycetales. Rhizobia form intimate associations in the form of root nodules with the root systems of plant species in the family Leguminosae - a large and diverse family of dicotyledonous plants, both herbaceous and woody, comprised of nearly 20,000 species (Allen and Allen 1981). Two subfamilies are predominantly woody trees and shrubs, viz., the Caesalpinioideae with about 180 genera and 2800 species and the Mimosoideae with about 65 genera and 2900 species. The third major subfamily, the Papilionoideae with over 500 genera and an estimated 14,500 species, both woody and herbaceous, includes most of the grain and forage legumes of importance to agriculture. Although considerably less inten-

NATO ASI Series, Vol. H17
Cell to Cell Signals in Plant, Animal and Microbial Symbiosis. Edited by S. Scannerini et al.

sively studied than in herbaceous species, root nodulation among the woody species shows lower frequency and is more variable. These associations in selected woody species are of particular interest and will be the subject of our explorations and comparisons.

One known exception to the rule that the Rhizobiaceae nodulate only members of the legume family is the special case of root nodulation of *Parasponia*, a woody tropical tree in the elm family (Ulmaceae) that forms N_2-fixing nodules in response to infection by *Bradyrhizobium* (Trinick 1973). This association is of particular interest since in many ways it represents an intermediate type of association both physiologically and structurally between legumes and actinorhizal plants.

Frankia is a widespread soil filamentous bacterium that, unlike the Rhizobia, infects a broad and diverse range of host families forming dinitrogen-fixing root nodules. The known hosts of *Frankia* include over 200 species in 24 genera distributed among 8 families, all of them woody dicotyledonous plants (Moiroud and Gianinazzi-Pearson 1984). The group, referred to collectively as actinorhizal plants, shows diversity in the expression of the microbial-root associations, paralleling in many ways the comparable associations between *Rhizobium* and legume host plants. Comparisons of the cellular interactions among these associations seem especially pertinent as we learn more of their structural and functional relationships.

These two major groups of symbiotic associations tend to be reviewed separately and an extensive literature exists for each group. Recent reviews of the legume-*Rhizobium* association include the comprehensive book on the Leguminosae by Allen and Allen (1981) and the structural, including ultrastructural, reviews by Newcomb (1980, 1981). The biology of *Frankia* in relation to actinorhizal plants has been reviewed recently by Moiroud and Gianinazzi-Pearson (1984), the physiology by Tjepkema et al. (1986), and ultrastructure by Newcomb and Wood (1987). An earlier comprehensive treatise on dinitrogen fixation edited by Hardy and Silver (1977) included discussions of nodulation both in the legumes and in actinorhizal plants. A more popular discussion of these topics is found in Sprent (1979). Recent reviews directed toward the specific interactions between host and microbial symbiont have been written for legume-*Rhizobium* symbioses (Vincent 1980, Halverson and Stacey 1986) and for actinorhizal-*Frankia* symbiosis (Rodriguez-Barrueco and Subramanian 1986).

Within the broad subject of cell-to-cell interactions in the microbial-woody dicotyledonous plant associations, I have selected topics which lend themselves to careful analysis at this time since considerable progress has been made in recent years in our understanding of the interactions. The topics will be discussed in the context of the continuing dialogue that goes on between the partners that begins with the first steps of recognition, proceeds through invasion, infection, cell proliferation and accommodation until the fully developed functional nodule is formed. I will emphasize within this time course of events the mode of entry and infection, and the mechanisms utilized by each partner in achieving oxygen protection for the enzyme nitrogenase.

INFECTION AND THE MODE OF ENTRY

According to Vincent (1980), the sequence of events in the development of nodule symbioses can be subdivided into a series of well defined stages: preinfection, infection and nodule formation and nodule function. Each of these stages may be subdivided into a number of steps, each of which may be influenced by the host plant genome, the bacterial genome or by the environment within which the events occur. Many of these steps have been identified by phenotypic codes for either host or micro-organism.

For the purposes of this review emphasis will focus on relatively narrow aspects of the second and third stages. The first stage will be considered elsewhere in this volume by Dazzo. In Table 1 the second stage of Vincent (1980) has been subdivided into sixteen descriptive events to allow us

Table 1. A summary of events in early infection and nodule development in nodulated plants.

Host genera	Early Infection Events									Nodule Development					
	1a	1b	2	3	4	5	6	7	8	9	10	11	12	13	14	15
Rhizobium - herbaceous legumes																
Trifolium repens	X	X	X	-	X	-	X	-	-	X	-	X	-	-	X	X
Arachis hypogeae	X	-	X	X	X	-	X	-	-	X	-	X	-	-	X	X
Stylosanthes	-	-	X	X	X	-	X	-	-	X	-	X	-	-	X	X
Rhizobium - *woody legumes*																
Andira spp.	?	?	X	?	X	-	-	X	-	-	X	X	-	-	X	?
Rhizobium - *nonlegumes*																
Parasponia	X	-	X	X	X	X	-	X	-	X	-	-	X	-	X	X
Frankia - Actinorhizal plants																
Elaeagnus	-	-	X	X	-	X	-	X	X	?	-	-	X	-	X	X
Alnus	X	X	X	-	X	X	-	X	X	X	-	-	X	-	X	X
Myrica	X	X	X	-	X	X	-	X	X	X	X	-	X	X	-	X
Casuarina	X	X	X	-	X	X	-	X	-	X	X	-	X	X	-	X

Key to Table 1.

Early Infection Events

1a. Root hair curling or deformation
1b. Root hair penetration
2. Bacterial thread formation
3. Intercellular penetration
4. Cortical cell proliferation
5. Lateral root initiation
6. Bacteroid formation
7. Bacteria persistent in thread
8. Endophyte vesicle formation

Nodule Development Events

9. Cell-to-cell infection
10. Infected cell walls modified
11. Vascular tissue peripheral
12. Vascular tissue central
13. Nodule root formation
14. Coralloid nodules
15. Hemoglobin formation

X = *Occurs*

\- = *Does not occur*

? = *Undetermined*

to make comparisons among a number of different symbiotic associations involving *Rhizobium* and *Frankia* root nodulation. Nine different host genera have been selected for comparative purposes to serve as the basis for analysis of similarities and differences in the mode of entry and the infection process itself. In some entries in Table 1 either the event does or does not occur; in other entries, alternative events are recorded.

The choice of host genera for analysis was determined first by the availability of pertinent information. For the most part, the nine genera represented in Table 1 have been well studied. A few question marks in the table need to be completed. The genera were selected in the second place because they represented clear examples of distinctive types with respect to the early infection events and early nodule development. They were chosen to illustrate the diversity of sequential steps in achieving a structure wherein the microorganism was able to function in dinitrogen fixation. Four groups are represented among the host genera: herbaceous legume-*Rhizobium* associations (three), woody legume-*Rhizobium* combinations (one), woody non-leguminous-*Rhizobium* symbiosis (one), and woody-actinorhizal-*Frankia* symbioses (four).

What is remarkable at the outset in studying Table 1 is to see the diversity of the interactive events involved in the different associations. Only one process is scored in common for all, i.e., bacterial thread formation, and even that must be qualified in that "threads" differ remarkably in time of occurrence, chemical nature, ultra-structure and persistence among the different symbioses. It does seem to be true, however, that host plants usually produce a polysaccharide sheath or encapsulation around the entering or entered bacteria that excludes them temporarily if not continuously from full entry into the host cytoplasm. A host plasmalemma outside this thread wall is a persistent structure in every case, and persists when the thread wall no longer is present, as for example when the bacterium is released into the host cell cytoplasm in some legume-*Rhizobium* combinations, forming bacteroids with peribacteroid membranes. In no other early infection or nodule development event do all the listed symbioses share!

Rhizobium - herbaceous legumes.

Looked at from the perspective of the hosts listed in Table 1, all are distinctive in some feature and were placed on the list because of their expressed differences. The first listed, Trifolium repens, is one of the most studied symbioses and represents a sequence considered typical of many herbaceous legume-Rhizobium symbioses. The sequence of events listed is perhaps the most familiar and typical, e.g. root hair curling, bacterial thread formation, cortical cell proliferation, cell-to-cell infection, bacteroid formation, peripheral vascular tissue development, single lobed or coralloid nodule formation and the synthesis of leghaemoglobin. Such events are normally associated with the development of effective dinitrogen-fixing root nodules in many herbaceous legumes.

Arachis hypogaea differs in that, although root hair deformation is observed when roots are inoculated with Rhizobium, entry is not via root hairs but involves inter-cellular penetration, i.e., entry via intercellular middle lamella dissolution at the epidermis, invasion of inter-cellular spaces and ultimately infection by penetration of cell walls within the root cortex of the host plant (Chandler 1978). Other events in Arachis are closely similar to those in Trifolium.

Early infection in Stylosanthes has been studied by Chandler et al. (1982). No root hair curling associated with Rhizobium inoculation was observed although some root hair branching occurred and root hairs were most frequent around sites of lateral roots. Rhizobia entered intercellular spaces at the base of lateral roots, penetrating into the root cortex by causing cell wall alterations in cortical cells, entering them and eliciting progressive cell collapse. Finally, bacteria were released as bacteroids into inner cortical cells which divided repeatedly to form the nodule tissue. This form of intercellular penetration involves collapse and compression of root cortical cells before the mutualism can be effectively established. In other characteristics root nodules in Stylosanthes differed only in minor detail from Arachis.

Rhizobium - woody legumes.

Root nodule development in woody species in the legumes has only recently begun to receive careful attention with respect to the details of structure, e.g., studies by Baird et al. (1985) on Prosopis, by Lawrie (1983) on Aotus, and by deFaria et al. (1986, 1987) on Andira spp. and other woody legumes especially of the subfamily Caesalpinioideae. While in Prosopis nodule development similar to herbaceous indeterminate type nodules was observed (Baird et al. 1985), Andira species represent a novel form of host-endophyte association (Sprent et al. 1986). The most striking and novel feature of this association is the lack of bacteroid formation within infected root nodule cells and instead the occurrence of infection threads within which Rhizobium cells are retained. This structural modification is similar to that seen in Parasponia. de Faria et al. (1986) observed persistent infection threads in seven species of Andira they studied. In a survey of root nodule structure in genera of tree species in the evolutionarily primitive sub-families of the legumes, notably members of the Caesalpinioideae, and of the Papilionoideae, de Faria et al. (1987) reported persistent thread formation in 12 genera studied, but none in the more advanced sub-family Mimosoideae. From their structural studies, these authors concluded that bacteria spread via intercellular spaces unconfined by threads, enter root cortical cells by cell wall penetration and thereafter are confined by host-produced cell wall material and membranes, forming persistent intracellular threads. No infection threads were observed to cross cell boundaries. In these anatomical studies, care was taken to examine nodules that had been shown to be active in acetylene reduction (i.e., were capable of fixing dinitrogen). The structural relationship between active nitrogenase and host cell membrane and wall modifications may well involve oxygen protection mechanisms similar to those believed to function in Parasponia discussed below.

Rhizobium - nonlegume.

A unique association that caused much interest when first reported by Trinick (1973) was the nodulation by Rhizobium of the roots of a tropical tree in the non-leguminous family Ulmaceae. The genus Parasponia (originally misidentified and reported as Trema) is comprised of several species from South Pacific islands including Java and Papua New Guinea, and in Indonesia and Malesia (Akkermans et al. 1978). Trinick (1973) demonstrated that the infective soil microorganisms isolated from root nodules of Parasponia were slow-growing strains of Rhizobium (now called Brady-rhizobium) and that the root nodules were capable of fixing dinitrogen. This unusual association has been studied in considerable detail with respect to the anatomy of the root nodules, their ultrastructure and in particular the stages of infection and early nodule development.

Trinick and Galbraith (1976) and Trinick (1979) in examining mature nodules pointed out that the nodules possessed central vascular strands with infected cells occupying swollen cortical cells in a crescent around the central bundle. In this type of structure, more reminiscent of modified lateral roots than legume nodules, they resembled coralloid-type nodules of the actinomycete-nodulated non-leguminous plants now referred to as actinorhizal plants. Infection threads were observed within infected cells and could be observed to pass from cell to cell. Trinick (1979) noted in his ultrastructural studies that rhizobial cells were not released from the infection threads, unlike the typical herbaceous legume nodules where release results in bacteroid formation.

More recently detailed ultrastructural studies of

initial infection events and early nodulation in *Parasponia rigida* have been reported (Lancelle and Torrey 1984, 1985, Price et al. 1984, and Smith et al. 1986). The distinctive features seen in *Parasponia* root nodules are noted in Table 1. Although root hair modification caused by *Rhizobium* is seen in *Parasponia* seedling roots (Lancelle and Torrey 1984), no root hair penetration occurs. Rather infection is via intercellular penetration at the base of induced multicellular root hairs. Cortical cell proliferation and lateral root induction like that observed in actinorhizal nodules occur, and the nodules formed are structurally closer to *Frankia*-induced nodules than *Rhizobium*-induced herbaceous legume nodules. Most distinctive is the persistence of bacteria within infection threads and the lack of bacteroid formation (Lancelle and Torrey 1985). Smith et al. (1986) have demonstrated the existence of striking chemical differences in cell walls of the invasive infection threads that traverse nodule cortical cells intercellularly and the intracellular threads that retain *Rhizobium* cells which are presumed to be the locus of nitrogen-fixing activity in the nodule. Thus the host seems to program two different chemical responses in relation to threads with different functions, concerned with invasion as opposed to dinitrogen fixation and maintenance.

Thus, *Parasponia* is a most interesting intermediate case, in many respects, between the legumes and the actinorhizal symbiotic associations. *Parasponia* is related to the legume-symbiosis by a common microbial partner. The same strain of *Bradyrhizobium* that nodulates *Parasponia* effectively nodulates host plants in the leguminous cowpea miscellany such as the herbaceous *Macroptilium* or *Stylosanthes* (Price et al. 1984). In fact a common series of bacterial genes function in nodule development in herbaceous legumes and in *Parasponia* (Marvel et al. 1985, 1987). However, the hosts behave very differently in the cell-to-cell response. In *Parasponia* modified lateral roots provide the structural site for bacterial invasion and occupancy as opposed to cortical cell proliferation forming a nodule structure **de novo.** Persistent infection thread formation provides an alternate host structure as opposed to bacterial release and bacteroid formation. Such thread wall modifications are construed by Smith et al. (1986) as providing a barrier to direct entry of molecular oxygen to the site of oxygen-labile nitrogenase. Parasponia root nodules have features in common with the primitive woody legume species such as Andira spp. and such relationships offer much material for comparison or for speculations concerning evolutionary origins as well as future potential for biotechnological engineering!

Frankia - actinorhizal plants.

Now let us turn to the Frankia-actinorhizal plant associations that lead to root nodule development and to symbiotic dinitrogen fixation. In Table 1 are listed four genera that represent the range of developmental behavior seen within this diverse group. A single soil bacterial genus, Frankia, is involved, the filamentous, non-motile actinomycete that occurs in a number of strains that we know to differ primarily with respect to the host with which they can associate. No species designations of Frankia strains have yet been given although it is clear that "cross-inoculation" groups exist, i.e., strains that nodulate one host species, genus, even family but fail to nodulate members of other actinorhizal groups. This specificity is analogous to cross-inoculation groups that occur in the Rhizobium-legume symbioses but is perhaps even less well understood.

Four actinorhizal genera have been selected for discussion because they represent some of the variation in cell-to-cell interactions observable in the group. Differences center on mode of infection, presence or absence of nodule roots, vesicle formation, occurrence of host cell wall modifications after infection, and formation of hemoglobin. In each of these differences host-microorganism interaction is expressed.

Until the reports by Miller and Baker (1985, 1986) the only mode of entry of Frankia into host plants was believed to be by root hair deformation and cell wall penetration. Such infections were well documented in *Alnus* (Angulo Carmona 1974, Lalonde 1977, Berry and Torrey 1983, Berry et al. 1986), in *Comptonia* (Callaham and Torrey 1977, Callaham et al. 1979), and in *Myrica* and *Casuarina* (Callaham et al. 1979). Root hair deformation accompanied by intimate association with the non-motile filamentous *Frankia* growing in the root environment allowed the cell-to-cell contact necessary to effect chemical dissolution of the root hair cell wall and invasion to occur. The host-produced polysaccharide encapsulation laid down around the invading filaments is perpetuated wherever the microsymbiont goes within the host and persists throughout the life of the association. The polysaccharide nature of the encapsulation remains to be precisely defined although it is probably both pectic in nature (Lalonde and Knowles 1975) and cellulosic (Berg, private communication).

Infection in *Elaeagnus* is different, not involving root hairs which may not even be deformed but occurring by direct intercellular penetration (Miller and Baker 1985). Filaments of *Frankia* at the root surface penetrate intercellular material between epidermal cells, dissolve the existing middle lamella and penetrate into the intercellular spaces of the root cortex. In some respects such direct intercellular penetration is reminiscent of the process observed in some herbaceous legumes invaded by *Rhizobium*, as seen in *Arachis* or *Stylosanthes*, and is not remarkably dissimilar from the early infection in the non-legume *Parasponia*.

Miller and Baker (1986) have shown that infection via root hairs versus intercellular penetration is controlled by the genome of the host rather than the bacterium. The same pure cultured strain of *Frankia* could be shown to infect *Elaeagnus* by intercellular penetration and *Myrica* by root hair deformation. In no case to date has a single host plant been shown to be subject to both modes of infection. At present the evidence suggests that members of the host family Elaeagnaceae are subject to infection by intercellular penetration, those of the Myricaceae, the Casuarinaceae and the Betulaceae studied to date show root hair penetration.

The ultimate host cell entry is not dissimilar in either case, i.e., the endophyte filament achieves attachment to a cell wall by chemical dissolution, enters through the host cell wall and is there encapsulated by vigorous activities within the host cytoplasm. The filamentous bacterium is retained within the capsule, elaborating its various structures including terminal vesicles and sometimes enlarged sporangia enclosed in the encapsulation within the host cytoplasm.

Progression of the infection from cell to cell may differ following these two types of infection. In Alnus, Myrica and Casuarina invasive filaments of Frankia, sometimes substantially larger in diameter than filaments occupying the cytoplasm (Newcomb et al. 1978, Berg and McDowell 1987a), penetrate host cell walls and traverse a longitudinal path through cortical tissues toward the nodule lobe apex. Such cell-to-cell invasions are best seen in longitudinal sections of nodule lobes. These invasive filaments in Frankia can be compared to the infection threads formed during Rhizobium infection of herbaceous legumes such as has been well documented by many workers. For example, in root nodules of Pisum, the infection thread passes from the root hair epidermal cell into outer root cortical cells and thence with frequent branchings from cell to cell through cell walls into inner cortical cells. This passage is well illustrated in the studies by Libbenga and Harkes (1973).

Such directional invasive infections are in contrast to those observed in associations involving intercellular penetrations. In Rhizobium-woody legumes such as observed in Andira (de Faria et al. 1986) infection threads were never seen to cross cell boundaries. Rather the bacteria spread via intercellular passages and infect successive individual cells. Similar distribution of the microsymbiont may occur in some of the herbaceous legumes, for example, in Arachis or Stylosanthes and needs further careful study.

In Elaeagnus where intercellular penetration occurs, the evidence is lacking as to whether, once infected cells are formed in modified lateral roots, there are invasive filaments of *Frankia* formed that invade successive cortical cells as the nodule develops. The presence of cell-to-cell infection should be checked as published papers on *Elaeagnus* nodule structure (Baker et al. 1980, Newcomb et al. 1987) fail to give evidence on this point.

Two morphological types of nodules are found among the actinorhizal plants (cf. Becking 1977). The *Alnus* type includes all those in which the multilobed nodule is comprised of usually numerous highly modified lateral roots formed together at a single site, forming a coralloid structure with each swollen lobe terminated by a papilla produced by the cessation of activity of a lobe meristem. In the *Myrica*/*Casuarina* type each nodule lobe in a cluster forms a nodule root which results from the continued activity of the nodule lobe meristem (Bowes et al. 1977). The past tendency by authors to refer this character to the host genome rather than to the effect of the microorganism has not been vigorously proved. Recent studies in the Casuarinaceae have added information in support of this view. *Allocasuarina lehmanniana* produces coralloid root nodules, lacking nodule roots (Zhang and Torrey 1985). A *Frankia* isolate from *A. lehmanniana* grown in pure culture used to reinoculate seedlings of this species produces coralloid nodules. The same *Frankia* isolate used as inoculum on seedling roots of *Casuarina* species produces nodules with vertically-upward growing nodule roots. This experiment supports the view that the host genome determines the morphological expression of the nodule.

Several characteristics listed in Table 1 that reflect interactions between the bacterial symbiont and the host cell concern interrelationships which bear on mechanisms by host or microsymbiont which tend to optimize conditions within the symbiosis to facilitate active nitrogen fixation by nitrogenase. These characteristic modifications develop fairly early in nodule development and peak during the most active period of nitrogen fixation.

Cellular modifications affording nitrogenase protection from molecular oxygen.

In all symbiotic associations involving nitrogenase, considerable attention has been paid to structural and physiological modifications that serve to provide protection of the oxygen-labile nitrogenase from direct exposure to molecular oxygen. In herbaceous legume nodules, modifications include nodule cortical differentiation, with reduction of intercellular spaces, bacteroid formation, and leghemoglobin production and function. In woody legume nodules, persistent thread formation with suberized walls may provide partial oxygen protection to nitrogenase within the bacteria.

In the case of actinorhizal root nodules, the problems of access of oxygen are somewhat different. *Frankia* cells grown in free-living culture in the absence of fixed nitrogen substrates develop terminal hyphal swellings termed vesicles within which nitrogenase is formed (Tjepkema et al. 1981). Under aerobic conditions, the vesicles possess a multilaminate envelope which provides protection of the N_2-fixing enzyme within the vesicle from direct access of molecular oxygen (Torrey and Callaham 1982). Thus *Frankia* is unique among actinomycetes in possessing the capability of fixing atmospheric nitrogen directly.

In actinorhizal root nodules *Frankia* typically differentiates vesicles in the symbiotic state as well, depending in part on the structural relationships of host and symbiont. In the Myricaceae, the *Frankia* vesicles are elongate and club-shaped as compared to the spherical or pear-shaped structures observed in *Alnus* or in *Elaeagnus*. The range of vesicle shapes and intracellular arrangements in actinorhizal plants has been discussed in some detail by Torrey (1985). *Frankia* vesicle shape within nodules is determined by the host. This conclusion was reached by Lalonde (1979) in experiments in which he inoculated seedlings of *Alnus* with a pure cultured *Frankia* strain, HFPCpI1 isolated from

Comptonia. Vesicles in Comptonia, like those in the Myricaceae, are club-shaped and arranged peripherally in infected cells. Vesicles in Alnus are peripheral in position but almost spherical in shape. Alnus seedlings formed root nodules when inoculated with the isolate from Comptonia but the vesicles were spherical, typical of Alnus. Thus the expression of Frankia vesicle differentiation depends on the genome of the host plant.

In only one group of actinorhizal plants does Frankia in the nodule fail to form vesicles. The unique group is the Casuarinaceae where Frankia remains filamentous in mature nodules. Correlated with the absence of vesicles in nodules of Casuarina is a specialized cell wall modification which has been intensively studied by Berg (1983) and Berg and McDowell (1987a, 1987b). Cells of Casuarina root nodules occupied by Frankia filaments are specially modified by cell wall thickening followed by deposition of lignin-like and suberin-like materials not found in uninfected cells. These modified cell walls stain differently with histological stains and show impermeability properties to water-soluble substances. Berg (1983) and Berg and McDowell (1987a) argue that these modifications also present physical barriers to gaseous diffusion into the infected cells and the site of the nitrogen-fixing enzyme nitrogenase. The evidence supports the view that the host cell walls provide at least partial protection of the nitrogenase from denaturation by molecular oxygen.

Similar cell wall modifications appear to occur in infected cells in Myrica gale and related Myrica species (Schaede 1939) but are not seen in host cells of Alnus, Elaeagnus and other hosts forming root nodules in which Frankia produces the more typical vesicle morphology. No comprehensive study of cell wall differences among different actinorhizal plants has been made but it is clear that different host plants respond differently to Frankia and that these differences in cell wall properties almost certainly affect the microsymbiont and its function in the symbiotic relationship.

Little or no evidence exists to show that infected cells of root nodules in Rhizobium-legume symbioses show modified cell walls that might influence microsymbiont function. However, there is some suggestion that specialized cell wall structures do develop in some legume root nodules (de Faria et al. 1986). In nodule development in most herbaceous legumes, central bacteroid-containing tissues are surrounded by a nodule cortex lacking intercellular air spaces, a specialization that reduces access of diffusible gases including oxygen to the interior nitrogen-fixing tissues of the nodule. These structural relations have been studied and described by Tjepkema (1983) who also compared legume and actinorhizal nodules.

A final entry in Table 1 is concerned with the occurrence of hemoglobins in plant-microbial symbioses. The surprising fact that hemoglobins, structurally similar to those produced by many animal, including mammalian, systems occurred in root nodules of legumes was first reported by Kubo (1939). These compounds have been studied extensively since then in relation to the nitrogen-fixing capacities of nodulated legumes and actinorhizal plants. Hemoglobins occur in all legume root nodules shown to be effective in dinitrogen fixation and function in an oxygen-carrier role essential for the N_2-fixing process in legume nodules (Appleby 1974). In legume nodules, hemoglobin is present only in infected cells in the host cytoplasm (Verma and Bal 1976). The protein component of hemoglobin is coded by plant genes and the heme component by the Rhizobium bacteroids (for a recent review cf. Verma and Nadler 1984). Thus, hemoglobin represents an important product of the plant-microbe interaction that serves a vital function in the symbiotic relationships.

The occurrence of hemoglobin has been reported in Parasponia root nodules where it is presumed to serve an oxygen-carrier role as in the Rhizobium-legume symbioses (Appleby et al. 1983). It was of great interest therefore to discover that hemoglobin occurs also in some actinorhizal root nodules but apparently not in others (Davenport 1960, Tjepkema 1983,

1984). Highest levels of hemoglobin in actinorhizal plants have been reported in Casuarina and Myrica root nodules (Tjepkema 1983). These genera have been shown to possess modified cell walls around cells infected with Frankia. Nodules in these genera also show highly modified forms of Frankia vesicles or total lack thereof. Other host genera showing Frankia-induced root nodules that contain little or no hemoglobin include Datisca and Ceanothus (not included in Table 1), both of which are quite active in dinitrogen fixation. Low amounts of hemoglobin were observed in Alnus and Elaeagnus species measured (Tjepkema 1983) correlated with spherical vesicles. Direct evidence for a role of hemoglobin in actinorhizal root nodules remains to be obtained although speculations of a possible role have been reviewed recently by Tjepkema et al. (1986). Actinorhizal nodules lacking hemoglobin such as Datisca show rates of dinitrogen fixation equivalent to hemoglobin-containing nodules such as Casuarina.

CONCLUSION

This brief review has been limited to the events of infection and early nodule development found in some of the woody plants that establish symbiotic relationships with soil microorganisms capable of forming nitrogenase. Establishment of these associations involves complex and subtle cellular events between host and microbial symbionts. The cell-to-cell interactions have evolved in diverse ways providing first for entry and then for establishment and cohabitation. Effectivity of the microbial potential for dinitrogen fixation depends upon a range of cellular modifications, involving both microorganism and higher plant partner, that result in restricted access of molecular oxygen to the site of the bacterial enzyme nitrogenase. Such structural modifications still provide the essential energy and reducing power to effect dinitrogen reduction. The complexities and subtleties of cell-to-cell interactions in symbiosis are no more dramatically demonstrated than in a comparative consideration of the effective dinitrogen-fixing symbioses of woody species.

ACKNOWLEDGEMENTS

The author expresses his continued indebtedness to research colleagues, associates and technical staff. His research in this field has been supported over many years by the Maria Moors Cabot Foundation for Botanical Research of Harvard University. Additional support has been provided by grants from the Department of Energy (DE-FG02-84-ER-13198) and the United States Department of Agriculture (83-CRCR-1-1285).

REFERENCES

Akkermans, A. D. L., Abdulkadir, S., Trinick, M. J. 1978. N_2-fixing root nodules in Ulmaceae: Parasponia or (and) Trema spp. Plant Soil 49: 711-715.

Allen, O. N., Allen, E. K. 1981. The Leguminosae. Univ. of Wisconsin Press, Madison, Wisconsin.

Angulo Carmona, A. F. 1974. La formation des nodules fixateurs d'azote chez Alnus glutinosa (L.) Vill. Acta Bot. Neerl. 23: 257-303.

Appleby, C. A. 1974. Leghemoglobin. In: A. Quispel (Ed.) The Biology of Nitrogen Fixation. Elsevier Publ. Co., Inc., New York, pp. 521-554.

Appleby, C. A., Tjepkema, J. D., Trinick, M. J. 1983. Hemoglobin in a non leguminous plant, Parasponia: possible genetic origin and function in nitrogen fixation. Science (Washington, D.C.) 220: 951-953.

Baird, L. M., Virginia, R. A., Webster, B. D. 1985. Development of root nodules in a woody legume, Prosopis glandulosa Torr. Bot. Gaz. 146: 39-43.

Baker, D., Newcomb, W., Torrey, J. G. 1980. Characterization of an ineffective actinorhizal microsymbiont, Frankia sp. EuIl (Actinomycetales). Can. J. Microbiol. 26: 1072-1089.

Becking, J. H. 1977. Dinitrogen-fixing associations in higher plants other than legumes. In: R. W. F. Hardy and W. S. Silver (Eds.) A Treatise on Dinitrogen Fixation. Sect. III. Biology. John Wiley & Sons, New York, pp 185-275.

Berg, R. H. 1983. Preliminary evidence for the involvement of suberization in infection of Casuarina. Can. J. Bot. 61: 2910-2918.

Berg, R. H., McDowell, L. 1987a. An unusual host-endophyte interaction in Casuarina actinorhizae. Planta (in press).

Berg, R. H., McDowell, L. 1987b. Endophyte differentiation in Casuarina actinorhizae. Protoplasma (in press).

Berry, A. M., Torrey, J. G. 1983. Root hair deformation in the infection process of Alnus rubra. Can. J. Bot. 61: 2863-2876.

Berry, A. M., McIntyre, L., McCully, M. E. 1986. Fine structure of root hair infection leading to nodulation in the Frankia - Alnus symbiosis. Can. J. Bot. 64: 292-305.

Bowes, B., Callaham, D., Torrey, J. G. 1977. Time-lapse photographic observations of morphogenesis in root nodules of Comptonia peregrina (Myricaceae). Am. J. Bot. 64: 516-525.

Callaham, D., Torrey, J. G. 1977. Prenodule formation and primary nodule development in roots of Comptonia (Myricaceae). Can. J. Bot. 55: 2306-2318.

Callaham, D., Newcomb, W., Torrey, J. G., Peterson, R. L. 1979. Root hair infection in actinomycete-induced root nodule initiation in Casuarina, Myrica and Comptonia. Bot. Gaz. 140 (suppl.): 81-89.

Chandler, M. R. 1978. Some observations on infection of Arachis hypogaea L. by Rhizobium. J. Exp. Bot. 29: 749-755.

Chandler, M. R., Date, R. A., Roughley, R. J. 1982. Infection and root-nodule development in Stylosanthes species by Rhizobium. J. Exp. Bot. 33: 47-57.

Davenport, H. E. 1960. Haemoglobin in the root nodules of Casuarina cunninghamiana Nature (Lond.) 186: 653-654.

Faria, S. M. de, McInroy, S. G., Sprent, J. I. 1987. The occurrence of infected cells with persistent infection threads in legume root nodules. Can. J. Bot. 65: 553-558.

Faria, S. M. de, Sutherland, J. M., Sprent, J. I. 1986. A new type of infected cell in root nodules of Andira spp. (Leguminosae). Plant Sci. 45: 143-147.

Halverson, L. J., Stacey, G. 1986. Signal exchange in plant microbe interactions. Microbiol. Rev. 50: 193-225.

Hardy, R. W. F., Silver, W. S. (Eds.). 1977. A treatise on dinitrogen fixation III. Biology. John Wiley & Sons,

New York.
Kubo, H. 1939. Über das Hämoprotein aus den Wurzelknöllchen von Leguminosen. Acta Phytochim. 11: 195-200.

Lalonde, M. 1977. Infection process of the Alnus root nodule symbiosis. In: W. Newton, J. R. Postgate and C. Rodriguez-Barrueco (Eds.) Recent Developments in Nitrogen Fixation. Academic Press, London, pp. 569-589.

Lalonde, M. 1979. Immunological and ultrastructural demonstration of nodulation of the European Alnus glutinosa (L.) Gaertn. host plant by an actinomycetal isolate from the North American Comptonia peregrina (L.) Coult. root nodule. Bot. Gaz. 140 (Suppl.): S35-S43.

Lalonde, M., Knowles, R. 1975. Ultrastructure, composition, and biogenesis of the encapsulation material surrounding the endophyte in Alnus crispa var. mollis root nodules. Can. J. Bot. 53: 1951-1971.

Lancelle, S. A., Torrey, J. G. 1984. Early development of Rhizobium-induced root nodules of Parasponia rigida. I. Infection and early nodule initiation. Protoplasma 123: 26-37.

Lancelle, S. A., Torrey, J. G. 1985. Early development of Rhizobium-induced root nodules of Parasponia rigida. II. Nodule morphogenesis and symbiotic development. Can. J. Bot. 63: 25-35.

Lawrie, A. C. 1983. Infection and nodule development in Aotus ericoides (Vent.) G. Don, a woody native Australian legume. J. Exp. Bot. 34: 1168-1180.

Libbenga, K. R., Harkes, P. A. A. 1973. Initial proliferation of cortical cells in the formation of root nodules in Pisum sativum L. Planta (Berl.) 114: 17-28.

Marvel, D. J., Kuldau, G., Hirsch, A., Richards, E., Torrey, J. G., Ausubel, F. M. 1985. Conservation of nodulation genes between Rhizobium meliloti and a slow-growing Rhizobium strain which nodulates a non-legume host. Proc. Nat. Acad. Sci. (U.S.) 82: 5841-5848.

Marvel, D. J., Torrey, J. G., Ausubel, F. M. 1987.

Rhizobium symbiotic genes required for nodulation of legume and non-legume hosts. Proc. Nat. Acad. Sci. (U.S.) 84: 1319-1323.

Miller, I. M., Baker, D. D. 1985. The initiation, development and structure of root nodules in Elaeagnus angustifolia L. (Elaeagnaceae). Protoplasma 128: 107-119.

Miller, I. M., Baker, D. D. 1986. Nodulation of actinorhizal plants by Frankia strains capable of both root hair infection and intercellular penetration. Protoplasma 131: 82-91.

Moiroud, A., Gianinazzi-Pearson, V. 1984. Symbiotic relationships in Actinorhizae. In: D. P. S. Verma and T. Holm (Eds.) Genes Involved in Microbe-Plant Interactions. Springer-Verlag, New York, pp. 205-223.

Newcomb, W. 1980. Control of morphogenesis and differentiation of pea root nodules. In: W. E. Newton and W. H. Orme-Johnson (Eds.) Nitrogen Fixation. Vol. II. Symbiotic Associations and Cyanobacteria. Univ. Park Press, Baltimore, MD, pp. 87-102.

Newcomb, W. 1981. Nodule morphogenesis and differentiation. Intern. Rev. Cytol. Suppl. 13: 247-298.

Newcomb, W., Baker, D., Torrey, J. G. 1987. Ontogeny and fine structure of effective root nodules of the autumn olive (Elaeagnus umbellata). Can. J. Bot. 65: 80-94.

Newcomb, W., Peterson, R. L., Callaham, D., Torrey, J. G. 1978. Structure and host-actinomycete interactions in developing root nodules of Comptonia peregrina. Can. J. Bot. 56: 502-531.

Newcomb, W., Wood, S. M. 1987. Morphogenesis and fine structure of Frankia (Actinomycetales): the microsymbiont of nitrogen-fixing actinorhizal root nodules. Intern. Rev. Cytol. (in press).

Price, G. D., Mohapatra, S. S., Gresshoff, P. M.. 1984. Structure of nodules formed by Rhizobium strain ANU 289 in the nonlegume Parasponia and the legume Siratro (Macroptilium atropurpureum). Bot. Gaz. (Chicago) 145: 444-451.

Rodriguez-Barrueco, Subramanian, C. 1986. Host-endophyte specificity in Frankia symbiosis. In: New Trends in Biological Nitrogen Fixation. N. S. Subba Rao, (Ed.). New Delhi: Oxford and IBH Publ. Co.

Schaede, R. 1939. Die Actinomyceten-Symbiose von Myrica gale. Planta 29: 32-46.

Smith, C. A., Skvirsky, R. C., Hirsch, A. M. 1986. Histochemical evidence for the presence of a suberin-like compound in Rhizobium-induced nodules of the nonlegume Parasponia rigida. Can. J. Bot. 64: 1474-1483.

Sprent, J. I. 1979. The Biology of Nitrogen-fixing Organisms. McGraw-Hill Book Co. (UK) Ltd., London.

Sprent, J. I, Sutherland, J. M., Faria, S. M. de 1986. Structure and function of root nodules from woody legumes. In: C. H. Stirton and J. L. Zarucchi (Eds.) Biology of the Leguminosae. Missouri Botanical Garden, St. Louis, MO., (in press).

Tjepkema, J. D. 1983. Hemoglobins in the nitrogen-fixing root nodules of actinorhizal plants. Can. J. Bot. 61: 2924-2929.

Tjepkema, J. D. 1984. Oxygen, hemoglobins, and energy usage in actinorhizal nodules. In: C. Veeger and W. E. Newton (Eds.) Advances in Nitrogen Fixation Research. Nijhoff, The Hague, Netherlands, pp. 467-473.

Tjepkema, J. D., Ormerod, W., Torrey, J. G. 1981. Factors affecting vesicle formation and acetylene reduction (nitrogenase activity) in Frankia sp. CpIl. Can. J. Microbiol. 27: 815-823.

Tjepkema, J. D., Schwintzer, C. R., Benson, D. R. 1986. Physiology of actinorhizal nodules. Ann. Rev. Plant Physiol. 37: 209-232.

Torrey, J. G. 1985. The site of nitrogenase in Frankia in free-living culture and in symbiosis. In: H. J. Evans, P. J. Bottomley and W. E. Newton (Eds.) Nitrogen Fixation Research Progress. M. Nijhoff Publ., Dordrecht, The Netherlands, pp. 293-299.

Torrey, J. G., Callaham, D.. 1982. Structural features of the vesicle of Frankia sp. CpIl in culture. Can. J.

Microbiol. 28: 749-757.

Trinick, M. J. 1973. Symbiosis between Rhizobium and the non-legume, Trema aspera. Nature (London) 244: 459-460.

Trinick, M. J. 1979. Structure of nitrogen-fixing nodules formed by Rhizobium on roots of Parasponia andersonii Planch. Can. J. Microbiol. 25: 565-578.

Trinick, M. J., Galbraith, J. 1976. Structure of root nodules formed by Rhizobium on the non-legume Trema cannabina var. scabra. Arch. Microbiol. 108: 159-166.

Verma, D. P. S., Bal, A. K. 1976. Intracellular site of synthesis and localization of leghaemoglobin in soybean root nodules. Proc. Nat. Acad. Sci. (U. S.) 73: 3843-3847).

Verma, D. P. S., Nadler, K. 1984. Legume-Rhizobium symbiosis: host's point of view. In: Verma D. P. S. and Th. Holm (Eds.) Genes Involved in Microbe-Plant Interactions. Springer-Verlag, New York, pp. 57-93.

Vincent, J. M. 1980. Factors controlling the legume Rhizobium symbiosis. In: W. E. Newton and W. H. Orme-Johnson (Eds.) Nitrogen Fixation. Vol. II. Symbiotic Associations and Cyanobacteria. Univ. Park Press, Baltimore, MD., pp. 103-129.

Zhang, Z., Torrey, J. G. 1985. Studies of an effective strain of Frankia from Allocasuarina lehmanniana of the Casuarinaceae. Plant Soil. 87: 1-16.

RECOGNITION MECHANISMS IN THE AZOLLA-ANABAENA SYMBIOSIS

M. Grilli Caiola and P. Albertano

Department of Biology, II University of Rome

v. O. Raimondo, 00173 Rome, Italy.

INTRODUCTION

The Azolla-Anabaena association shows the following characteristic features:

a) it is formed by more than two organisms;

b) it cannot be disassociated into its separate component organism, nor can the symbiosis be reconstituted;

c) the Anabaena symbiont is not free-living in nature.

Numerous studies have been carried out on this particular association during recent decades by many research groups with the aim of studying the physiology of the symbiosis and its potential application. There has been emphasis on physiological, ecological and applied studies, so that we know little about the interactions between the symbiotic components and very little about the recognition process between Azolla and its symbionts. This is partly because of the difficulty of separating the individual components so that their reassociation cannot be studied.

With the aim of understanding the relationship between host and symbionts, this paper describes some research on the association by a variety of microscopical techniques.

NATO ASI Series, Vol. H17
Cell to Cell Signals in Plant, Animal and
Microbial Symbiosis. Edited by S. Scannerini et al.

Before doing this, it is first necessary to consider each component of the association: the *Azolla* host, the *Anabaena* phycobiont, the bacteria, the hair cells and the envelope.

AZOLLA. It is a genus of aquatic ferns belonging to the Salviniaceae, comprising six extant species and about forty fossil species. *Azolla* morphology is characterized by small leaves (from 2 to 2.5 mm) arranged in distichous order along the branches which stem from a rhizome from which also roots depart. During their development, the leaves fold in their edges to form a foliar cavity communicating with the outside by a pore. A detailed study of *Azolla* leaf cavity development has been reported by Konar and Kapoor (1974) and Peters et al. (1978). These authors correlated the leaf cavity development with the endophyte evolution. The mature leaf cavity, 0.1-0.4 mm in diameter, is the functional unit containig the phycobiont, *Anabaena*, bacteria, hairs and the envelope (Fig. 1).

ANABAENA AZOLLAE Strasb. It is not free-living in nature, unless the Fjerdingstad's (1976) hypothesis that *A. azolla* was a status of *Anabaena variabilis* can be accepted. In the youngest leaves near the meristematic apex, *Anabaena* is formed by short trichomes of vegetative cells, but at the level of the third or fourth leaf, heterocysts are differentiated (Fig. 2). *Anabaena* heterocyst frequency (HF) varies according to the *Azolla* species, whereas the vegetative cell dimensions always increase from the apex toward the base and their shape changes from roundish to elongated. Vegetative cells, and peculiarly also heterocysts, are always autofluorescent in symbiosis. Akinetes are rare inside the leaves, but common inside micro- and megaspores (Konar and Kapoor 1974; Grilli Caiola and Moretti 1985).

BACTERIA. Bacteria in the *Azolla* leaf cavities were reported by Grilli (1963) and aftewards they were isolated and their number determined in *A. caroliniana* Willd. by Gates et al. (1980). The result was 1900-7400 bacterial cells in each

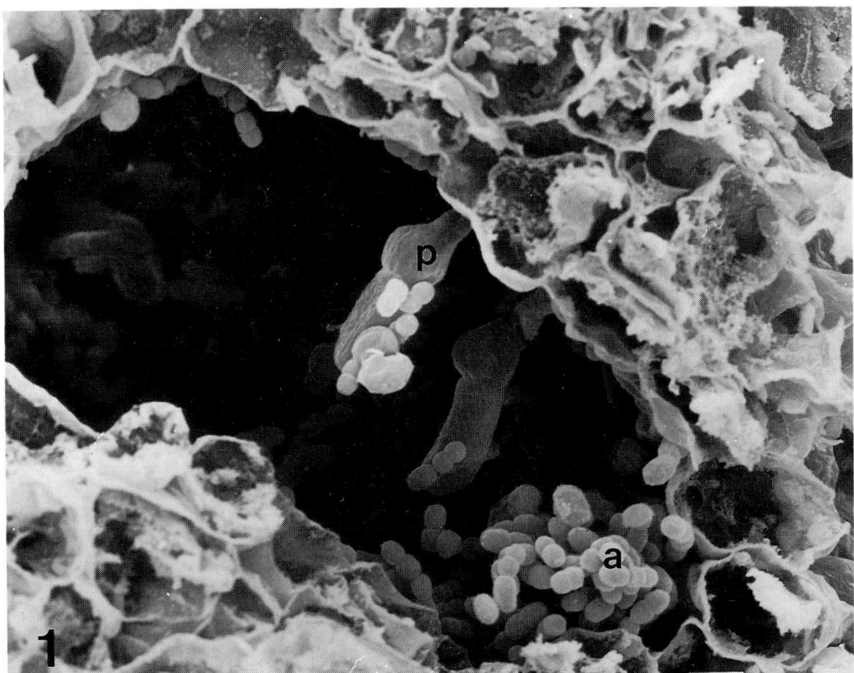

Fig. 1 - Leaf cavity of Azolla pinnata in which the phycobiont and two simple hairs are evident. 1145x.

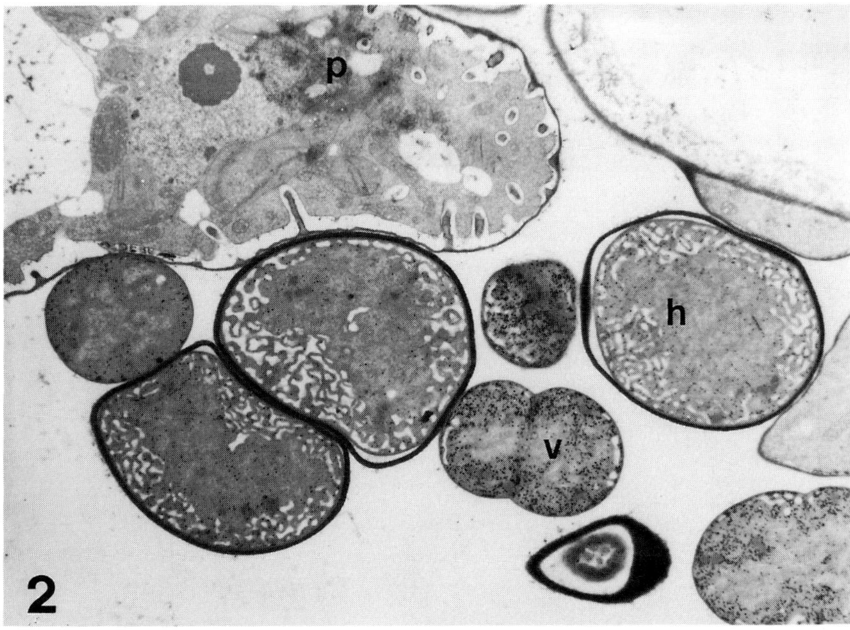

Fig. 2 - Vegetative cells and hetercysts of Anabaena azollae in the apical leaves of A. pinnata after PA-TSC-SP. 5775x.

cavity, with a ratio bacteria/*Anabaena* cell of 1:1. Increasing interest in identificatying the bacteria has occurred over the last ten years, but different authors obtained different results. Newton and Herdman (1979) found *Alcaligenes faecalis* and *Caulibacter fusiformis*; Gates et al. (1980) found *Corynebacterium* and more recently Wallace and Gates (1986) identified *Arthrobacter globiformis.* The same determination was achieved by Grilli Caiola et al. (in progress). The bacteria found inside the cavity of *Azolla* are rod shaped, sometimes cocci (Fig. 3).

HAIRS. The hairs protruding in the leaf cavity of *Azolla* are epidermal structures (Fig. 1, 2 and 4) whose ontogeny and role have been studied by Calvert et al. (1985) in *A. caroliniana*. In the young leaves without a cavity, and in conjiunction with foliar traces, two branched hairs occur, formed by a stalk terminating in apical cells with wall invaginations typical of transfer cells. Many simple hairs also occur randomly spread in the mature leaf cavities. Branched hairs do not seem to be present in the leaves of all *Azolla* species (Neumüller and Bergman 1981; personal data).

ENVELOPE. It is formed by mucilagenous material occurring in the leaf cavity in which the endophyte, bacteria and hairs are embedded (Fig. 5). In the upper leaves the envelope fills the cavity, but in the basal ones it is mainly localized at the periphery, leaving a central part free from structured material. The composition, origin and function of the envelope are not know so far.

Cell-to-cell recognition in *Azolla*

The presence in the *Azolla* of so numerous and different cells makes it difficult to ascertain the roles of each component in the association, and to gain evidence of any cell-to-cell recognition processes. This problem has been

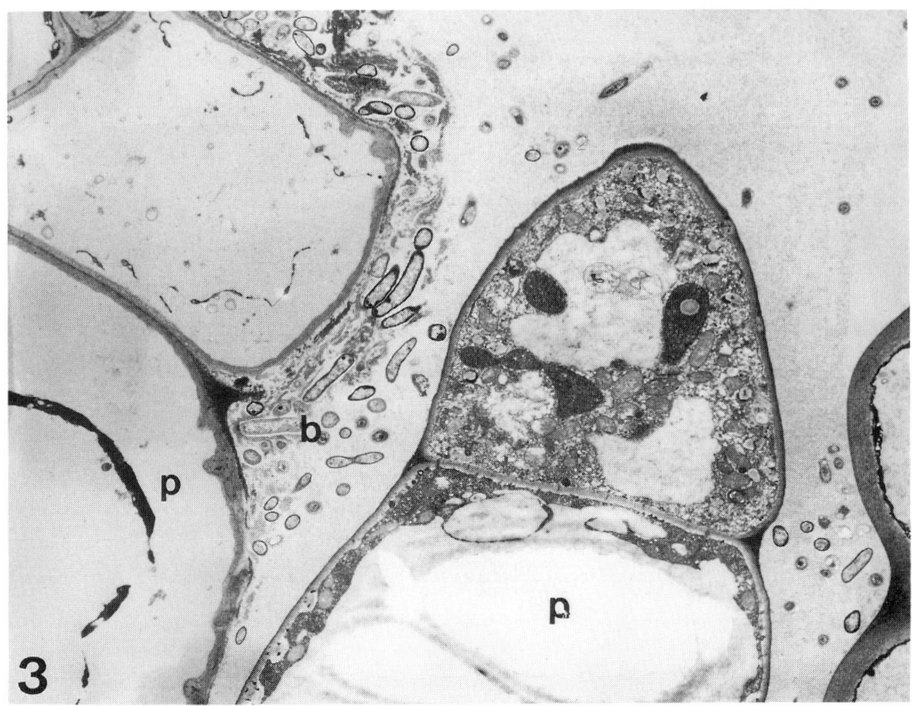

Fig. 3 - Apical leaf of A. carolíniana in which many bacteria surround the hair cells (p).

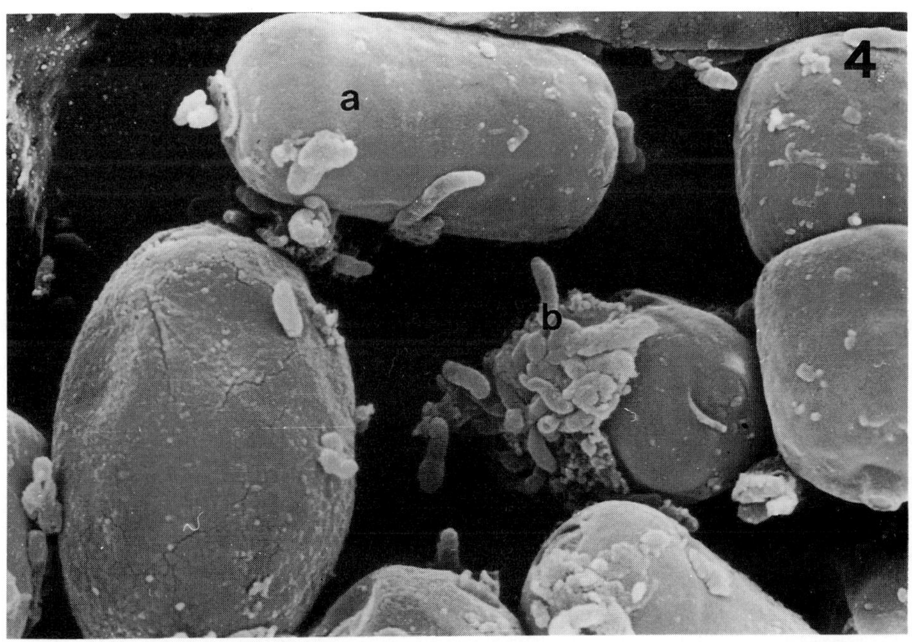

Fig. 4 - Bacteria (b) adhering to heterocysts (h) in a leaf cavity of A. microphylla. 1050x.

encountered in other symbiotic systems, where some researches have involved the identification of structures and lectins, and the use of immunological techniques. Such approaches have not so far been used to study interactions of the *Azolla* association with all its components: *Anabaena*, bacteria, hairs, envelope.

Structural recognition mechanisms: glycocalyx, fimbriae, pili.

Structural interactions between the cells have been found in many symbiotic or parasitic associations including cyanobacteria and fungi in lichens (Dick and Stewart 1980), and cyanobacteria and bacteria (Grilli Caiola and Pellegrini 1984). They are fibrillar structures appearing on the cell wall surface as glycocalyx, fimbriae or pili, capable of interacting with the host cells. We have tried to find evidence of similar structures in *Anabaena azollae* living in *Azolla caroliniana* Willd., *A. filiculoides* Lam., *A. mexicana* Presl., *A. microphylla* Kaulf. and *A. pinnata* R. Br., by means of SEM observations carried out on algal packets inside the leaf cavities, or by TEM observations either of ultrathin sections treated with specific staining or algal packets freshly isolated and negatively stained by PTA. In no case could we observe on the *Azolla, Anabaena* or bacterial cells, structures such as pili, fimbriae or glycocalyx. However, our researches reveal the nature of the adhesion mainly between the bacteria and the hair cells of *Azolla,* and bacteria and *Anabaena* heterocysts. Such adhesion is often of the polar type as seen in Rhizobia-legume associations (Fig. 6, 7), but were not observed structural mechanisms between the cells. The adhesion to heterocysts could be related to the particular composition of the envelope of these cells, which differs from the wall of the vegetative cells.

Recognition mechanisms mediated by lectins

The presence of lectins in the *Azolla-Anabaena* association has been noted by some authors (Mellor et al. 1981; Kobiler et al. 1981, 1982). Mellor found

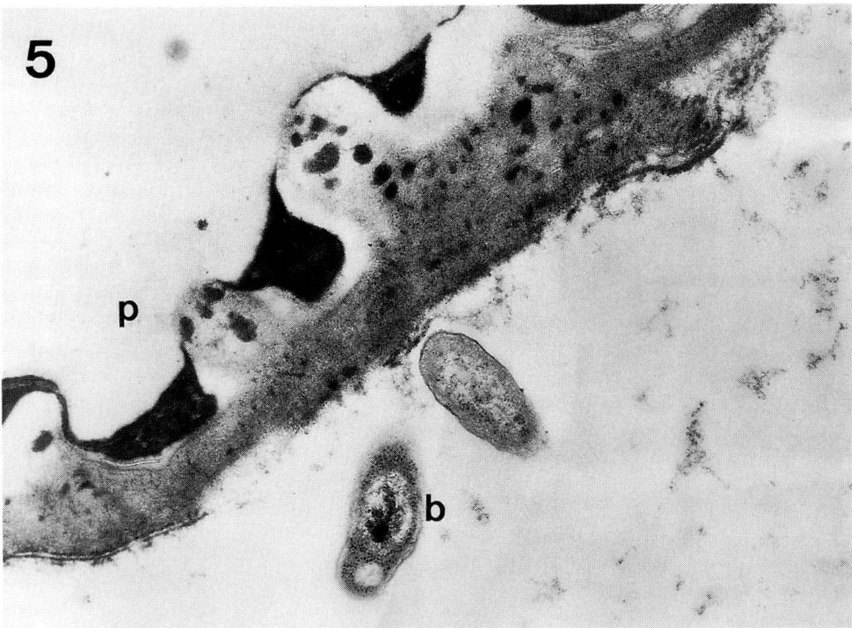

Fig. 5 - Cell wall invaginations in a hair cell of A. caroliniana and bacteria. Membranous structures and electrodense material are visible in the cell wall by tannic acid. 25500x.

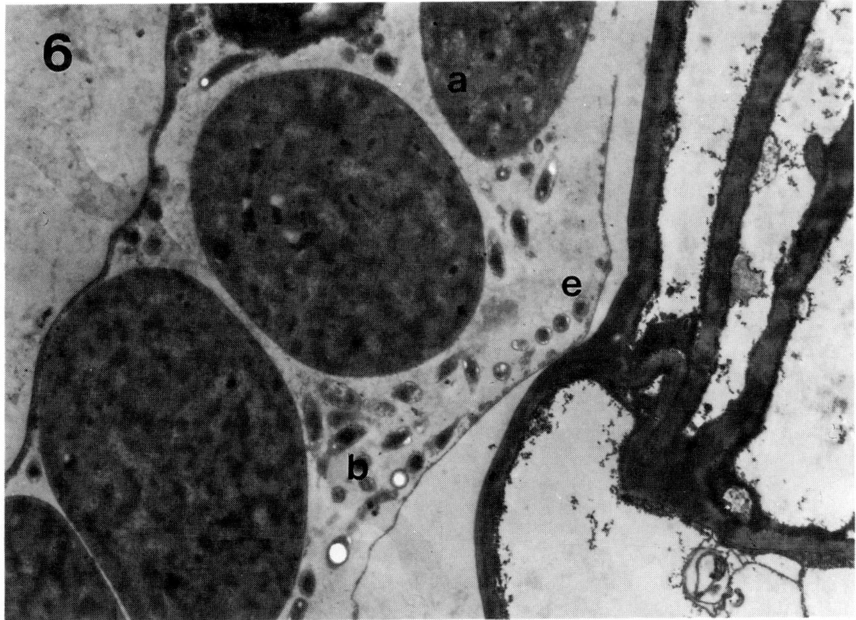

Fig. 6 - Leaf cavity of A. mexicana with Anabaena azollae and bacteria evidently embedded in the envelope. 7875x.

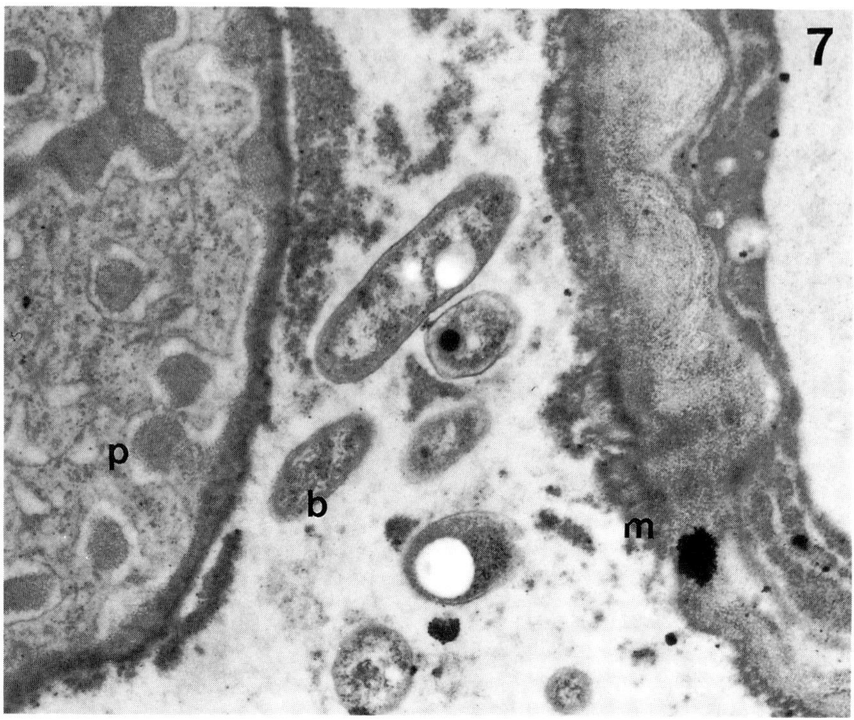

Fig. 7 - Hair cell (p), bacteria (b) and epidermal cell wall (on the right) of Azolla caroliniana. On the cell wall surface of the hair and epidermis, electrodense granular material is visible, and it seems to contribute to the envelope formation. The hair cell shown typical wall invaginations as transfer cells. 25500x.

a, Anabaena azollae; b, bacteria; e, envelope; h, heterocyst; m, mucilage; p, hair cell; v, vegetative cell.

haemoagglutinatinin activity in extracts of cell-free *Azolla-Anabaena*. Such activity was mainly localized in *Azolla* and was influenced by the presence of nitrate in the medium. This compound would affect both lectin synthesis and development of *Anabaena*, consequently influencing lectin production in *Azolla*. Kobiler, on the other hand, identified lectins on the cyanobacterial cell surface and attributed the adhesion between host and symbiont cells to recognition process mediated by these lectins. The same author localized in the leaf cavity a glycoconiugate receptor capable of binding the lectins produced by *Anabaena azollae*.

Recognition mechanisms involving antigen-antibody binding

By using fluorescent antibodies some authors (Gates et al. 1980; Ladha and Watanabe 1982, 1984; Arad et al. 1985) demonstrated that *A. azollae* isolated from leaf cavities had antigenic activity against antibodies prepared from the same cells, but these antibodies did not recognize either *Anabaena azollae* isolated from *Azolla* and grown in culture or other free-living or cultured cyanobacteria. Fluorescent antibodies against *A. azollae* symbiont indicated that all *A. azollae* freshly isolated from *Azolla* showed identical highly specific antigens, but no antibody cross-reacted with free-living cyanobacteria *or* *A. azollae* in cultures. These negative results obtained from these experiments indicated either that *A. azollae* isolated strains were not the true symbionts of *Azolla*, or that *A. azollae* antigenic capacity could change during isolation and growth in culture. A third explanation could be that a recognition mechanism is unnecessary because *A. azollae* passes directly from the sporophyte to gametophyte of *Azolla*, so that new infection of *Azolla* by *Anabaena* need never occur. The numerous reports on the impossibility of culturing *A. azollae* without *Azolla*, could support this hypothesis.

Different results were obtained by us using fluorescent antibodies against an *A. azollae* strain isolated from *A. caroliniana* and grown in culture, a strain not used by the other authors previously cited. Our immunological experiments carried out on five species of Azolla revealed that heterocysts behave differently from vegetative cells. The fluorescent antibodies always reacted with vegetative cells of A. azollae in all the Azolla species, but never with the heterocysts. This result is related to the presence of a layer covering the heterocyst cell wall, the heterocyst envelope, which covers the outer membrane. Furthermore we always noted an intense aspecific reaction of A. azollae and the leaf envelope to antibodies against cultured A. azollae and also to pre-immune rabbit serum. Antigenic activity either of freshly isolated A. azollae packets and the envelope from the same leaf cavity, was not eliminated by concanavalin A treatment. No responses were observed at the bacterial level.

In contrast to reports of other authors, Arad et al. (1985) found by quantitative and immunological determinations and ELISA, the identity of antigens inside the A. azollae from A. filiculoides and A. caroliniana, but not from A. pinnata.

Other adhesion mechanisms: the role of the envelope.

A different adhesion mechanism between Azolla and Anabaena cells was shown by Robins et al. (1986) when studying the envelope of the leaf cavity by low temperature SEM. These authors believe that the mucilage of the envelope might be produced by the cyanobacteria in order to adher to the Azolla cell surface. In this way, the adhesion between host and symbiont would not depende on specific interactions involving lectins.

Concerning the origin and role of the envelope, cytochemical studies have been carried out on thin and ultrathin sections of apical leaves of different Azolla species. The results indicate that Anabaena, bacteria and Azolla might be involved in the envelope production, as demonstrated by the presence of material reacting to Alcian blue, tannic acid and ruthenium red around epidermal and hair cells of the Azolla, and bacteria.

In conclusions, there is some evidence for the existence of lectin mediated recognition signals between cells in the Azolla-Anabaena symbiosis, but the complexity of the system, involving many types of cells, is such that the results obtained by the various methods used are not sufficient to characterize the nature of the recognition mechanisms.

All the species of Azolla examined have shown a common morphology and ultrastructure of the phycobiont. Consequently, it appears that A. azollae Strasb. is the only symbiont present in all the species considered. As well as morphology, the development of the Anabaena from the apical to the basal leaf of the fern appears to be similar in all species.

The role of the bacteria needs to be clarified: they are always found inside the leaf cavities of Azolla species and could contribute to forming the envelope or to producing growth regulating substances which determine the leaf curving and the ensuing development of the leaf cavity.

The envelope in the apical zone could bind Anabaena and bacteria to the host and could be the site of metabolic exchanges with the Azolla hairs.

Finally it can be hypothized that adhesion mechanisms mediated by the envelope could make recognition unnecessary in the apex of Azolla, where Anabaena and bacteria are constantly present. Furthermore, the host-symbiont adhesion could be reinforced by the metabolic dependance of partners, varying from the apex to the base.

Research work supported by CNR, Italy. Special grant I.P.R.A. - Sub-project 1. Paper N. 1473.

References

Arad, H., Keysari, A., Tel-Or, E., Kobiler, D. (1985). A comparison between cell antigens in different isolates of Anabaena azollae. Symbiosis 1: 195-204.

Calvert, H.E., Pence, M.K., Peters, G.A. (1985). Ultrastructural ontogeny of leaf cavity trichomes in Azolla implies a functional role in metabolite exchange. Protoplasma 129: 10-27.

Dick, H., Stewart, W.D.P. (1980). The occurrence of fimbriae on a N_2 fixing cyanobacterium which occurs in a lichen symbiosis. Arch. Microbiol. 124: 107-109.

Fjerdingstad, E. (1976). Anabaena variabilis status azollae. Arch. Hydrobiol., Suppl., 49, Algological Studies 17: 377-381.

Gates, J.E., Fisher, R.W., Candler, R.A. (1980). The occurrence of coryneform bacteria in the leaf cavity of Azolla. Arch. Microbiol. 127: 163-165.

Gates, J.E., Fisher, R.W., Goggin, T.W., Azrolan, N.I. (1980). Antigenic differences between Anabaena azollae fresh from the Azolla fern leaf cavity and free-living cyanobacteria. Arch. Microbiol. 128: 126-129.

Grilli, A. (1964). Infrastrutture di Anabaena azollae vivente nelle foglioline di Azolla caroliniana. Ann. Microbiol. 14: 69-90.

Grilli Caiola, M., Moretti, A. (1985). Impiego degli azotofissatori in agricoltura. Il sistema Azolla-Anabaena. CNR Monografia n. 7, Progetto finalizzato IPRA, p. 53-69.

Grilli Caiola M., Pellegrini, S. (1984). Lysis of Microcystis aeruginosa (Kütz.) by Bdellovibrio-like bacteria. J. Phycol. 20: 471-475.

Kobiler, D., Cohen-Sharon, A., Tel-Or, E. (1981). Recognition between the N_2-fixing Anabaena and the water fern Azolla. FEBS Lett. 133: 157-160.

Kobiler, D., Cohen-Sharon, A., Tel-Or, E. (1982). Lectins involved in the recognition between Anabaena and Azolla. Israel J. Botany 31: 324-328.

Konar, R.N., Kapoor, R.K. (1974). Embryology of Azolla pinnata. Phytomorphology 24: 228-261.

Ladha, J.K., Watanabe, I. (1982). Antigenic similarity among Anabaena azollae separated from different species of Azolla. Biochem. Biophys. Res. Comm. 109: 675-682.

Ladha, J.K., Watanabe, I. (1984). Antigenic analysis of Anabaena azollae and the role of lectin in the Azolla-Anabaena symbiosis. New Phytol. 98: 295-300.

Mellor, R.B., Gadd, G.M., Rowell, P., Stewart, W. D.P. (1981). A phyto-haemoagglutinin from the Azolla-Anabaena symbiosis. Biochem. Biophys. Res. Comm. 99: 1348-1353.

Neumüller, M. Bergman, B. (1981). The ultrastructure of Anabaena azollae in Azolla pinnata. Physiol. Plant. 51: 69-76.

Newton, J.W., Herman, H.I. (1979). Isolation of cyanobacteria from the aquatic fern Azolla. Arch. Microbiol. 120: 161-165

Peters, G.A., Toia Jr., R.E., Levine N.J., Raveed, D. (1978). Azolla-Anabaena azollae relationship. VI. Morphological aspects of the association. New Phytol. 80: 583-593.

Robins, R.J., Hall, D.O., Shi, D.J., Turner, R.J., Rhodes, M.J.C. (1986). Mucilage acts to adher cyanobacteria and cultured plant cells to biological and inert surfaces. FEMS Microbiol. Lett. 34: 155-160.

Wallace, W.H., Gates, J.E. (1986). Identification of eubacteria isolated from leaf cavities of four species of the N_2-fixing Azolla fern as Arthrobacter Conn. and Dimmick. Appl. Environ. Microbiol. 52: 425-429.

THE FUNCTIONAL MORPHOLOGY OF CELL-TO-CELL INTERACTIONS IN LICHENS

Rosmarie Honegger
University of Zürich
Institute of Plant Biology
CH-8008 Zürich Switzerland

INTRODUCTION

Lichen mycobionts are a large, polyphyletic and taxonomically heterogenous group comprising about 21% of all fungi (Hawksworth and Hill 1984). 98% of lichen-forming fungi are Ascomycetes, a few belong to the Basidiomycetes (Aphyllophorales and Agaricales), and the rest are Deuteromycetes. Only those mycobiont-photobiont associations are considered as lichens in which the fungal partner is the exhabitant in the sense of Law and Lewis (1983). Mutualistic symbioses in which fungi are inhabitants of multicellular algae are traditionally not regarded as lichens (Hawksworth 1988). Of the 46 orders of Ascomycetes accepted by Eriksson and Hawksworth (1986) 10 comprise and 6 consist entirely of lichen-forming taxa. Phycobionts of 20 chlorophycean genera (5 orders) and cyanobionts of 12 cyanobacterial genera (4 orders) have been reported as lichen photobionts. However, in an immense number of lichens the photobiont has never been identified, not even on a generic level.

The present study aims to summarise the different types of mycobiont-photobiont interactions in lichens, to review published and to present new data on the functional morphology of Lecanorales with stratified thalli containing phycobionts of the genus *Trebouxia* s. lat. (taxonomic problems related to the distinction of *Trebouxia* Puymaly and *Pseudotrebouxia* Archibald are discussed by Gärtner 1985), and to discuss the significance of these findings for our understanding of cell-to-cell signals in the lichen symbiosis.

MYCOBIONT - PHOTOBIONT INTERACTIONS IN LICHENS

Considering the taxonomic diversity of the organisms involved and the supposed great phylogenetic age of some of the lichen-forming taxa it is not surprising that various modes of mycobiont-photobiont interactions have developed ranging from loose associations between fungal hyphae and photobiont cells to

NATO ASI Series, Vol. H17
Cell to Cell Signals in Plant, Animal and
Microbial Symbiosis. Edited by S. Scannerini et al.

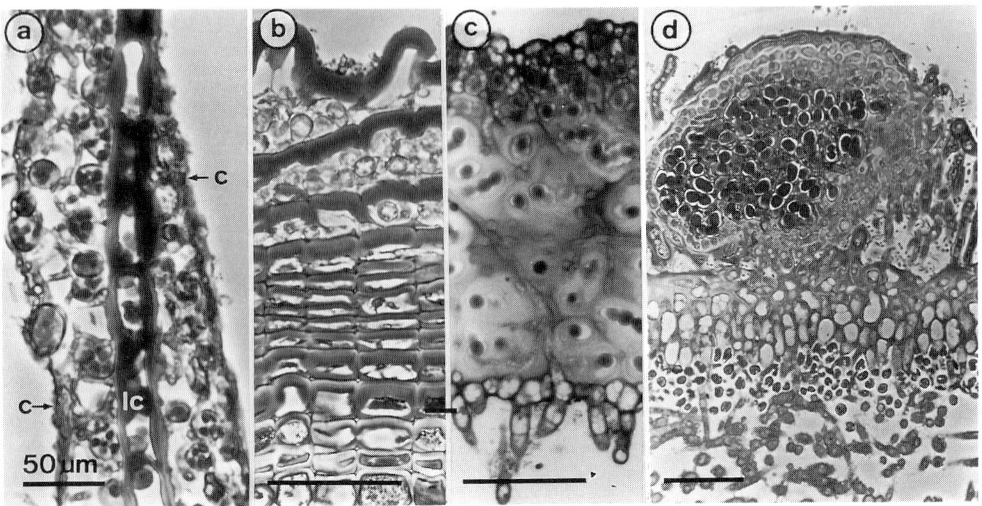

Figs. 1a-d. Taxonomic and morphological diversity in lichens. a: Dimerella lutea (Gyalectales) developing underneath the cuticle (c) of the liverworth Frullania tamarisci whose leaf cells (lc) are penetrated by the mycobiont. b: Graphis elegans (Graphidales), endophleodal thallus on Ilex europaeus. d: Leptogium saturninum (Lecanorales), dorsiventrally organised, homoeomerous gelatinous thallus with no gas-filled internal space. d: Peltigera leucophlebia (Peltigerales), dorsiventrally organised, foliose thallus with Coccomyxa phycobionts in the algal layer and Nostoc sp. as cyanobiont in the external cephalodium. Semithin sections of metacrylate-embedded specimens.

highly evolved lichens with very complex stratified thalli (reviewed by Hawksworth 1988) in which the photobiont cells are positioned in a distinct layer and controlled by the mycobiont (Figs. 1a-d; Greenhalgh and Anglesea 1979; Hill 1985; Honegger 1985, 1987). Correlations have been found between thallus morphology, phycobiont cell wall structure and composition, and the type of mycobiont-phycobiont interaction in a large number, but by no means in all lichens with green algal symbionts (Tschermak 1941; Plessl 1963; Honegger 1985, 1986a). No such correlations seem to occur in lichens with cyanobacterial photobionts.

Three different types of cell walls have so far been detected in lichen phycobionts: first, tripartite walls with a trilaminar outermost wall layer containing a highly resistant, sporopollenin-like biopolymer (Coccomyxa, Elliptochloris; Honegger and Brunner 1981; Brunner 1985; Brunner and Honegger 1985); second, cellulosic cell walls with a fibrous inner and an amorphous outer part, the latter containing acid polysaccharides and proteins (Trebouxia s. lat., Trentepohlia; Bubrick and Galun 1980b; Honegger 1984; Brunner 1985; Brunner and Honegger 1985); and third, an intermediate type with tripartite cell wall and outermost trilaminar layer, but without a resistant biopolymer

(Dictyochloropsis, Myrmecia, Pseudochlorella; Brunner 1985; Brunner and Honegger 1985).

Extensive comparative light microscopy investigations on the mycobiont-photobiont interface have been performed by Tschermak (1941) and Plessl (1963) Their findings were largely confirmed and complemented by scanning and transmission electron microscopy studies of various investigators (summarised by Honegger 1984, 1985, 1986a). Simple wall-to-wall apposition (Fig.2a) appears to be the predominant mode of mycobiont-phycobiont interaction in crustose and foliose asco- and basidiolichens with enzymatically non-degradable biopolymer in the cell wall of the phycobiont. The only exception was found in Micarea prasina whose thallus consists entirely of goniocysts. The cells of its Elliptochloris phycobiont are densely ensheathed by mycobiont hyphae which penetrate the algal cell wall at always the same point (Brunner 1985). Other goniocyst-inhabiting phycobionts with sporopollenin-like biopolymer are not penetrated by fungal haustoria (Honegger and Brunner 1981; Oberwinkler 1984). Intracellular haustoria (Fig. 2e) are the predominant type of interaction in a range of lichens with crustose, non-stratified thallus and phycobionts with either cellulosic, or intermediate cell walls. Different types of intraparietal haustoria (Figs. 2f-h) occur in distantly related squamulose, foliose and fruticose lichens whose phycobionts have either cellulosic, or intermediate cell walls. Intragelatinous fungal protrusions (Fig. 2a) appear to be the predominant type of mycobiont-cyanobiont interaction in diverse, distantly related groups of ascolichens (Jacobs and Ahmadjian 1973; Peveling 1973; Englund 1977; Spector and Jensen, 1977; Boissière 1982; Honegger 1982, 1985). Intragelatinous fungal protrusions forming wall-to-wall appositions at the cyanobiont cell wall surface (Fig. 2b) have been reported in numerous Lecanorales and Lichinales (Paran et al. 1971; Tschermak-Woess et al. 1983; Büdel 1987). Very peculiar fungal sheaths covering the surface of Scytonema colonies and forming intracellular haustoria throughout the trichomes except heterocysts (Fig. 2c) were found in thelephoracean basidiolichens of the genus Dictyonema (reviewed by Tschermak-Woess 1983; Oberwinkler 1984).

Based on the observation of intracellular haustoria, especially in artificially resynthesised lichens Ahmadjian and Jacobs (1983) considered lichen mycobionts as algal parasites. However, although many species of lichen phycobionts can be found in a free-living state (Tschermak-Woess 1978; Bubrick et al. 1984) most of them are ecologically very much more successful in symbiosis with lichen mycobionts than when apart. Therefore the system can be considered as a mutualistic symbiosis (Law and Lewis 1983; Smith and Douglas 1987). Nevertheless, the biology of a large number of mainly crustose lichen

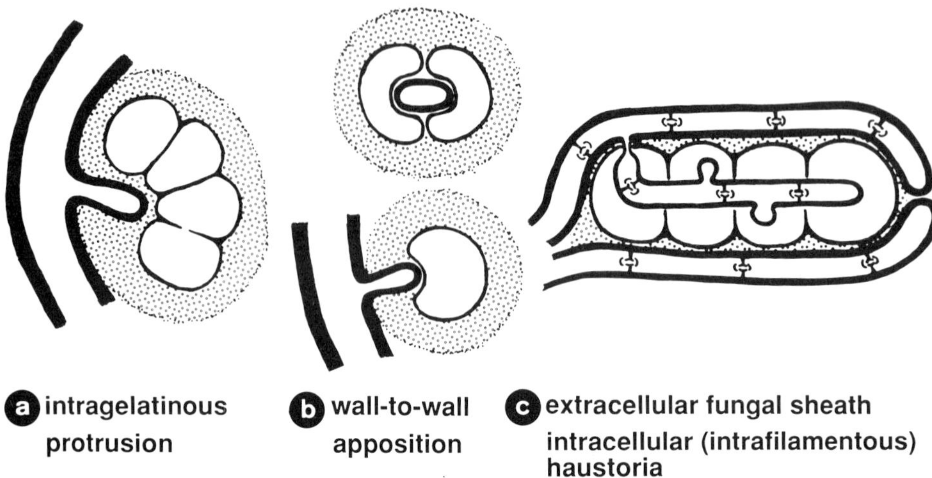

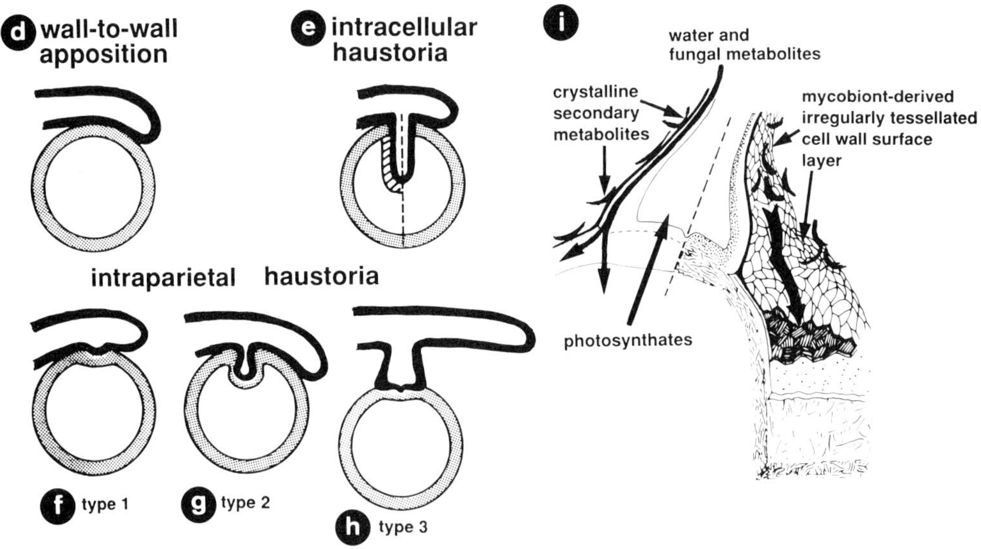

Figs. 2a-h: Diagrammatic interpretation of different types of mycobiont-photobiont relationships in lichens. Fig. 2i: detail of the mycobiont-phycobiont interface in intraparietal haustoria of type 3, and possible routes of transfer (arrows) of water and fungal metabolites on the one, and of photosynthates on the other hand. a after Honegger, 1985; c after Oberwinkler, 1984; d-h after Honegger 1986a; i after Honegger 1985, 1986b.

species has never been investigated. It is essential to keep in mind the enormous diversity of symbionts and types of interactions, and our rather rudimentary knowledge of the biology of lichens when discussing symbiotic relationships in these systems.

THE FUNCTIONAL MORPHOLOGY OF LECANORALES WITH STRATIFIED THALLUS AND PHYCOBIONTS OF THE GENUS TREBOUXIA s. lat.

Lecanoralean Ascomycetes with stratified squamulose, foliose or fruticose thalli are among those lichens which produce the most complex vegetative thalli or stromata in the fungal kingdom. The majority of physiological, ecophysiological and biochemical investigations have been performed with members of this group. These thalli contain a zone of conglutinated hyphae bound together by an extracellular gelatinous material, and a system of aerial hyphae some of which contact the phycobiont cells. In dorsiventrally organised squamulose and foliose taxa the conglutinated zone covers the thallus as a cortical layer either on the upper surface only (e.g. Cladoniaceae), or on both upper and lower surfaces (e.g. Parmeliaceae). The complex thallus morphology and anatomy in combination with intraparietal haustoria of either type 2 (Cladoniaceae; Fig. 2g) or type 3 (Parmeliaceae; Fig. 2h) reflect the high level of evolutionary development of these symbiotic systems which are supposed to be the result of a prolonged co-evolution of both symbionts.

The morphologist Goebel (1926a, 1926b) was the first to distinguish between hydrophilic "Schwellhyphen" (swelling hyphae) of the cortical layer of lichens with stratified thallus and "Lufthyphen" (aerial hyphae) of the medullary layer whose water repellency he supposed to be due to the mycobiont-derived secondary products crystallising at the cell wall surface. He noticed that the aerial hyphae were readily wetted by water after incubation in solvents such as ether, chloroform, benzene etc. Goebel recognised the significance of both conglutinated cortical and water-repellent aerial hyphae for the uptake and translocation of water in the lichen thallus. He proposed capillary water translocation along the cell wall surfaces of hyphae, cilia and rhizinae outside the cortical layer, then water uptake either by the gelatinous material, or the terminal fungal cells of the cortical layer and subsequent translocation within the cell wall (which he termed "membrane") of the aerial hyphae of the algal and medullary layer. "Capillary translocation of water (along the cell wall surface of aerial hyphae of the medulla) is completely out of question" (Goebel 1926a). Unfortunately his observations were largely forgotten.

Until recently capillary water translocation along the surface of medullary hyphae was supposed to be the rule (summarised by Blum 1973; Jahns 1984). Water saturation would then mean completely wet cell wall surfaces in the medullary and algal layer, a less than optimal situation for gas exchange by the photobiont cells.

Ultrastructural studies show that the cell wall surfaces of aerial hyphae in lichen thalli are not only covered to varying degrees by crystalline extracellular secondary metabolites, but primarily by a thin, outermost cell wall layer which reveals a certain water repellency of its own (Fig. 2i; Honegger 1984, 1985, 1986b). Besides forming a haustorial apparatus at the mycobiont-phycobiont interface lecanoralean mycobionts spread their thin outermost cell wall layer over the cell wall surface of the algal cell. This happens at the very first contact between the growing fungal hypha and the young auto- or aplanospores, respectively, when the latter are still ensheathed by the degrading mother cell wall (Honegger 1984, 1985). Consequently the structure and composition of the cell wall surface of Trebouxia phycobionts is different in symbiosis than in the cultured, non-symbiotic state (Honegger 1984). Differences in the cell wall surface composition between cultured and symbiotic Trebouxia phycobionts of the teloschistalean Xanthoria parietina have been demonstrated by Bubrick and coworkers (Bubrick and Galun 1980b; Bubrick et al. 1982) by means of histochemical and immunological methods. The cell wall surface layers of aerial hyphae in Lecanorales are recognised on the ultrastructural level as a somewhat rigid zone with an irregularly tessellated pattern in freeze-fracture preparations, or as a thin, electron-opaque layer in ultrathin sections (Figs. 4d-e; Honegger 1984, 1985, 1986b). Comparable cell wall surface layers revealing a distinct rodlet pattern were first detected in aerial hyphae of non-lichenised fungi, especially in hyphomycetous conidiophores and conidia (reviewed by Aronson 1981), or spreading as a continuous layer over the apical ends of the cells forming the hymenial surface of basidiocarps (McLaughlin 1982). Proteins and lipids have been identified as the main components of this rodlet layer in conidia (Aronson 1981; Cole & Pope 1981).

The irregularly tessellated surface layer of aerial hyphae and phycobiont cell wall surfaces in Lecanorales have not been chemically analysed, but in histochemical studies a positive reaction was observed with Coomassie blue, a protein stain, and with the fluorochrome Auramin O (Honegger 1986b) which binds to lipids and acidic waxes (Gahan 1984) and is also used as a Feulgen stain for hydrolysed fungal nuclei (Lemke et al. 1976). Some mycobiont-derived secondary metabolites crystallise within and on this cell wall surface layer of the aerial hyphae, others in the gelatinous material of the cortical layer (Peveling, 1970; Honegger, 1986b). Considerable differences between lichen species in the water repellency of the aerial hyphae were observed (Goebel, 1926a, 1926b). Such differences are obvious during the preparative procedures for electron microscopy, when thallus fragments are to be infiltrated by the fixing solution. Some species are distinctly more water-repellent than others and sink into the fixing solution only after prolonged (up to 40 min) evacuation with the water aspirator. Such differences might be either due to the presence of more or chemically different secondary metabolites at the wall surface, or to differences in the structure or composition of the cell wall surface layer. This problem was tentatively investigated with histochemical methods using the parmeliacean Hypogymnia physodes with fairly water-repellent aerial hyphae, and Cladonia caespiticia, a Cladoniaceae forming minute thallus squamules with only an upper cortical layer and with extremely water-repellent aerial hyphae.

The results listed in Tab.1 and Figs. 3a-f, 4a-f are tentatively interpreted as follows. The water-repellency of the aerial hyphae is primarily due to the cell wall surface layer and its composition. Extracellular lichen products may enhance this effect. The highly water-repellent aerial hyphae of Cladonia caespiticia are not covered by crystalline secondary products, but by their irregularly tessellated surface layer (Fig. 4e) which forms knobs and saccules (Fig. 4f) and resists extraction procedures with solvents. Therefore this surface layer can also be seen in ultrathin sections of specimens dehydrated with acetone (Fig. 4e; Honegger 1986a, 1987). The lipids on the cell wall surface layer of aerial hyphae in Hypogymnia physodes were not fully extracted, but large amounts were removed by acetone and the chloroform-methanol mixture (Tab. 1; Fig. 3c'). This could be the reason why this surface layer is less obvious in ultrathin sections of acetone dehydrated specimens of the Parmeliaceae (Fig. 3f; Honegger 1984, 1986a, 1986b). The differences between Hypogymnia physodes and Cladonia caespiticia in the resistance of their cell wall surface lipids during extraction procedures might be due to formation of complexes with proteins in the latter species. Protein-lipid complexes resist these extraction procedures (Gahan 1984). Chemical analyses of the outer and inner parts of the cell walls of lichenised Ascomycetes are required.

Preliminary comparative histochemical and ultrastructural studies (unpubl.) indicate the presence of a more or less water-repellent cell wall surface layer on the aerial hyphae in numerous, distantly related lichenised Ascomycetes with stratified thalli. Proteins, as inferred from the staining reaction with Coomassie blue, are apparently always present. Lipids are present in varying amounts as shown by the intensity and colouration of the fluorescence with Auramin O. So far, high lipid contents are correlated with an irregularly tessellated pattern of the cell wall surface in the Lecanorales and in the peltigeralean Lobaria pulmonaria, whereas peltigeralean species with faint Auramin O reaction (e.g. P. leucophlebia) revealed a distinct rodlet pattern at the cell wall surface of medullary hyphae (Honegger, 1982). Intrathalline differences in the Auramin O fluorescence were noted in some species. In the peltigeralean Sticta fuliginosa the medullary hyphae give a faint, but the tomentum and cyphellae of the lower thallus surface an intense reaction with Auramin O. Sticta fuliginosa is like other species of this genus, devoid of crystalline fungal products (Geyer 1985).

Table 1. Histochemical investigation of the cell wall surface layer of aerial hyphae and Trebouxia phycobionts in the stratified thalli of Hypogymnia physodes and Cladonia caespiticia

Preparative procedure	Hypogymnia physodes *	Cladonia caespiticia *
Coomassie blue (1)	phy: cell wall surface blue my : cell wall surface of medullary hyphae blue	(phy ensheathed by mycobiont hyphae) my : cell wall surfaces blue
Phase contrast (2)	my : crystalline lichen products more or less abundant at the surface of aerial hyphae (physodic and physodalic acids)	my : no crystalline lichen products
Autofluorescence (2) at 405 nm (3)	phy: chloroplast red (chlorophyll) my : no autofluorescence	phy: chloroplast red my : blue fluorescence in the central part of the cortical layer
Auramin O (4) at 405 nm (3)	phy: chloroplast red, cell wall surface golden yellow my : surface of medullary hyphae intense golden yellow	phy: chloroplast red my : surface of aerial and cortical hyphae intense greenish yellow
Auramin O after removal of lipids (5) at 405 nm (3)	phy: chloroplast red; walls of damaged cells bluish my : thin surface layer of aerial hyphae golden yellow (intensity diminished) bluish fluorescence of damaged cells	phy: chloroplast red my : surface of aerial and cortical hyphae intense greenish yellow (unchanged)

* 14 µm thick, freshly prepared cryostat sections of freshly collected material

1 0.25% Coomassie blue in a 10 : 3 : 87 mixture of methanol, acetic acid and water (Heslop-Harrison, 1983), 30 sec sections were mounted in water

2 Unstained sections mounted in water

3 Violet light excitation (405 nm) was used instead of the recommended blue light excitation (450 - 490 nm) in order to avoid a strong, interfering white to yellow autofluorescence of the hyphae

Photographs were taken with a Zeiss Photomikroskop II equipped with a III RS condensor (epifluorescence)

4 0.01% Auramin O in Tris-HCl, pH 7.4, 10 min (Heslop-Harrison, 1983). Sections were mounted in Tris-HCl

5 Subsequent incubation in 96% acetone and a 2 : 1 (v/v) mixture of chloroform and methanol, 5 min each, staining with Auramin O (10 min) and mounting in Tris-HCl

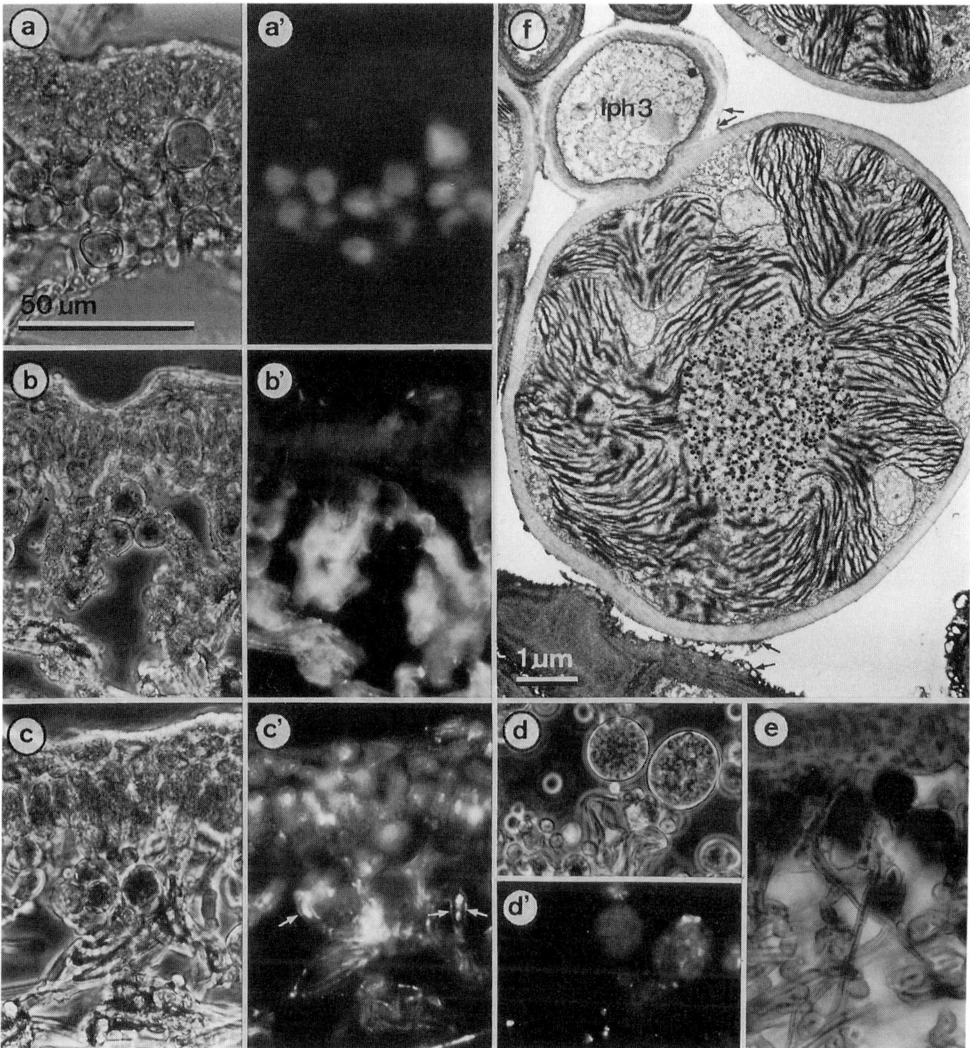

Figs. 3a-f. Histochemical and ultrastructural studies in Hypogymnia physodes. a, a': phase contrast and autofluorescence of an unstained cryostat section. Red chlorophyll fluorescence at 405 nm. b, b': phase contrast and fluorescence at 405 nm after staining with Auramin O. Medullary hyphae and the surface of the phycobiont cells appear bright golden yellow. c, c': phase contrast and fluorescence at 405 nm after removal of the lipids and staining with Auramin O. A thin, golden yellow cell wall surface layer is recognisable (arrows). Damaged fungal and algal cells are bluish white. d, d': phase contrast and fluorescence at 405 nm of the cultured phycobiont after staining with Auramin O. Red chlorophyll fluorescence and yellow intracellular lipid droplets are seen, but no fluorescent cell wall surface layer. This is formed by the mycobiont in symbiosis. e: bright field micrograph of a section stained with Coomassie blue. f: TEM micrograph of chemically fixed, dehydrated and epoxy resin embedded material (methods see Honegger, 1986b). iph 3: intraparietal haustorium of type 3; arrows point to the very thin outermost cell wall layer. Same magnification in a-e.

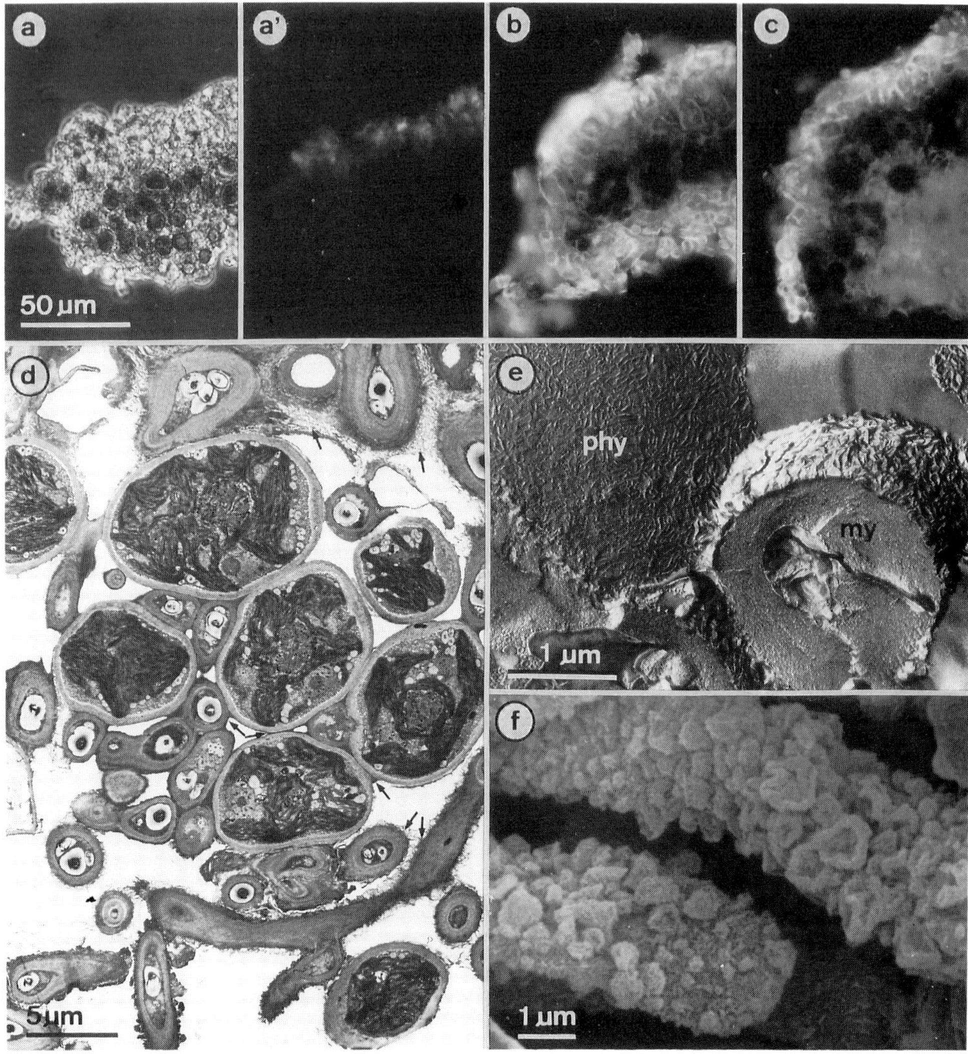

Figs. 4a-f: Histochemical and ultrastructural studies in Cladonia caespiticia. a,a': phase contrast and autofluorescence at 405 nm. Intense blue fluorescence in the central part of the cortical layer. b: fluorescence at 405 nm after staining with Auramin O. The cell wall surfaces or aerial and cortical hyphae appear bright greenish yellow. c: fluorescence at 405 nm after removal of the lipids and staining with Auramin O. The bright greenish to yellow fluorescence of the cell wall surface layer is unchanged. d : ultrathin section of chemically fixed, dehydrated end epoxy resin embedded material. The electron-dense cell wall surface layer of aerial and cortical hyphae and at the mycobiont-phycobiont interface is retained (arrows). d: freeze-fracture preparation of the mycobiont-phycobiont interface in chemically fixed, glycerinated material. The irregularly tessellated surface layer is spreading from the mycobiont (my) over the phycobiont (phy) cell wall surface. f: scanning electron micrograph of the surface of aerial hyphae in freeze-dryed material. No crystalline extra-cellular lichen products, but knobs and saccules of the cell wall surface layer are seen. Preparative procedures for TEM: see Honegger (1986b).

DISCUSSION

Based on an improved knowledge of the mycobiont-photobiont interface and of the functional morphology of stratified lichen thalli, more precise models can be developed for the translocation not only of water and dissolved salts, eventually of pollutants and naturally occurring metals, but also of fungal metabolites and of photobiont-derived compounds.

The haustorial complex and the different types of haustoria and intragelatinous fungal protrusions are with high probability the site of mobilisation and uptake of photosynthates (e.g. polyols) and other photobiont-derived compounds (e.g. substrates and/or precursors for cell wall synthesis in young, developing photobiont cells, and enzymatically degraded cell wall or cell sheath components of mature photobionts; Hill 1972; Honegger 1985).

The present data on the functional morphology of stratified thalli are complemented by A. Beckett's low temperature scanning electron microscope (LTSEM) studies of native, chemically untreated thalli of the foliose _Parmelia sulcata_ under either water saturated, or completely dry conditions (Brown et al. 1987). As shown in these LTSEM micrographs, the phycobiont cells and aerial hyphae of the mycobiont shrink very drastically when dry, but recover rapidly after rewetting, the phycobionts returing to their globose shape. Interestingly, no water (it would be seen as ice crystals in this type of preparation) was detected on the surface of either aerial hyphae or phycobiont cells in water saturated thalli (capillary water was supposed to fill the intercellular space in this species; Jahns 1984). These data confirm earlier reports by Goebel (1926a, 1926b) and Honegger (1984, 1985, 1986b) on water translocation in the outer part of the cell wall of aerial hyphae in lichen-forming Ascomycetes.

It is very likely that not only water and dissolved salts, but also extracellular fungal metabolites are transported in the passively moving water flow along the fungal cell wall, and in this way reach the photobiont cells. Mycobiont-derived urea and phenolic secondary metabolites are supposed to play a role in cyclic regulatory processes in lecanoralean lichens (Vicente 1985; Legaz 1985; summarised by Richardson 1985; Honegger 1986b). Phenolic secondary metabolites have been shown to reduce photosynthesis and growth of cultured phycobiont cells in vitro (Kinraide and Ahmadjian 1970) and to occur on and within phycobiont cells in the lichen thallus (Ahmadjian and Jacobs 1985; Honegger 1986b, 1987; Lines, in Smith and Douglas 1987; Avalos and Vicente 1987). The role of these phenolics in the control and regulation of symbiotic photobionts and in processes related to growth pattern formation in the complex lichen thalli require further investigation.

Cell wall surface layers probably playing a role in translocation seem to occur not only in aerial hyphae of various fungi, but also in fungus-higher plant cell interactions and even in higher plants (e.g. pollen-stigma interface; see below). The structure and composition of these surface layers may vary, but their function is apparently always the same: on the one hand directing the flow, on the other reducing the loss of solutions on their way to structures in need of these compounds. Sealing of cells which are donors of required substances appears to be important in symbiotic systems. The cementing material between host and fungal cells in the Hartig net of ascomycetous and basidiomycetous ectomycorrhizae (e.g. Scannerini & Bonfante-Fasolo, 1983) might be functionally comparable to the cell wall surface layer at the mycobiont-photobiont interface in lichens. The cells of the root cortex and photobionts of lichens require water and salts, but they are also the donors of carbohydrates. At least in lichens these carbohydrates are not leaking out of the photobiont cells into the intercellular space as supposed in earlier investigations (summarised by Honegger 1985), but are, with high probability, mobilised in the central part of the haustorial complex, whereas water is brought in at the periphery of the haustorium (Figs. 2i, 3f; Honegger 1985, 1986b; Smith and Douglas 1987).

Particularly interesting is the situation at the pollen-stigma interface. Exine-derived lipids spread over the stigma surface where they apparently undergo chemical changes (Ferrari et al. 1985) and subsequently canalise the water flow from the papilla to the dry pollen grain. The rapidly spreading pollen oils provide a first, unspecific binding of the pollen grain to the stigma surface. Recognition processes take place after the release of exine-derived proteinaceous fractions by the hydrated pollen grain (summarised by Ferrari et al. 1985). Similar events might happen at the first contact between hyphae of lichen-forming fungi and algal cells. In resynthesis experiments aerial mycobiont hyphae were found to adhere to compatible and incompatible algal cells (Ahmadjian and Jacobs 1983), even when these had completely different cell walls from the compatible phycobiont (e.g. sporopollenin-containing _Coccomyxa_ cells bound to the hyphae of _Xanthoria parietina_ whose _Trebouxia_ phycobionts have cellulosic walls; Bubrick et al. 1985). After a first unspecific binding of algal cells by lipids of the fungal cell wall surface, a whole series of events including the specific action of algal binding proteins (Bubrick & Galun 1980a; Bubrick et al. 1981, 1985) might be triggered which finally lead to the recognition of compatible partners. These most fascinating aspects of the lichen symbiosis require further investigations.

Lichen-forming fungi and their photobionts provide interesting model systems for the study of certain aspects of fungus-host cell interactions. Or, as demonstrated by Sir David Smith (1978): Lichens can tell us much about "real fungi", especially about the biotrophic ones.

Acknowledgements. My sincere thanks are due to Professor Sir David Smith for encouraging comments and for improving the English style of this manuscript, and to Mrs. V. Kutasi for doing all the darkroom work.

REFERENCES

Ahmadjian V, Jacobs JB (1983) Algal-fungal relationships in lichens: recognition, synthesis, and development. In: LJ Goff (ed): Algal Symbiosis. Cambridge: Cambridge University Press, pp. 147-172

Ahmadjian V, Jacobs JB (1985) Artificial re-establishment of lichens. IV. Comparison between natural and synthetic thalli of Usnea strigosa. Lichenologist 17: 149-166

Aronson JA (1981) Cell wall chemistry, ultrastructure, and metabolism. In: GT Cole, B Kendrick (eds): Biology of Conidial Fungi. New York: Academic Press, pp. 459-507

Avalos A, Vicente C (1987) The occurrence of lichen phenolics in the photobiont cells of Evernia prunastri. Plant Cell Reports 6: 74-76.

Blum OB (1973) Water relations. In: V Ahmadjian, ME Hale (eds): The Lichens. New York: Academic Press, pp. 381-400

Boissière M-C (1982) Cytochemical ultrastructure of Peltigera canina: some features related to its symbiosis. Lichenologist 14: 1-28

Brown DH, Rapsch S, Beckett A, Ascaso C (1987) The effect of desiccation on cell shape in the lichen Parmelia sulcata. New Phytol 105: 295-299

Brunner U (1985) Ultrastrukturelle und chemische Untersuchungen an Flechtenphycobionten aus 7 Gattungen der Chlorophyceae (Chlorophytina) unter besonderer Berücksichtigung sporopollenin-ähnlicher Biopolymere. Zürich: Juris Druck & Verlag, Dissertation.

Brunner U, Honegger R (1985) Chemical and ultrastructural studies on the distribution of sporopollenin-like biopolymers in 6 genera of lichen phycobionts. Can J Bot 63: 2221-2230

Bubrick P, Galun M (1980a) Proteins from the lichen Xanthoria parietina which bind to phycobiont cell walls. Correlation between binding patterns and cell wall cytochemistry. Protoplasma 104: 167-173

Bubrick P, Galun M (1980b) Symbiosis in lichens: differences in cell wall properties of freshly isolated and cultured phycobionts. FEMS Microbiol Lett 7: 311-313

Bubrick P, Galun M, Frensdorff A (1981) Proteins from the lichen Xanthoria parietina which bind to phycobiont cell walls. Localization in the intact lichen and cultured mycobiont. Protoplasma 105: 207-211

Bubrick P, Galun M, Ben-Yaacov M, Frensdorff A (1982) Antigenic similarities and differences between symbiotic and cultured phycobionts from the lichen Xanthoria parietina. FEMS Microbiol Lett 13: 435-438

Bubrick P, Galun M, Frensdorff A (1984) Observations on free-living Trebouxia de Puymaly and Pseudotrebouxia Archibald, and evidence that both symbionts from Xanthoria parietina (L.)Th.Fr. can be found free-living in nature. New Phytol 97: 455-462

Bubrick P, Frensdorff A, Galun M (1985) Selectivity in the lichen symbiosis. In: DH Brown (ed): Lichen Physiology and Cell Biology. New York: Plenum, pp. 319-334

Büdel B (1987) Zur Biologie und Systematik der Flechtengattungen Heppia und Peltula im südlichen Afrika. Bibl Lichenol vol 23. Berlin: J. Cramer

Cole GT, Pope LM (1981) Surface wall components of Aspergillus niger conidia. In: G Turian, HR Hohl (eds): The Fungal Spore. Morphogenetic controls. London: Academic Press, pp. 195-215

Englund B (1977) The physiology of the lichen Peltigera aphthosa, with special reference to the blue-green phycobiont (Nostoc sp.). Physiol Plant 41: 298-304

Eriksson O, Hawksworth DL (1986) Outline of the Ascomycetes - 1986. Systema ascom 5: 185-324

Ferrari TE, Best V, More TA, Comstock P, Muhammad A, Wallace H (1985) Intercellular adhesions in the pollen-stigma system: pollen capture, grain binding and tube attachments. Am J Bot 72: 1466-1474

Gahan PB (1984) Plant Histochemistry and Cytochemistry. London: Academic Press

Gärtner G (1985) Taxonomische Probleme bei den Flechtenalgengattungen Trebouxia und Pseudotrebouxia. Phyton 25: 101-111

Geyer M (1985) Hochdruck-Flüssigkeits-Chromatographie (HPLC) von Flechten-Sekundärstoffen. Dissertation Universität Essen

Goebel K von (1926a) Die Wasseraufnahme der Flechten. Ber deutsch Bot Ges 44: 158-161

Goebel K von (1926b) Morphologische und biologische Studien. Ein Beitrag zur Biologie der Flechten. Annls jardin bot Buitenzorg 36: 1-83

Greenhalgh GN, Anglesea D (1979) The distribution of algal cells in lichen thalli. Lichenologist 11: 283-292

Hawksworth DL (1988) The variety of fungal-algal symbioses, their evolutionary significance, and the nature of lichens. Bot J Linn Soc: in press

Hawksworth DL, Hill DJ (1984) The lichen-forming fungi. Glasgow: Blackie

Heslop-Harrison J, Heslop-Harrison Y (1983) Pollen-stigma interaction in the Leguminosae: the organisation of the stigma in Trifolium pratense L. Ann Bot 51: 571-585

Hill DJ (1972) The movement of carbohydrate from the alga to the fungus in the lichen Peltigera polydactyla. New Phytol 71: 31-39

Hill DJ (1985) Changes in photobiont dimensions and numbers during co-development of lichen symbionts. In: DH Brown (ed): Lichen Physiology and Cell Biology. New York: Plenum, pp. 303-317

Honegger R (1982) Cytological aspects of the triple symbiosis in Peltigera aphthosa. J Hattori Bot Lab 52: 379-391

Honegger R (1984) Cytological aspects of the mycobiont-phycobiont relationship in lichens. Lichenologist 16: 111-127

Honegger R (1985) Fine structure of different types of symbiotic relationships in lichens. In: DH Brown (ed): Lichen Physiology and Cell Biology. New York: Plenum Press, pp. 287-302

Honegger R (1986a) Ultrastructural studies in lichens. I. Haustorial types and their frequencies in a range of lichens with trebouxioid phycobionts. New Phytol 103: 785-795

Honegger R (1986b) Ultrastructural studies in lichens. II. Mycobiont and photobiont cell wall surface layers and adhering crystalline lichen products in four Parmeliaceae. New Phytol 103: 797-808

Honegger R (1987) Questions about pattern formation in the algal layer of lichens with stratified (heteromerous) thalli. In: E Peveling (ed): Progress and Problems in Lichenology in the Eighties. Bibl Lichenol 25:

59-71. Berlin: J. Cramer
Honegger R, Brunner U (1981) Sporopollenin in the cell wall of Coccomyxa and Myrmecia phycobionts of various lichens: an ultrastructural and chemical investigation. Can J Bot 59: 2713-2734
Jacobs JB, Ahmadjian V (1973) The ultrastructure of lichens. V. Hydrothyria venosa, a freshwater lichen. New Phytol 72: 155-160
Jahns HM (1984) Morphology, Reproduction and water relations - a system of morphogenetic interactions in Parmelia saxatilis. Nova Hedwigia Beih 79: 715-737
Kinraide WTB, Ahmadjian V (1970) The effects of usnic acid on the physiology of two cultured species of the lichen alga Trebouxia Puym. Lichenologist 4: 234-247
Law R, Lewis DH (1983) Biotic environments and the maintenance of sex - some evidence from mutualistic symbioses. Biol J Linn Soc 20:249-276
Legaz ME (1985) The regulation of urea biosynthesis. In: C Vicente, DH Brown, ME Legaz (eds): Surface Physiology of Lichens. Madrid: Complutense University Press, pp. 57-73
Lemke PA, Ellison JR, Marino R, Morimoto B, Arons E, Kohman P (1975) Fluorescent Feulgen staining of fungal nuclei. Exptl Cell Res. 96: 367-373
McLaughlin DJ (1982) Basidial and basidiospore development. In: K Wells, E Wells (eds): Basidium and basidiocarp. New York: Springer, pp. 37-74
Oberwinkler F (1984) Fungus-alga interactions in basidiolichens. Nova Hedwigia Beih 79: 739-774
Paran N, Ben-Shaul Y, Galun M (1971) Fine structure of the blue-green phycobiont and its relation to the mycobiont in two Gonohymenia lichens. Arch. Mikrobiol. 76: 103-113
Peveling E (1970) Die Darstellung von Oberflächenstrukturen von Flechten mit dem Raster-Elektronenmikroskop. Ber dtsch Bot Ges NF 4: 89-101
Peveling E (1973) Vesicles in the phycobiont sheath as possible transfer structures between the symbionts in the lichen Lichina pygmaea. New Phytol 72: 343-345
Plessl A (1963) Ueber die Beziehungen von Pilz und Alge im Flechtenthallus. Oest Bot Z 110: 194-269
Richardson DHS (1985) The surface physiology of lichens with particular reference to carbohydrate transfer between the symbionts. In: C Vicente, DH Brown, ME Legaz (eds): Surface Physiology of Lichens. Madrid: Complutense University Press, pp. 25-55
Scannerini S, Bonfante-Fasolo P (1983) Comparative ultrastructural analysis of mycorrhizal associations. Can J Bot 61: 917-943
Smith DC (1978) What can lichens tell us about real fungi? Mycologia 70: 915-934
Smith DC, Douglas AE (1987) The Biology of Symbiosis. London: Edward Arnold
Spector DL, Jensen TE (1977) Fine structure of Leptogium cyanescens and its cultured phycobiont Nostoc commune. Bryologist 80: 445-460
Tschermak E (1941) Untersuchungen über die Beziehungen von Pilz und Alge im Flechtenthallus. Oest Bot Z 90: 233-307
Tschermak-Woess E (1978) Myrmecia reticulata as a phycobiont and free-living - free-living Trebouxia - the problem of Stenocybe septata. Lichenologist 10: 69-79
Tschermak-Woess E (1983) Das Haustorialsystem von Dictyonema kennzeichnend für die Gattung. Plant Syst Evol 143: 109-115
Tschermak-Woess E, Bartlett J, Peveling E (1983) Lichenothrix riddlei is an ascolichen and also occurs in New Zealand - light and electron microscopical investigations. Plant Syst Evol 143: 109-115
Vicente C (1985) Surface physiology in lichens: facts and concepts. In: C Vicente, DH Brown, ME Legaz (eds): Surface Physiology in Lichens. Madrid: Complutense University Press, pp. 11-24.

Host - fungus interactions in ectomycorrhizae

Y. Piché (1), R.L.Peterson (2), and H.B. Massicotte (2)

1. Centre de Recherche en Biologie Forestière, Département des Sciences Forestières Foresterie et Géodésie, Université Laval, Québec, Canada G1K 7P4
2. Department of Botany, University of Guelph, Guelph, Ontario, Canada N1G 2W1

I. Introduction

Ectomycorrhizae are symbiotic associations between fungi, primarily members of the Basidiomycotina and a few Ascomycotina, and roots of most important forest species in temperate regions and a few forest species in tropical ecosystems (Malloch et al., 1980). The conservative estimate of 5000 mycobiont species involved in ectomycorrhizal associations (Malloch et al., 1980) will no doubt be altered as research into this important symbiotic association is extended, particularly into tropical regions. If one considers that strains of a particular fungal species may differ in the way they interact with a host species (Kropp et al., 1987) and the number of hybrids and cultivars of host species in existence, then the number of possible ectomycorrhizal types is enormous.

The interaction of ectomycorrhizal fungi with plant roots is

NATO ASI Series, Vol. H17
Cell to Cell Signals in Plant, Animal and Microbial Symbiosis. Edited by S. Scannerini et al.

characterized structurally by the formation of a mantle of fungal hyphae encasing the root surface and a network of intercellular hyphae, the Hartig net, which forms most of the interface between fungus and root tissue (see Plate 1). Hyphae emanating from the outer mantle form the extraradical mycelium, an absorbing interface with the soil. Some of these hyphae may form sclerotia, complex aggregates of hyphae organized into storage structures, and sporocarps (sporophores), reproductive structures forming spores, which may be epigeous or hypogeous. Since ectomycorrhizal fungi are intimately associated with roots and form an interface between root tissue and the soil, nutrients from the soil pass through the fungal layer before reaching root cells (Harley, 1978). Because of the prevalence and importance of ectomycorrhizae in the vast tracts of coniferous and angiosperm forests, it is essential to gain a better understanding of the interaction between host roots and their mycosymbionts, particularly if these associations are to be manipulated to provide maximum growth responses of trees.

As pointed out by Gianinazzi- Pearson (1984), before a mycorrhizal organ can form, an appropriate fungus and plant must form a compatible pair, which implies that there may be a certain degree of recognition and specificity involved. This topic has been discussed in several recent reviews (Harley and Smith, 1983; Gianinazzi - Pearson, 1984; Harley, 1985) and it is evident that compared, for example, to the Rhizobium - legume symbiosis, there is very little firm evidence concerning recognition, specificity, and compatibility in ectomycorrhizal associations. Although it is not our intent to restate all the arguments for and against these phenomena being operative in ectomycorrhizae, a few salient points need to be made. One of the striking features of ectomycorrhizae is that most of the mycobionts involved are not host specific, at least as revealed by laboratory synthesis experiments. More specificity might exist in field situations but this has been poorly documented. One exception to lack of host specificity is shown by the gasteromycete, Alpova diplophloeus which appears to be specific to the genus, Alnus (Molina, 1979), but when one considers the total number of fungal species able to form ectomycorrhizae, there is still, overall, a very low specificity involved in this association.

It is of interest that although there appears to be low specificity among ectomycorrhizal fungi, the host involved may control the type of development which follows root colonization. For example, many species forming typical ectomycorrhizae with members of the Pinaceae form arbutoid mycorrhizae with ericaceous species (Molina and Trappe, 1982). Also, structure and physiology of a particular host - fungus combination may depend on environmental factors such as light, temperature, pH and water status in which the mycorrhiza develops, indicating that there is some degree of plasticity in the final morphology.

The understanding of changes in both symbionts during the establishment of an ectomycorrhiza relies on a suitable method of synthesis. The method must provide root material at various stages of development so that the interaction between symbionts at both the cellular and subcellular level can be determined. The growth pouch technique (Fortin et al., 1980) has been very useful in studies of both gymnosperm and angiosperm ectomycorrhizae; some of the results relevant to events involved in root colonization will be summarized in this paper. Also, some recent work (Kropp et al., 1987), comparing the formation of ectomycorrhizae using either monokaryotic or dikaryotic strains of Laccaria bicolor, will be discussed.

II. Synthesis of ectomycorrhizae in growth pouches

The advantages of using the growth pouch system for developmental studies of ectomycorrhizae include the ease of establishing seedlings and colonizing them with pure cultures of ectomycorrhizal fungi. Over one hundred host-fungus combinations have been tested to date (Godbout and Fortin, 1983; Samson and Fortin, 1986; Massicotte et al.,1987). The ease in documenting initial contact between hyphae and the root surface and subsequent stages in ectomycorrhiza formation is another advantage of the growth pouch technique, and is essential when studying recognition and compatiblity. The non-sterility of this technique makes it unsuitable, however, for the identification of substances (signals) involved in recognition events since contaminants such as bacteria and algae are likely to be present. A considerable amount of information concerning ectomycorrhiza morphology (Figure 1), topography of the mantle using scanning electron microscopy (Figure 2), and basic features of ectomycorrhizae using light microscopy (Figure 3) and transmission electron microscopy (Figure 4) has been obtained for several host-fungus combinations (see Massicotte et al., 1986, 1987; Melville et al., 1987b, for detailed examples). Information obtained from two host-fungus combinations pertinent to the topics of recognition and compatibility is summarized briefly in the following sections.

III. *Pinus strobus* - *Pisolithus tinctorius*

One of the earliest responses of both root cells and fungal hyphae in this interaction is the production of substances that stain positively by the Thiery reaction for polysaccharides (Piche et al., 1983a, b). These substances may be similar to the fibrillar material produced by hyphae of the ectomycorrhizal fungus, *Terfezia leptoderma* as they approach the root surface of Helianthemum salicifolium (Gianinazzi - Pearson, 1984). Whether these substances in either case are involved in recognition, or simply attachment of hyphae to the root surface, has not been determined. A striking feature of ectomycorrhiza establishment, including Pinus strobus - Pisolithus tinctorius, is the extensive colonization of short lateral roots once fungal hyphae contact the root surface, suggesting that these roots may be releasing exudates conducive to hypha proliferation (see Piche et al., 1983a).

Concurrent with or subsequent to hyphal proliferation on the root surface, penetration of the root by inner mantle hyphae occurs, initiating an intercellular Hartig net (Piche et al., 1983). Gianinazzi - Pearson (1984) argues that a form of recognition occurs at this stage since the fungus usually penetrates between root cells that are at a particular stage of development. Whether cells in this zone have an altered metabolism which favors fungal growth or a middle lamella consisting of substances which can be utilized readily as substrates for fungal growth is not known.

Pinus root cells do not show pronounced structural changes when interfaced with Hartig net hyphae of Pisolithus tinctorius, they are undoubtedly affected physiologically but this has not been determined. Short lateral roots of Pinus strobus, like roots of other pine species, respond to ectomycorrhizal fungi by dichotomizing (see Piche et al., 1982), a developmental event that can be triggered by treatment of roots with exogenous auxin or ethephon, a source of ethylene (Rupp and Mudge, 1985). Repeated dichotomies can occur resulting in clusters of roots and an increase in absorbing surface.

IV. Alnus crispa - Alpova diplophloeus

The gasteromycete, Alpova diplophloeus, forms ectomycorrhizae only

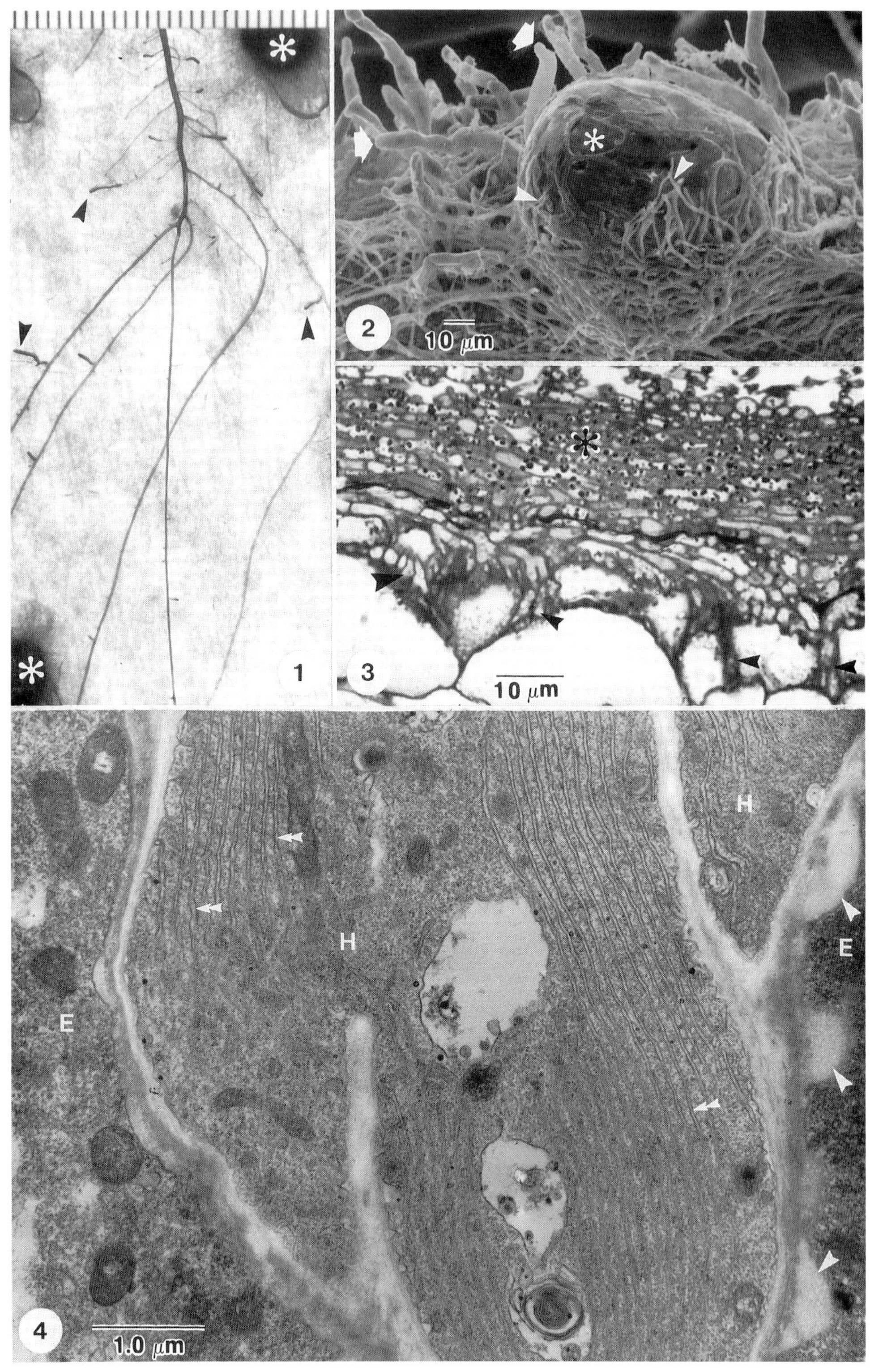
1
2
10 μm
3
10 μm
H
H
E
E
4
1.0 μm

Plate 1. Figures 1-4. Synthesis of ectomycorrhizae in growth pouches.

Figure 1. Ectomycorrhizae synthesized between *Eucalyptus pilularis* and *Pisolithus tinctorius*. Arrowheads indicate mycorrhizae. Fungal plug inocula (*) are evident. Scale in mm.

Figure 2. Scanning electron micrograph of early stage in the colonization of *Betula alleghaniensis* with *Paxillus involutus*. Fungal hyphae (arrowheads) are attached onto the root surface. Root hairs are indicated by arrows.

Figure 3. Light micrograph of longitudinal section of an *Alnus rubra* - *Alpova diplophloeus* ectomycorrhiza. A well developed mantle (*) and a paraepidermal Hartig net (arrowheads) have developed.

Figure 4. Transmission electron micrograph of a transverse section of an *Alnus crispa* - *Alpova diplophloeus* ectomycorrhiza showing the interface between epidermal cells (E) and Hartig net hyphae (H). Wall ingrowths (arrowheads) are evident in the epidermal cells while numerous cisternae of rough endoplasmic reticulum (double arrowheads) are present in the hyphae.

with species in the genus Alnus (Molina, 1981) and it has been of interest therefore, to study the interaction between the symbionts Alnus crispa - Alpova diplophloeus to determine if this level of specificity results in any particular specialization at the cellular level. A very detailed account of the ontogeny of this association has been published previously (Massicotte et al., 1986) and several conclusions can be drawn from this work. Cells of both symbionts in the interface region respond by altering their cytoplasm; Hartig net hyphae change dramatically by synthesizing large amounts of rough endoplasmic reticulum. Root cells respond by synthesizing additional wall material in the form of ingrowths along cell walls interfacing with Hartig net hyphae which themselves branch irregularly. Hartig net formation in general may involve several recognition steps as exemplified in recent studies of this process. Blasius et al. (1986) and Melville et al. (1987a) have shown that hyphae in this region of the root change from a linear type of growth to a less polarized growth which results in a complex, interdigitated system.

An important feature of the Alnus crispa - Alpova diplophloeus ectomycorrhiza is that senescence, as determined by changes in cellular organelles in both symbionts, and the deposition of secondary wall material over the wall ingrowths, occurs very quickly. This implies that perhaps only a restricted portion of the mycorrhizal organ is involved in transport of nutrients.

V. New approaches in studying ectomycorrhizae

Although a detailed understanding of the genetics of many higher fungi, particularly genera such as Neurospora and Sordaria (Kendrick, 1985) exists, this is not the case for ectomycorrhizal fungi. Genetic studies on ectomycorrhizal species belonging to the Basidiomycotina have been impeded by the difficulties encountered in germinating basidiospores. A recent method for basidiospore germination has, however, been developed by Fries (1983). Basidiospores are germinated on agar containing charcoal and mycelium of the same fungal species as the basidiospores to be germinated. Since a high germination rate can be obtained with this method, it is now possible to study mating systems of ectomycorrhizal fungi (Kropp et al., 1987). These authors have reported that monokaryons and reconstituted dikaryons of Laccaria bicolor are capable of forming ectomycorrhizae, but there is considerable variation in the type of association formed. Some isolates are vigorous colonizers while others had lost the ability to form ectomycorrhizae. This is an important observation and this approach provides a method to explore intraspecific variability in ectomycorrhizal fungal species and the possibility that specific genes are involved in the control of the establishment of the ectomycorrhizal association.

The expression of specific genes results in the synthesis of new proteins and therefore it should be possible to monitor changes in protein profiles at various stages in the establishment of an ectomycorrhiza. Some preliminary results by Hilbert et al. (1987) are very encouraging. These authors have shown that in the ectomycorrhiza formed between Eucalyptus globulus and Paxillus involutus, a group of new polypeptides (ectomycorrhizins) appeared that were not present in either symbiont growing in isolation. It should now be possible, using those ectomycorrhizal associations in which all the events leading to a mature ectomycorrhiza have been documented carefully, to monitor changes in proteins at each step. This might provide some insight into the genetic control of this complex symbiotic association.

VI. Discussion

Since evidence is lacking that a single strain of ectomycorrhizal fungus is specific for a single host genotype, it is probably premature at this time to think of ectomycorrhizal associations in terms of the molecular events which occur, for example during the establishment of the *Rhizobium* - legume symbiosis. In spite of this, there are some observations concerning host - ectomycorrhizal fungus interactions that may be relevant to recognition and compatibility. At present, compatibility in this symbiotic association can be discussed in terms of structural events only, since there is a paucity of information correlating structural features with physiological events. The term 'compatibility' when used in reference to ectomycorrhizal associations usually implies the attainment of a particular structural integration of the two symbionts involving inner mantle and Hartig net hyphae and associated root cells.

Using this criterion of compatibility, a general picture of compatibility and the converse, incompatibility responses of host-fungus combinations has been established. Evidence is based on microscopical observations similar to those summarized for *Pinus* and *Alnus* mycorrhizae in preceding sections. The events leading to the establishment of a compatible ectomycorrhizal association are represented schematically in Figure 5, which is a modified version of an excellent discussion on plant-pathogen interactions (Heath, 1981). In attempting to unravel the events which occur during mycorrhizal establishment it may be useful, and indeed, essential, to think in terms of discrete steps that might be amenable to experimental manipulation.

For a successful interaction to occur, there must be accomodation of the fungus and host at the metabolic level (Heath, 1981) so that a series of events can take place leading to the formation of the functioning ectomycorrhiza. The first step, the attraction and attachment of fungal hyphae to the root surface, may involve the secretion of substances by both symbionts (Piche et al., 1982; 1983a, b), some of which may repress the production of anti-fungal compounds which would inhibit further development of the ectomycorrhizal fungal hyphae. Harley and Smith (1983) have suggested that a non-specific binding of fungal hyphae to root cell walls may be mediated by lectins; although this is an attractive hypothesis, there is no evidence for it. Any discussion of the involvement of specific signal molecules which precede the attachment phase would be purely speculative at this time. It is obvious from the many ectomycorrhizae studied (Massicotte et al., 1987) that subsequent to the attachment of the first hyphae onto the surface of short roots there is a rapid proliferation of hyphae. The fungal hyphae on the root surface may release substances that induce root cells to leak sugars onto the surface which are then absorbed by hyphae. The proliferation of hyphae would, in turn, lead to increased nutrient uptake for the developing root. Fungal proliferation is organized and varies with host-fungus combination (Massicotte et al., 1987; Melville et al., 1987b), to form a mantle and Hartig net typical for each ectomycorrhiza.

Since there appears to be lack of strict specificity in ectomycorrhizal systems, a condition as simple as the presence of the appropriate exudates for fungal proliferation may be operative in determining whether a particular host - fungus combination is successful. It would, therefore, be of considerable interest to have more information concerning the nature of root exudates released by a variety of woody plant species. Gianinazzi - Pearson (1984) points out that although a host-fungus combination may develop a mantle and Hartig net, structural features that indicate they are compatible, they may in fact lack functional compatibility i.e. they may fail to show a physiological relationship that enhances plant growth.

In order for a functional ectomycorrhiza to form, it is likely that alterations in both symbionts must occur. The most pronounced alterations at the cellular and subcellular levels occur in the interface established between Hartig net hyphae (and perhaps inner mantle hyphae) and root cells. Structural changes in root cells may either be very subtle and involve changes in organelle number or structure, or in some cases, may be more evident, as in the elaboration of host cell wall in the form of ingrowths that increase the absorbing surface of the cell (Ashford and Allaway, 1982; Massicotte et al., 1986). Gross morphological changes occur infrequently but in pine ectomycorrhizae, root apices dichotomize repeatedly. Hartig net hyphae branch extensively to form a labyrinthine system; whether this is a requirement for exchange of substances between the symbionts is not known, but it does seem to characterize a number of ectomycorrhizal systems that have been studied (Blasius et al., 1986 ; Melville et al., 1987a), and it may turn out to be a reliable indicator of compatibility between host - fungus partners. As more information related to interface formation is obtained, it may become possible to predict, based on the structure of the two symbionts in this region, which host - fungus combinations show compatibility.

The final step in the formation of an ectomycorrhiza is the senescence of both symbionts. This has not received much attention in the literature but from the detailed study of *Alnus crispa* - *Alpova diplophloeus* ectomycorrhizae (Massicotte et al., 1986), it is apparent that this step involves more than simply a degradative process. In this ectomycorrhiza, senescence is characterized by the synthesis of additional host cell wall material to cover the wall ingrowths induced by the presence of the fungus.

In addition, changes in cellular organelles of both symbionts indicate that the association may become less functional.

One can assume that if the interaction between a particular host species and ectomycorrhizal fungus is unsuccessful then the block could occur at any of the steps shown in Figure 5. Responses of host root cells that have been interpreted to indicate incompatibility with a fungal species include lignification and subsequent disorganization of cortical cells (Molina and Trappe, 1982), accumulation of phenolic compounds in root cells (Molina, 1981; Molina and Trappe, 1982; Malajczuk et al., 1984) and failure to form a mantle or Hartig net (Duddridge, 1985; Kropp et al., 1987). Manipulation of the medium in which mycorrhizae are synthesized by either adding or deleting glucose can also induce alterations in the interface established, some of which are similar to incompatibility responses (Duddridge, 1985).

Research concerning the control of each of the steps in the establishment of an ectomycorrhiza will only advance once the genetics of ectomycorrhizal fungi is better understood. It should be possible then to produce mutants which lack the genetic information for one or more of the steps involved in ectomycorrhiza establishment and thereby determine the role played by the mycosymbiont in each developmental event.

Acknowledgements

We thank Lewis Melville for help with the manuscript, Dr. Anne Ashford for useful discussions, Melanie Chapple for proof-reading the manuscript, and the Natural Sciences and Engineering Research Council of Canada for financial support.

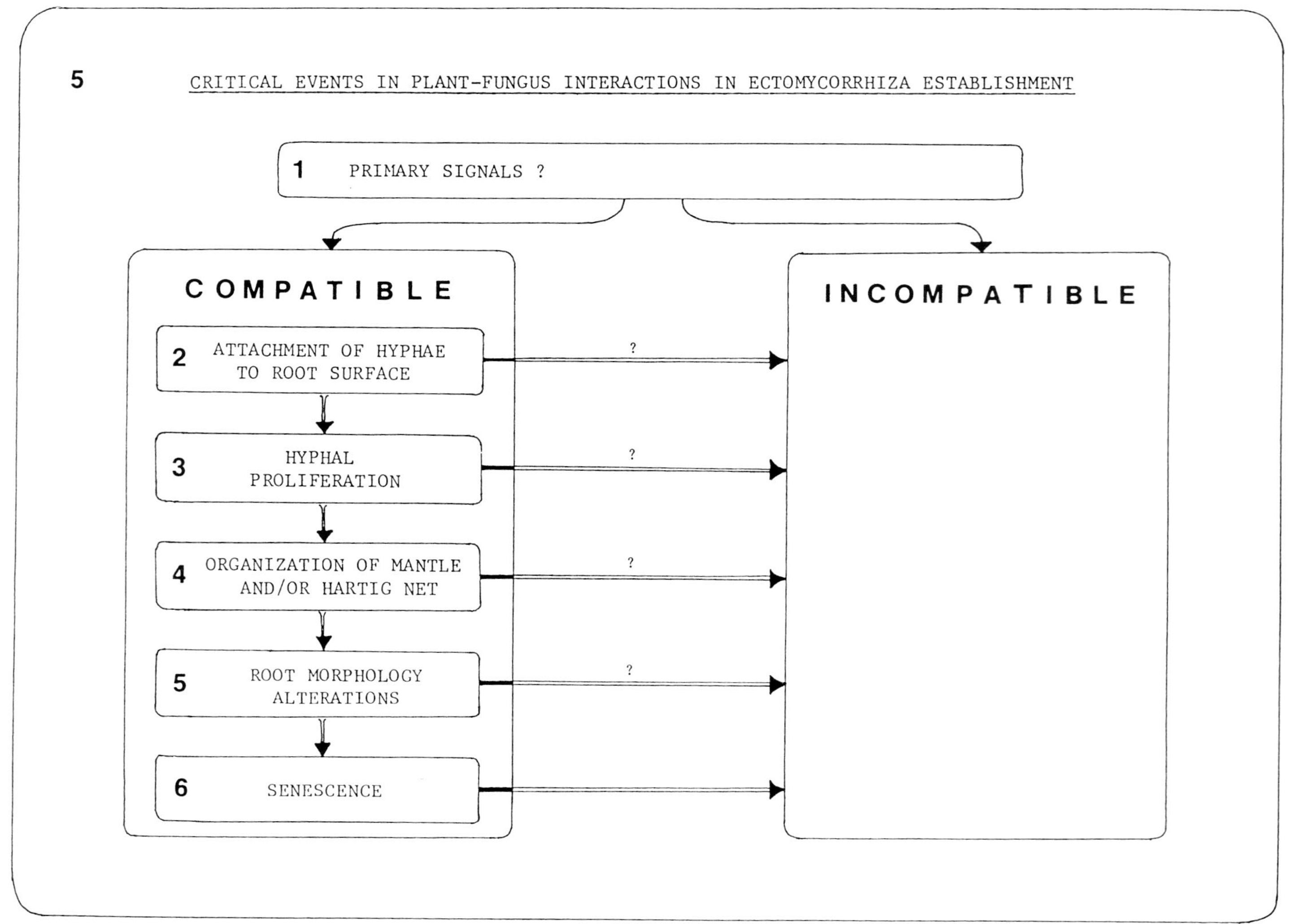
5
CRITICAL EVENTS IN PLANT-FUNGUS INTERACTIONS IN ECTOMYCORRHIZA ESTABLISHMENT
1 PRIMARY SIGNALS ?
COMPATIBLE
INCOMPATIBLE
2 ATTACHMENT OF HYPHAE TO ROOT SURFACE
3 HYPHAL PROLIFERATION
4 ORGANIZATION OF MANTLE AND/OR HARTIG NET
5 ROOT MORPHOLOGY ALTERATIONS
6 SENESCENCE
?
?
?
?

References

ASHFORD, A.E., and W.G. ALLAWAY. 1982. A sheathing mycorrhiza on Pisonia grandis R. Br. (Nyctaginaceae) with development of transfer cells rather than a Hartig net. New Phytol. 90: 511-517.

BLASIUS, D., W. FEIL, I. KOTTKE, and F. OBERWINKLER. 1986. Hartig net structure and formation in fully ensheathed ectomycorrhizas. Nord. J. Bot. 6: 837-842.

DUDDRIDGE, J.A.. 1985. A comparative ultrastructural analysis of the host-fungus interface in mycorrhizal and parasitic associations. In Developmental biology of higher fungi. Edited by D. Moore, L.A. Casselton, D.A. Wood, and J.C. Frankland. Cambridge University Press, Cambridge. pp. 141-173.

FORTIN, J.A., Y. PICHE, and M. LALONDE. 1980. Technique for the observation of early morphological changes during ectomycorrhiza formation. Can. J. Bot. 58: 361-365.

FRIES, N. 1983. Spore germination, homing reaction and intersterility groups in Laccaria laccata (Agaricales). Mycologia 75: 221-227.

GIANINAZZI-PEARSON, V. 1984. Host-fungus specificity, recognition and compatibility in mycorrhizae. In Genes involved in plant microbe interactions. Edited by D.P.S. Verma and T.H. Hohn. Springer-Verlag, Wien, New York. pp. 225-253.

GODBOUT, C., and J.A. FORTIN. 1983. Morphological features of synthesized ectomycorrhizae of Alnus crispa and A. rugosa. New Phytol. 94: 249-262.

HARLEY, J.L.. 1978. Nutrient absorption by ectomycorrhizae. Physiol. Veg. 16: 533-545.

HARLEY, J.L. 1985. Specificity and penetration of tissues by mycorrhizal fungi. Proc. Indian Acad. Sci. (Plant Sci.) 94: 99-109.

HARLEY, J.L. and S.E. SMITH. 1983. Mycorrhizal Symbiosis. Academic Press, New York. 483p.

HEATH, M.C. 1981. A generalized concept of host-parasite specificity . Phytopathol. 71: 1121-1123.

HILBERT. J.L., V. GAUDIN, F. MARTIN, and F. LAPEYRIE. 1987. Identification of "ectomycorrhiza-specific" proteins (ectomycorrhizins) involved in the development of Paxillus-Eucalyptus symbiosis. Proc. 7th N. Am. Conf. on Mycorrhizae. Gainesville, Florida.(in press).

KENDRICK, B. 1985. The Fifth Kingdom. Mycologue publications. Waterloo, Canada. 363p.

KROPP, B.R., B.J. McAFEE, and J.A. FORTIN. 1987. Variable loss of ectomycorrhizal ability in monokaryotic and dikaryotic cultures of Laccaria bicolor. Can. J. Bot. 65: 500-504.

MALAJCZUK, N., R. MOLINA, and J.M. TRAPPE. 1984. Ectomycorrhiza formation in Eucalyptus. II. The ultrastructure of compatible and incompatible mycorrhizal fungi and associated roots. New Phytol. 96: 43-53.

MALLOCH, D.W., K.A. PIROZYNSKI, and P.H. RAVEN. 1980. Ecological and evolutionary significance of mycorrhizal symbioses in vascular plants (a review).Proc. Natl. Acad. Sci. USA. 77: 2113-2118.

MASSICOTTE, H.B., R.L. PETERSON, C.A. ACKERLEY, and Y. PICHE. 1986. Structure and ontogeny of Alnus crispa- Alpova diplophloeus ectomycorrhizae. Can. J. Bot. 64: 177-192.

MASSICOTTE H.B., L.H. MELVILLE, and R.L. PETERSON. 1987. Scanning electron microscopy of ectomycorrhizae - potential and limitations. Scanning Electron Microscopy. (In press).

MELVILLE, L.H., C.A. ACKERLEY, H.B. MASSICOTTE, and R.L. PETERSON. 1987a. Morphogenesis and model of Hartig net in two ectomycorrhizal systems.

Proc. 7th N. Am. Conf. on Mycorrhizae (May 2-8). Gainesville, Fl. (in press).

MELVILLE, L.H., H.B. MASSICOTTE, and R.L. PETERSON. 1987b. Morphological variations in developing ectomycorrhizae of Dryas integrifolia and five fungal species. Scanning Electron Microscopy (In press).

MOLINA R. 1979. Pure culture synthesis and host specificity of red alder mycorrhizae. Can. J. Bot. 57: 1223-1228.

MOLINA R. 1981. Ectomycorrhizal specificity in the genus Alnus. Can. J. Bot. 59: 325-334.

MOLINA, R. and J.M. TRAPPE. 1982. Lack of mycorrhizal specificity by ericaceous hosts Arbutus menziesii and Arctostaphylos uva-ursi. New Phytol. 90: 495-509.

PICHE, Y., J.A. FORTIN, R.L. PETERSON, and U. POSLUSZNY. 1982. Ontogeny of dichotomizing apices in mycorrhizal short roots of Pinus strobus. Can. J. Bot. 60: 1523-1528.

PICHE, Y., R.L. PETERSON, and C.A. ACKERLEY. 1983a. Early development of ectomycorrhizal short roots of pine. Scanning Electron Microscopy, III, 1467-1474.

PICHE, Y., R.L. PETERSON, M.J. HOWARTH, and J.A. FORTIN. 1983b. A structural study of the interaction between the ectomycorrhizal fungus, Pisolithus tinctorius and Pinus strobus roots. Can. J. Bot. 61: 1185-1193.

RUPP, L.A., and K.W. MUDGE. 1985. Ethephon and auxin induce mycorrhiza-like changes in the morphology of root organ cultures of mugo pine. Physiol. Plant. 64: 316-322.

SAMSON, J., and J.A. FORTIN. 1986. Ectomycorrhizal fungi of Larix laricina and the interspecific and intraspecific variation in response to temperature. Can. J. Bot. 64: 3020-3028.

MORPHOLOGICAL INTEGRATION AND FUNCTIONAL COMPATIBILITY BETWEEN SYMBIONTS IN VESICULAR ARBUSCULAR ENDOMYCORRHIZAL ASSOCIATIONS

V. Gianinazzi-Pearson and S. Gianinazzi
Laboratoire de Phytoparasitologie, Station d'Amélioration des Plantes, INRA
BV 1540, 21034 Dijon Cédex, France

INTRODUCTION

Mutualistic symbiotic associations between plant roots and soil fungi, mycorrhizae, commonly occur throughout the plant kingdom. Mycorrhizae can be divided into morphologically distinct groups whose structural and functional diversity is determined by the plant and fungal taxa involved (see Harley and Smith, 1983 ; Gianinazzi-Pearson, 1984). It is, nevertheless, remarkable that the large majority of plant species form the same type of mycorrhizal association, vesicular-arbuscular (VA) endomycorrhizae (Harley and Harley, 1987) and this raises the question of whether compatibility systems have been specifically developed towards the fungi involved. Investigations of symbiont interactions in this type of mycorrhizal association are hampered by the fact that the fungal associates, belonging to four genera in the Endogonales (Trappe, 1982), cannot be grown in pure culture so that, apart from a preliminary study on axenically infected clover (Pons, 1984), most information comes from electron microscope observations made on non-axenic associations.

Host-fungus relationships are particularly complex in VA endomycorrhizae compared to other types of mycorrhizae (Scannerini and Bonfante-Fasolo, 1983 ; Gianinazzi-Pearson, 1984). Contact between the fungal hyphae and host cells leads to a sequence of interactions which result in the establishment of an important endocellular biotrophic phase during which cellular relationships between the symbionts are functionally compatible. These are characterised by bidirectional nutrient exchange between the mycorrhizal associates : the plant acts as a carbon source for the fungus, whilst the fungus releases to the plant certain mineral nutrients, particularly phosphorus, that it has absorbed and translocated from the soil (Gianinazzi-Pearson and Gianinazzi, 1983 ; Harley and Smith, 1983).

NATO ASI Series, Vol. H17
Cell to Cell Signals in Plant, Animal and Microbial Symbiosis. Edited by S. Scannerini et al.

INFECTION MORPHOLOGY

Initial cell to cell contact between VA endomycorrhizal symbionts appears to be fortuitous and anything resembling adhesion or attachment of fungal hyphae to host cell walls, similar to that occurring in ericoid endomycorrhiza (Gianinazzi-Pearson et al., 1986) has never been observed, even under axenic conditions (Pons, 1984). However, when cell contact is established, fungal penetration of the host root is preceded by the formation of a more or less well-defined haustorium indicating that some kind of recognition phenomenon occurs at this early stage in mycorrhiza formation.

The VA endomycorrhizal fungi only develop in the unsuberised tissues of the host root ; fungal colonisation is restricted to epidermal and cortical host tissue and the hyphae never enter meristematic or vascular regions of the root. Although infection patterns can vary slightly depending on the fungus or host species involved (Abbott, 1982 ; Bonfante-Fasolo, 1984), the distribution of the hyphal system within roots is largely determined by the type of host tissue colonised (Figure 1).

Hyphal penetration of the outer cell layers of the host root can be intercellular or intracellular ; in the latter case fungi colonise the epidermal or outermost cortical cells often forming a simple, unbranched intracellular coil. After this initial infection stage, intercellular hyphae usually form with increasing frequency as the fungi spread into the inner cortex, although in certain plant species hyphal development may remain confined uniquely to an intracellular position (Kinden and Brown, 1975 ; Jacquelinet-Jeanmougin, 1986). A common feature of all VA endomycorrhizal infections, however, is the intense intracellular fungal development in the innermost layers of the cortex where the fungi form complex much-branched haustoria, which are called arbuscules, within the parenchymal host cells. Arbuscules are important structures since they are assumed to be involved in preferential nutrient transfer between the symbionts (Gianinazzi-Pearson and Gianinazzi, 1983 ; Harley and Smith, 1983). The mechanisms responsible for this induction of arbuscule formation in cortical parenchyma cells are unknown, but it is evident that any signal emission or signal receptor molecules that may be involved must not only specifically occur in these tissues, but also be common to a large number of plant species. After a few days, the arbuscule degenerates, hyphae empty and the walls collapse ; again, the reason for fungal senescence, whether it is provoked by the host cell or not, is not known although in this context it is interesting that VA endomycorrhizal infections can elicit phytoalexin production in roots (Morandi and Gianinazzi-Pearson, 1986).

These variations in the morphological integration between the VA endomycorrhizal symbionts are accompanied by changes in the nature of their cellular relationship. The ultrastructural aspects of several VA endomycorrhizal associations have already been described in detail (see Carling and Brown, 1982 ; Bonfante-Fasolo, 1984 ; Gianinazzi-Pearson, 1984 for references) ; the aim of this paper is to interpret the ultracytological events occurring at the different host-fungus interfaces in relation to the functional relationships that may exist between the symbionts. For this purpose, comparisons are also made with relevant interactions occurring in roots of non-hosts and in biotrophic associations between plants and fungal pathogens.

CELLULAR RELATIONSHIPS BETWEEN SYMBIONTS

Changes in the morphology of VA endomycorrhizal fungi as they develop in host tissues are accompanied by modifications in their wall metabolism. There is a thinning out of the hyphal wall with a simplification of both its structure and composition, which begins in the hyphal coils and accentuates with hyphal proliferation to reach a maximum in the haustorial arbuscule (Bonfante-Fasolo and Gianinazzi-Pearson, 1986 ; Gianinazzi-Pearson, 1986). This aspect of symbiont interaction is treated in detail elsewhere in this volume by P. Bonfante-Fasolo.

Host cells show no evident response to the presence of hyphae of the fungal symbiont in the intercellular spaces of the root tissue, but as soon as a hypha breaches the cell wall and comes into contact with the host protoplast, the host plasmalemma elongates and actively deposits wall material containing proteins and polysaccharides around the invading fungus (Dexheimer, Gianinazzi and Gianinazzi-Pearson, 1979 ; Holley and Peterson, 1979 ; Scannerini and Bonfante-Fasolo, 1979 ; Gianinazzi-Pearson <u>et al.</u>, 1981). This wall building activity appears to result from a host reaction to accomodate the increased surface area of its host protoplast, similar to that which occurs in elongating cells or protoplasts, and the underlying mechanisms may well be similar. This initial response to fungal colonization is common to all infected host cells ; the nature of the host-fungus interface varies, however, in function of the type of cellular relationship which is established between macro and microsymbiont.

Host-fungus interfaces in outer root tissues

In the case of intracellular coils developing in the outer cell layers of the root, the fungal wall is still relatively thick and chitinous and it is continuously separated from the host plasmalemma by the wall material of host origin. The latter forms a physical barrier separating the fungus from the host protoplast and in this respect the host-fungus interface resembles very much that established with intercellular hyphae. The protoplasmic contents of these host cells change little in response to fungal invasion ; only a thin layer of cytoplasm surrounds the hypha and starch grains remain intact. In this type of root cell, Mg.ATPase activity is characteristically associated with the peripheral plasmalemma (unpublished observations on sycamore, onion and leek roots) ; this persists in an infected cell but no, or very weak, ATPase activity can be revealed bound to the host membrane extending around the fungus (Figure 1). These features of the host-fungus interface do not indicate a special involvement in nutrient transfer from fungal to host cells. On the contrary, the fungal plasmalemma of both hyphal coils and intercellular hyphae can possess Mg. ATPase activity, so that an active transport system for nutrient absorption by the fungus seems to exist during these phases of the mycorrhizal association. The persistence of starch grains in the colonised host cells suggests, however, that the carbon demands of the fungal symbionts are probably limited in these cells.

Host-fungus interfaces in parenchymal cortical cells

This situation is very different from that found in the parenchymal host cells of the inner root cortex, where the endomycorrhizal fungi form arbuscules. Here, profound changes occur in the host protoplast : cytoplasmic contents increase several-fold (Cox and Tinker, 1976 ; Toth and Miller, 1984) with a concomitant decrease in the cell vacuole, starch grains disap-

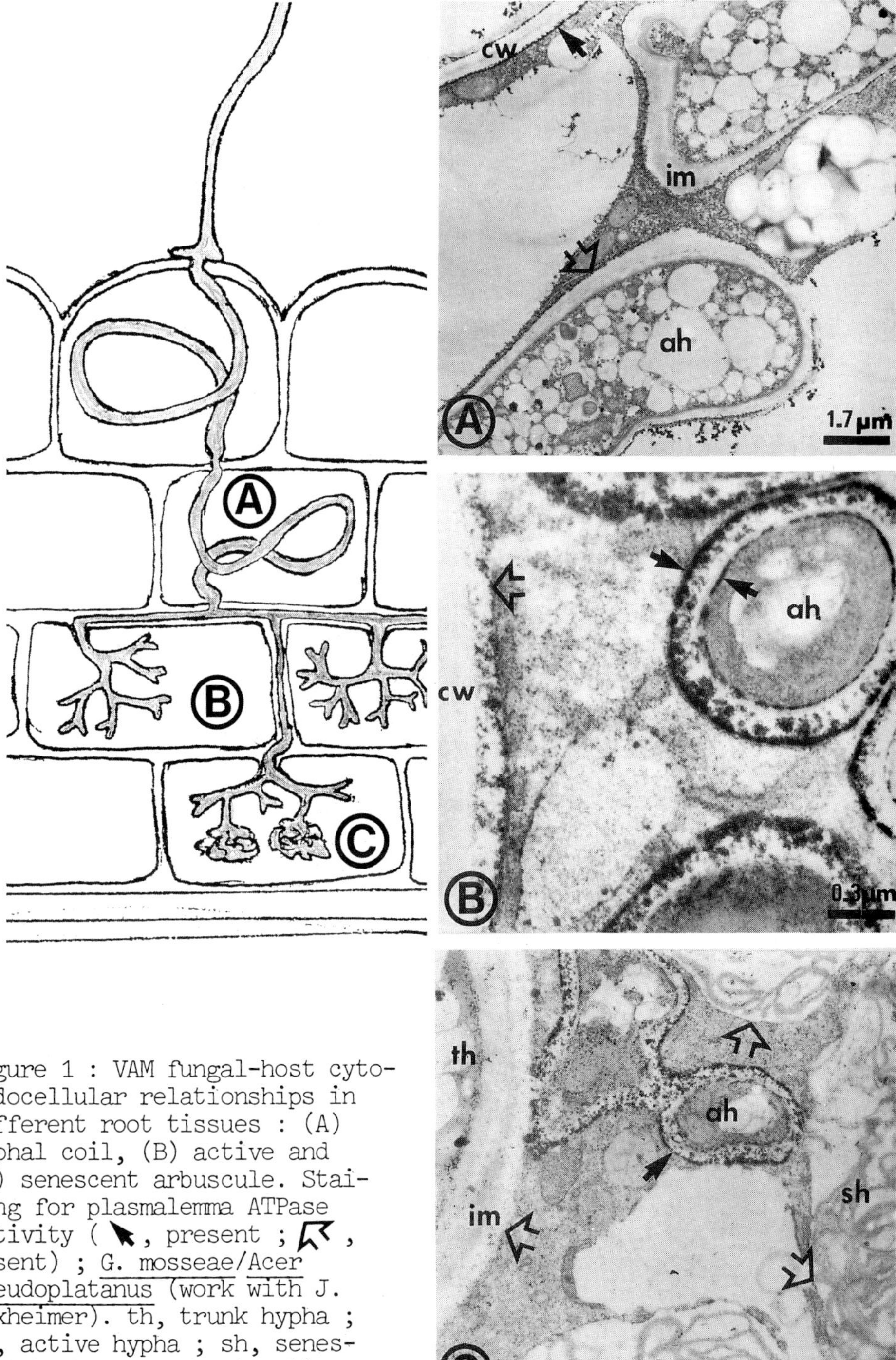

Figure 1 : VAM fungal-host cyto-endocellular relationships in different root tissues : (A) hyphal coil, (B) active and (C) senescent arbuscule. Staining for plasmalemma ATPase activity (↖, present ; ⇖, absent) ; G. mosseae/Acer pseudoplatanus (work with J. Dexheimer). th, trunk hypha ; ah, active hypha ; sh, senescent hypha ; cw, host wall ; im, interfacial material.

pear, cell organelles become more numerous, membrane systems increase and the host plasmalemma proliferates to closely surround the branching hyphae, creating an important surface of cellular contact between the symbionts (Figure 1). These aspects of the infected host cell are all signs of a metabolically active condition and recent observations indicate that they are associated with an intense transcriptional activity in the host nuclei (Berta et al., 1986).

As the fungal hyphae ramify deep into the host protoplast, the host-fungus interface is modified and becomes a physiologically complex zone. The fungal wall remains extremely thin and simplified over all the arbuscule ; it is tempting to interpret this specialised haustorium as a much extended hyphal tip (see P. Bonfante-Fasolo, this volume) over which nutrient absorption will be facilitated. Moreover, the amount of wall material deposited by the host cell against the invading fungus greatly decreases in the interfacial matrix around the fine arbuscule branches, so that the apoplast between fungal and host plasmalemma is much reduced and a close relationship is established between the cell surfaces of the symbionts. The change in the cell wall building activity of the host around the arbuscule does not appear to result from adverse alterations in the newly synthesized host plasmalemma, which shows the same morphology and cytochemical properties as the peripheral membrane (Dexheimer et al., 1979, 1982 ; Gianinazzi-Pearson et al., 1984 ; Gianinazzi-Pearson, 1986). Furthermore, neutral phosphatases, characteristic of sites of polysaccharide synthesis in animal and plant cells, become specifically localised along the host plasmalemma of the arbuscule interface (Gianinazzi et al., 1983 ; Jeanmaire et al., 1985), and plasmalemmasomes, interpreted as being involved in cell wall synthesis, are particularly frequent in the interfacial matrix around the fine arbuscule branches (Dexheimer et al., 1985). Both persist with hyphal senescence when host material accumulates again around the fungal remains. It therefore seems reasonable to conclude that the host plasmalemma surrounding arbuscular hyphae has a cell wall building activity but that the processes of wall deposition and organisation are somehow affected by the actively growing fungal hyphae. Diversion of wall precursors or localised hydrolysis of wall material (Jacquelinet-Jeanmougin, 1986) could provide an important supplementary source of carbohydrate to the endomycorrhizal fungus (Dexheimer et al., 1982, 1986 ; Harley and Smith, 1983). Mg.ATPase activity is very frequently localised along the plasmalemma of the living arbuscule branches, indicating the possibility of active transmembrane transport of solutes into the fungus.

Contrary to the outer cell layers of the root, the plasmalemma of uninfected parenchymal cells characteristically possesses little or no Mg.ATPase activity (unpublished observations on sycamore, onion, leek and soybean roots). When an endomycorrhizal fungus infects these cells, the peripheral plasmalemma remains inactive but the newly synthesized host plasmalemma surrounding the fine arbuscule branches becomes the site of intense Mg.ATPase activity (Figure 1b) (Dexheimer et al., 1982, 1986 ; Marx et al., 1982 ; Gianinazzi-Pearson, 1986). Inhibition of this enzyme activity by diethylstilbestrol (Marx et al., 1982) and vanadate (unpublished results) confirms that it is due to a plasmalemma-bound specific ATPase. This ATPase activity, contrary to the neutral phosphatase activity, can only be detected around living fungal hyphae of the arbuscule and it disappears from the membrane surrounding senescent hyphal branches (Figure 1c). A high level of ATPase activity on a membrane indicates a high transporting capacity, even though the solute(s) transported may not be known (Smith and Smith, 1986). It is therefore highly significant that the host cell activates its plasmalemma around the fungal haustorium where phosphate transfer to the host cell is considered to preferentially occur (Gianinazzi-Pearson and Gianinazzi, 1983). The cause and mechanisms controlling release of phosphate by the internal mycelium of VA endomycorrhizal fungi are not known, but the phosphate generating system probably depends on hydrolysis of polyphosphate in the fungal vacuoles under the influence of a suitable signal (Harley and Smith, 1983 ; Dexheimer et al., 1986 ; Smith and Smith, 1986). Possible transport mechanisms at the host-fungus interface and the sort of physiological or biochemical problems that these may pose are discussed in more detail by Harley and Smith (1983), and Smith and Smith (1986).

DISCUSSION

The cellular modifications occurring in both VA endomycorrhizal symbionts indicate that complex interactions exist between their genomes in which signal molecules must be involved. The occurrence of such cell to cell communication is revealed by the fact that the fungal wall is greatly simplified, that arbuscule formation only occurs in parenchymal cells and that the newly formed host plasmalemma surrounding arbuscular hyphae specifically acquires neutral phosphatase and ATPase activities.

The molecular mechanisms determining host-fungus compatibility in endomycorrhizal associations remain obscure. VA endomycorrhizal fungi can develop within root tissues of non-hosts but they have only been found in dead plant cells (Figure 2) (Glenn et al., 1985) and arbuscule formation does not usually occur (Hirrel et al., 1978 ; Ocampo et al., 1980 ; our unpublished observations). Extracellular material is often associated with the intracellular hyphae in non-host cells but its origin and significance have not been determined. It has been suggested that the lack of development of VA endomycorrhizal fungi within living non-host cells may be due to the absence of necessary stimuli (signals ?) rather than the production of an inhibitory factor by the cells (Glenn et al., 1985), but this has yet to be proven.

It appears from studies not only of VA endomycorrhizae but also ericoid endomycorrhizae (Gianinazzi-Pearson et al., 1986), that events occurring at the host plasmalemma are probably of fundamental importance in determining the type of biotrophic relationship established between the interacting organisms (Smith, 1979). This is further underlined by comparisons with haustorial infections by biotrophic pathogens, characterised by nutrient transport in one direction only, from plant to fungus (Manners and Gay, 1984). Here, the living host-fungus interface is morphologically quite similar to that in cells containing arbuscules, with fungal and plant plasmalemma separated by wall material of both symbionts. An important difference, however, lies in the fact that in plant-fungal pathogen interactions, the extrahaustorial host plasmalemma can be considerably altered as compared to the normal peripheral plasmalemma. In particular, changes have been observed in membrane structure and staining properties and these are associated with a specific lack of ATPase activity along the host plasmalemma surrounding the haustoria in the diseased tissue (Spencer-Phillips and Gay, 1982 ; Manners and Gay, 1983 ; Woods and Gay, 1987).

CONCLUSIONS

Although host-fungus cellular relationships in VA endomycorrhizae are very similar in many host species, they do differ in detail between cell

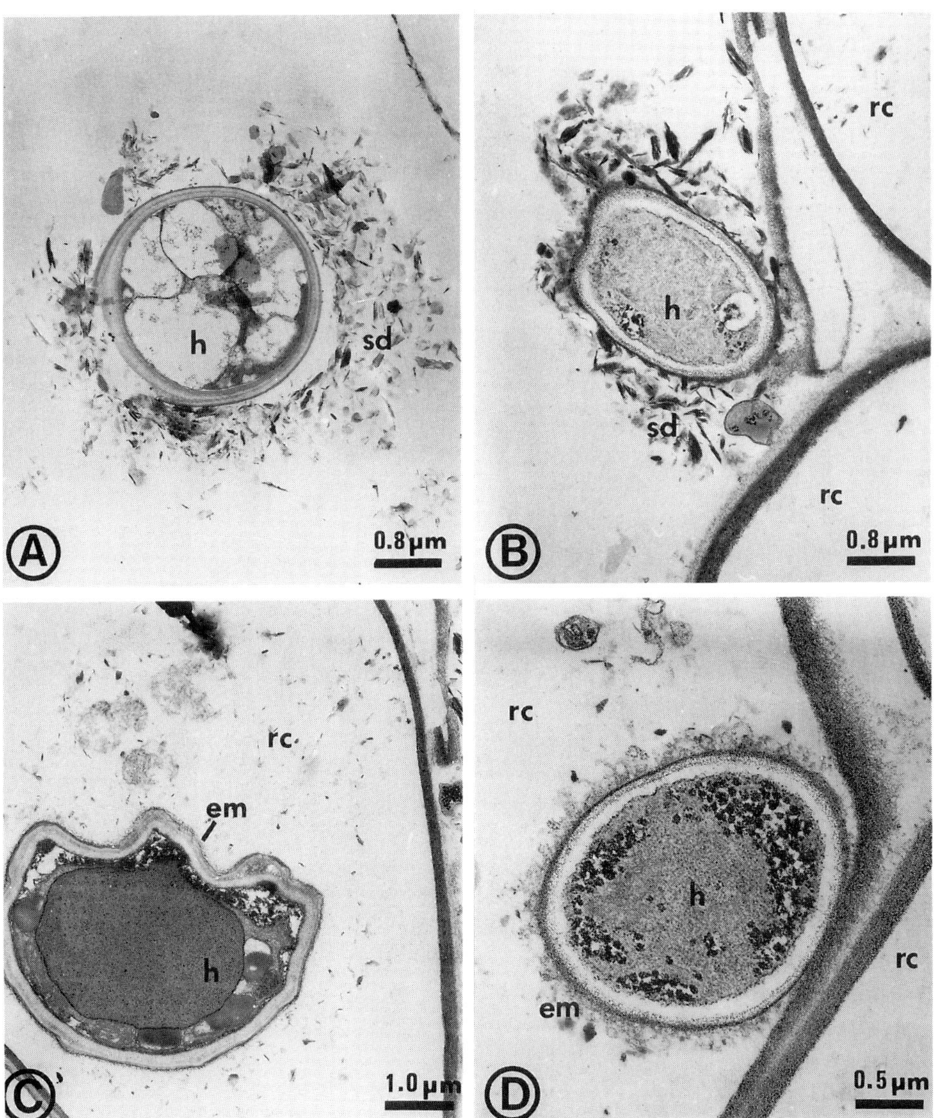

Figure 2 : Hyphae of Glomus E_3 associated with living roots of the non-host Beta vulgaris. A, B : external thick- and thin-walled hyphae surrounded by soil debris. C, D : endocellular hyphae within moribund root cells in the outer cortex and coated with diffuse material. h, hypha ; rc, root cell ; sd, soil debris ; em, extracellular material.

types in the same root. From an ultracytological point of view, and compared to intercellular hyphae or intracellular coils, the arbuscule appears to be a specialised haustorial organ which is well adapted as a site of nutrient exchange between fungus and host. Here, the important surface of cellular contact between the symbionts, the much reduced apoplast between their plasmalemmas and the concentration of energy generating enzyme systems along these membranes, together create a highly favourable situation for bidirectional movement of solutes across the host-fungus interface. The overall consequence of arbuscule formation in VA endomycorrhizal associations is, therefore, to increase the efficiency of the plant's nutrient uptake system by conferring on inner root cortical cells a new role of nutrient absorption from the soil via the hyphae of the fungal symbiont.

REFERENCES

Abbott, L.K. 1982. Comparative anatomy of vesicular-arbuscular mycorrhizas formed on subterranean clover. Aust J Bot 30: 485-499

Berta, G., Fusconi, A., Sgorbati, S., Trotta, A., Scannerini, S. 1986. Preliminary results on the ploidy and fine structure of the nuclei of the host cells in a VA mycorrhiza. Giorn Bot Ital 120: 84-86

Bonfante-Fasolo, P. 1984. Anatomy and morphology of VA mycorrhizae. In : Powell, C.L., Bagyaraj, D.J. (eds) VA mycorrhiza. CRC Press, Inc, Boca Raton Florida, pp 5-33

Bonfante-Fasolo, P., Fontana, A. 1985. VAM fungi in Ginkgo biloba roots : their interactions at the cellular level. Symbiosis 1: 53-57

Bonfante-Fasolo, P., Gianinazzi-Pearson, V. 1986. Wall and plasmalemma modifications in mycorrhizal symbiosis. In : Gianinazzi-Pearson, V., Gianinazzi, S. (eds) Physiological and genetical aspects of mycorrhizae. INRA Press, Paris, pp 65-73

Carling, D.E., Brown, M.F. 1982. Anatomy and physiology of vesicular-arbuscular and non mycorrhizal roots. Phytopathology 72: 1108-1114

Cox, G., Tinker, P.E. 1976. Translocation and transfer of nutrients in vesicular-arbuscular mycorrhizas I The arbuscule and phosphorus transfer : a quantitative ultrastructural study. New Phytol 77: 371-378

Dexheimer, J., Gianinazzi, S., Gianinazzi-Pearson, V. 1979. Ultrastructural cytochemistry of the host-fungus interfaces in the endomycorrhizal association Glomus mosseae/Allium cepa. Z Pflanzenphysiol 92: 191-206

Dexheimer, J., Gianinazzi-Pearson, V., Gianinazzi,S. 1982. Acquisitions récentes sur la physiologie des mycorhizes VA au niveau cellulaire. Coll de l'INRA 13: 61-73

Dexheimer, J., Marx, C., Gianinazzi-Pearson, V., Gianinazzi, S. 1985. Ultracytological studies of plasmalemma formations produced by host and fungus in vesicular-arbuscular mycorrhizae. Cytologia 50: 461-471

Dexheimer, J., Kreutz-Jeanmaire, C., Gerard, J., Gianinazzi-Pearson, V., Gianinazzi, S. 1986. Approche cellulaire du fonctionnement des endomycorhizes à vésicules et arbuscules : les plasmalemmes de l'interface. In : Gianinazzi-Pearson, V., Gianinazzi, S. (eds) Physiological and genetical aspects of mycorrhizae. INRA Press, Paris, pp 277-283

Gianinazzi, S., Gianinazzi-Pearson, V., Dexheimer, J. 1983. Role of the host-arbuscule interface in the VA mycorrhizal symbiosis : ultracytological studies of processes involved in phosphate and carbohydrate exchange. Plant Soil 71: 211-215

Gianinazzi-Pearson, V. 1984. Host-fungus specificity, recognition and compatibility in mycorrhizae. In : Verma DPS, Hohn, Th. (eds) Genes involved in microbe-plant interactions. Springer-Verlag, Vienna New York, pp 225-253

Gianinazzi-Pearson, V. 1986. Cellular modifications during host-fungus interactions in endomycorrhizae. In : Bailey, J.A. (ed) Biology and molecular biology of plant pathogen interactions. NATO ASI Series, vol. H1, Springer-Verlag, Berlin, pp 29-37

Gianinazzi-Pearson, V., Gianinazzi, S. 1983. The physiology of vesicular-arbuscular mycorrhizal roots. Plant Soil 71: 197-209

Gianinazzi-Pearson, V., Bonfante-Fasolo, P., Dexheimer, J. 1986. Ultrastructural studies of surface interactions during adhesion and infection by ericoid endomycorrhizal fungi. In : Lugtenberg, B. (ed) Recognition in microbe-plant symbiotic and pathogenic interactions. NATO ASI Series, Vol H4, Springer-Verlag, Berlin pp 273-282

Gianinazzi-Pearson, V., Dexheimer, J., Gianinazzi, S., Jeanmaire, C. 1984. Plasmalemma structure and function in endomycorrhizal symbiosis. Z Pflanzenphysiol 114: 201-205

Gianinazzi-Pearson, V., Morandi, D., Dexheimer, J., Gianinazzi, S. 1981. Ultrastructural and ultracytochemical features of a Glomus tenuis mycorrhiza. New Phytol 88: 633-639

Jacquelinet-Jeanmougin, S. 1986. Les endomycorhizes de Gentiana lutea L. : Détermination des champignons symbiotes ; aspects physiologiques et ultrastructuraux de ces associations. Thèse de doctorat, Dijon University, France, 131 pages

Jeanmaire, C., Dexheimer, J., Marx, C., Gianinazzi, S., Gianinazzi-Pearson, V. 1985. Effect of vesicular-arbuscular mycorrhizal infection on the distribution of neutral phosphatase activities in root cortical cells. J Plant Physiol 119: 285-293

Kinden, D.A., Brown, M.F. 1975. Electron microscopy of vesicular-arbuscular mycorrhizae of yellow poplar I Characterisation of endophytic structures by scanning electron microscopy. Can J Microbiol 21: 989-993

Harley, J.L., Harley, E.L. 1987. A check-list of mycorrhiza in the British flora. New Phytol Suppl 105: 1-102

Harley, J.L., Smith, S.E. 1983. Mycorrhizal symbiosis. Academic Press, London New York, 483 pages

Hirrel, M.C., Mehravaran, H., Gerdemann, J.W. 1978. Vesicular-arbuscular mycorrhizae in the Chenopodiaceae and Cruciferae : do they occur ? Can J Bot 56: 2813-2817

Holley, J.D., Peterson, R.L. 1979. Development of a vesicular-arbuscular mycorrhiza in bean root. Can J Bot 57: 1960-1978

Manners, J.M., Gay, J.L. 1983. The host-parasite interface and nutrient transfer in biotrophic parasitism. In : Callow, J.A. (ed) Biochemical plant pathology. John Wiley & Sons Ltd, Chichester New York, pp 163-195

Marx, C., Dexheimer, J., Gianinazzi-Pearson, V., Gianinazzi, S. 1982. Enzymatic studies on the metabolism of vesicular-arbuscular mycorrhiza IV Ultracytoenzymological evidence (ATPase) for active transfer processes in the host-arbuscule interface. New Phytol 90: 37-43

Morandi, D., Gianinazzi-Pearson, V. 1986. Influence of mycorrhizal infection and phosphate nutrition on secondary metabolite contents of soybean roots. In : Gianinazzi-Pearson, V., Gianinazzi, S. (eds) Physiological and genetical aspects of mycorrhizae. INRA Press, Paris, pp 787-791

Ocampo, J.A., Martin, J., Hayman, D.S. 1980. Influence of plant interactions on vesicular-arbuscular mycorrhizal infections I Host and non-hosts grown together. New Phytol 84: 27-35

Pons, F. 1984. L'endomycorhization VA in vitro de quelques espèces herbacées et ligneuses : aspects ultrastructuraux de l'association. 3ème cycle thesis, Dijon University, France, 116 pages

Scannerini, S., Bonfante-Fasolo, P. 1979. Ultrastructural cytochemical demonstration of polysaccharides and proteins within the host-arbuscule interfacial matrix in endomycorrhizas. New Phytol 83: 87-94

Scannerini, S., Bonfante-Fasolo, P. 1983. Comparative ultrastructural analysis of mycorrhizal associations. Can J Bot 61: 917-943

Smith, D.C. 1979. From extracellular to intracellular : the establishment of a symbiosis. Proc Royal Soc London B204: 115-130

Smith, F.A., Smith, S.E. 1986. Movement across membranes : physiology and biochemistry. In : Gianinazzi-Pearson, V., Gianinazzi, S. (eds) Physiological and genetical aspects of mycorrhizae. INRA Press, Paris, pp 75-84

Spencer-Phillips, P.T.N., Gay, J.L. 1981. Domains of ATPase in plasmamembranes and transport through infected plant cells. New Phytol 89: 393-400

Toth, R., Miller, R.M. 1984. Dynamics of arbuscule development and degeneration in a Zea mays mycorrhiza. Amer J Bot 71: 449-460

Trappe, J.M. 1982. Synoptic keys to the genera and species of zygomycetous mycorrhizal fungi. Phytopathology 72: 1102-1108

Woods, A.M., Gay, J.L. 1987. The interface between haustoria of Puccinia poarum (monokaryon) and Tussilago farfara. Physiol Mol Plant Pathol 30: 167-185

CELL TO CELL INTERACTIONS IN INSECT ENDOCYTOBIOSIS

P. NARDON
Biologie 406 - Institut National des Sciences Appliquées - 69621 Villeurbanne Cedex - France

From the evolutionary point of view, symbiosis represents the union of two (or more) divergent genomes into a new coevolutionary unit. The more sophisticated association is endocytobiosis, where the microorganism partner lives inside specialized cells of the host, and is maternally inherited. These intracellular symbioses, or endocytobioses, are found in numerous insect species (Brooks 1963, Buchner 1965, Koch 1967, Louis 1980, Schwemmler 1980, Dasch et al 1984). Their establishment and their maintenance under selection pressure suppose both mutual tolerance and mutual exchanges between the partners. In insects, the status of endocytobiotes is not yet clear. They are either yeast-like or bacterium-like. Many observations have been made on insects, but most of them are very old, and, in the absence of ultrastructural pictures it is often very difficult to proffer a worthwhile opinion. Moreover, biochemical, and in particular genetical studies are still limited to only a few models : cockroaches, some beetles and some Homoptera (Aphids and Leafhoppers). Nevertheless, despite the rare investigations at the cellular and molecular levels, we may assume that endocytobiotes are perfectly integrated into the host metabolism and homeostasis. Genetic exchange between insect and microorganisms has been hypothesized (Nardon 1978, Schwemmler 1980), but no convincing evidence has yet been reported and we must confess that we have no precise idea concerning the processes involved in such a transfer.

A great diversity of endocytobiosis types exists in insects. The literature documenting this diversity is extensive and no attempt will be made to exhaustively review it (see in Buchner 1965, Houk and Griffiths, 1980, Schwemmler 1987). Nevertheless, the most important features have to be reported. Furthermore I want to point out some characteristics related to the exchanges between the partners, so as to appreciate the nature of their relationships.

I - Endocytobiote morphology

The problem of symbiote taxonomy will not be considered here, and we will refer only to "yeasts" and "bacteria", this latter being used in the broad sense to include Eubacteria, Rickettsia or Chlamydia and Mollicutes.

A) Yeast symbiotes

Endocytobiotic yeasts are found in several families, particularly in Coleoptera (Anobiidae, Cerambycidae) and Homoptera (Coccidae) (Buchner 1965). Some of them have

NATO ASI Series, Vol. H17
Cell to Cell Signals in Plant, Animal and
Microbial Symbiosis. Edited by S. Scannerini et al.

been successfully cultivated *in vitro* and identified as *Candida* species in Cerambycidae and *Torulopsis* species in Anobiidae (Buchner, 1965 ; Bismanis 1976).

B) Bacterium-like symbiotes

A definite identification is very difficult since these endocytobiotes are not culturable *in vitro* . The classification proposed by Louis (1980), which I personally find very interesting, has not been retained in the last edition of Bergey's manual (Dasch et al, 1984). Generally, the symbiotic bacteria are non-sporulating, non-ciliated, and Gram-negative, as in beetles (Musgrave et al 1962, Nardon et al 1985), in Aphids (Hinde 1971b), in the bedbug *Cimex lectularius* (Chang and Musgrave 1973) and in cockroaches (Gharagozlou, 1966 ; Brooks, 1970). Gram-positive bacteria have been described in the Scolytid *Xyleborus ferrugineus* (Peleg et Norris 1973) and in the human body Louse *Pediculus* (Eberle and Mc Lean, 1983).

The ultrastructural study of symbiote membranes is of particular interest since they necessarily regulate the exchanges with the host. As was said above, most species are Gram-negative and the presence of a peptidoglycan sheath, more or less thick, has been confirmed (Daniel and Brooks 1967, Houk et al 1977). In some insects, the outer membrane is lacking as in the Heteroptera *Arocatus roeselii* (Louis 1980), and probably in the "t" endocytobiote of the leafhopper *Euscelis plebejus* (Louis et al 1976), and in "g" and "t" endocytobiotes of another leafhopper *Helochara communis* (Chang and Musgrave 1972).

The two symbiotes of *Euscelis plebejus* ("a" and "t") have been studied by electron microscopy in ultra thin sectioned and freeze-etched material (Louis and Nicolas 1976, Louis et al 1976). The density of the particles associated with the cytoplasmic membrane differed widely between the two symbiotes. Their density on the P faces (against the bacteria cytoplasm) is approximately ten fold greater in "a" than in "t" symbiotes, while on the E faces (facing the host cytoplasm) these densities are almost equal. The total number of particles is much lower in "t" (750 to 1 200 per μm^2) than in "a" symbiotes (1 900 to 3 350). Such a difference probably reflects a physiological peculiarity, and, from comparison with other structures poor in particles, the authors suggest that the "t" symbiote could play the role of a kidney in regulating the osmotic pressure. In a previous paper (Schwemmler, 1974) it has been claimed that the symbiote of *E. plebejus* could play a role in regulating osmotic pressure and pH.

C) Monosymbiosis and multisymbiosis

As it was seen above, *E. plebejus* is a disymbiotic insect. Multisymbioses are known in other insects. The most complex situation exists in Fulgoroid and Membracid Homoptera where as many as six different symbiotes can be found, each being precisely located (Buchner, 1965). It is often difficult to decide whether symbiotes are a single pleomorphic bacterium or several species. For instance, the bedbug *Cimex lectularius* is now known to harbour three procaryotes (Louis, 1980). It is very exciting to imagine how such complex associations have been acquired during evolution, and how the various genomes are able to cooperate.

II- Endocytobiote location and interaction with host

A) Diversity

In most insect species the symbiote-harbouring cells, bacteriocytes or mycetocytes, are grouped in compact organs, paired or unpaired : bacteriomes or mycetomes. Nevertheless, in some insects the host cells are scattered, as in the cockroaches, where they lie among fat cells and urate cells (Brooks, 1963 ; Milburn, 1966 ; Philippe, 1986). The symbiotic organs are principally associated with the intestine, connected (Anobiidae, Cerambycidae) or unconnected (most Curculionids) with its lumen (Buchner, 1965). They are also frequently located in the body cavity (lice, leafhoppers). They can be found also in fat tissue (some Curculionid Apioninae) or in Malpighian tubules (in the Scolytid *Coccotrypes* and in some Apioninae). Moreover, the symbiotes invade the ovaries in numerous insect species at the time of reproduction.

B) Correlation with host development and host cells. Pleomorphy

In the Cerambycid beetle, *Criocephalus rusticus*, morphological diversity of the yeasts are related with host cell seasonal changes (Riba and Chararas, 1976). Pleomorphy is widespread in yeasts (Buchner, 1965). In bacteria two forms are frequently described : "vegetative form", located in the larval bacteriome, and "infection form" whose role is to infect the transmission organ. Such different morphological types have been extensively studied in the leafhopper *Euscelis plebejus* where the differentiation of vegetative forms always occurs with entry into a new host cell, as is shown in the remarkable work of Körner (1972, 1974). Migratory forms of "a" symbiote show an increase of cytoplasmic density, a deposit of granular material at the cell wall periphery, cell division (Louis and Laporte, 1969), and asymmetric distribution of the ribosomes (Körner, 1972). The migratory forms leave the bacteriomes (which lie in the body cavity), reach the neighboring ovaries via the hemolymph, and penetrate into the posterior pole of oocytes. The infection form persists during the early embryogenesis (Körner, 1972) until the "a" symbiotes are captured by A1-bacteriocytes and then by A2-bacteriocytes. The "t" symbiotes also occur in an infection form (but with a lower density), but only for a short time (Körner, 1974), and they enter specific host cells. Finally an unpaired transitory bacteriome is formed which later divides into the definitive paired organ. In immature stages each kind of endocytobiote is separately located in its specific cells. But, at the time of reproduction, infection forms "a" and "t" both penetrate into the same special follicle cells (Hamon, 1971 and Körner, 1978). Such observations of strict correlation between host and symbiote differentiation have been reported in other insects (see in Buchner, 1965) and offer good evidence of a reciprocal control (see below).

According to Körner (1978) the correlation between differentiation of the insect and its symbiotes represent "a sort of host-symbiont-cycle", perfectly synchronized. What is important to our purpose is to remember that *the symbiote location varies according to the developmental stage of the host and also frequently with the sex* . This implies mechanisms of cell recognition.

C) Correlation with host sex. Transmission

In numerous insects, the symbiote location differs between male and female. In species where oocytes are infected with true endocytobiotes, the latter only persist in the female germ cells of progeny and are eliminated from the testes during embryogenesis. In the cockroach *Blattella germanica* , all males possess rudimentary ovaries with a few small oocytes. In a very elegant study, Brooks and Kurtti (1972) have shown that the oocytes of the males are infected by the symbiotic bacteroids, while the male germ cells are not. They conclude that the infection of a cell type necessitates a particular stimulus and suggest that the symbiotes are attracted by the yolk, whose deposition would be induced by mesoderm.

Another example is found in Cerambycidae where the symbiotes are yeasts. Shortly before pupation the gut mycetomes disappear in both sexes and yeasts are dispersed into the digestive lumen. While in the male symbiotes are eliminated with the feces, in the female they multiply in the hindgut and invade the transmission devices of the ovipositor (see in Buchner 1965). In the body louse *Pediculus* , before the third molting, the symbiotes leave the larval unpaired bacteriome only in females, to infect ovarial ampullae and then the oocytes (Aschner and Ries 1933). In males the bacteria degenerate and are not necessary to insect survival, contrary to the females. Furthermore male fertility is not affected by the extirpation of larval bacteriome while aposymbiotic females are unfertile and soon die (Aschner 1934). The relationship between host and symbiote varies according to the sex. These observations and others not reported here, lead us to suppose the existence of different signals in male and female since the bacterium population is identical in both sexes.

In the almond moth *Ephestia cautella* (Kellen et al, 1981), in mosquitoes (Wright and Barr, 1980) and in the leafhopper *Nephotettix cincticeps* (Mitsuhashi and Kono, 1975), non-pathogenic Rickettsia are found in great number in testes, but we do not consider this kind of association as a true endocytobiosis (no host specialized cells).

In all symbiotic insects so far studied, true endocytobiotes are always transmitted to the offspring by females only. Two main mechanisms exist : the symbiotes are inherited either inside the egg (following oocyte infection, as we have seen above in lice and leafhoppers), or by egg smearing as in Anobiidae or Cerambycidae (review in Buchner, 1965 ; Koch, 1967 ; Carayon, 1952). The symbiotes deposited on the egg surface come from either special devices (intersegmental tubules, vaginal pockets) of the egg-laying apparatus, or from the intestine. The emerging larvae are infected on devouring a part of the eggshell.

D) Symbiotes in host cells

Endocytobiotes, with or without a differentiated migratory form, are able to survive outside their host cell (*Pediculus*). How they penetrate into bacteriocyte or mycetocyte is not yet known in detail and a phagocytosis process is generally believed to occur. In *Helochara communis*, Chang and Musgrave (1975) describe a translocation of symbiotes by means of auxiliary cells which are able to phagocytose them.

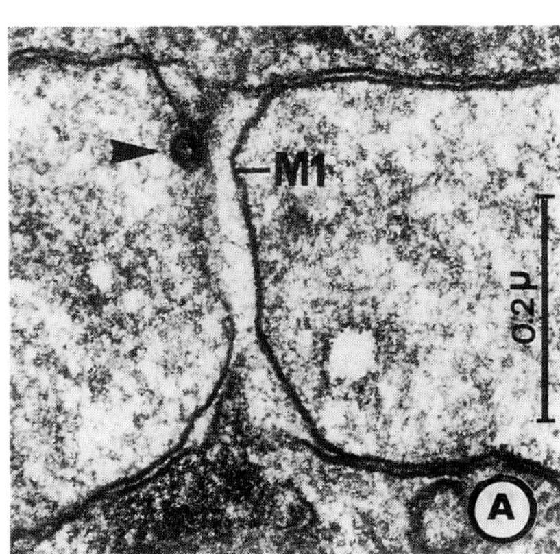

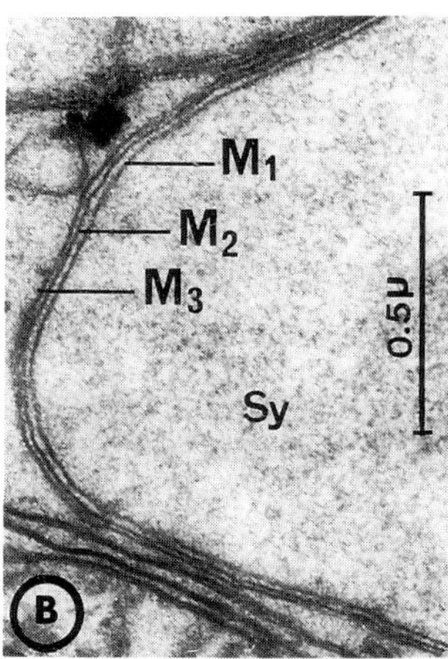

Figure 1 - Ultrastructure of endocytobiote envelope. Potassium permanganate fixation. A : *Metamasius hemipterus*, dividing symbiote. M1 : plasmic membrane. Arrow indicates a mesosome. B : "a" symbiote of *Euscelis plebejus* (Sy), limited by its cytoplasmic membrane (M1), a cell wall (M2) and a host membrane (M3). (From courtesy of C. Louis).

In insects, the connection of symbiote with the host cell is only known from morphological studies. One of the most documented is that of Louis (1980) who also investigated pathogenic procaryotes. From his observations and those of other authors (Malke and Bartsch, 1966 ; Gharagozlou, 1966 ; Milburn, 1966 ; Daniel and Brooks, 1972 ; Hinde, 1971 ; Körner, 1972, 1974 ; Chang and Musgrave, 1972, 1973) we can summarize the three kinds of association. In some rare instances, the symbiote, always a bacterium-like microorganism in that case, lies free in the host cytoplasm : in the Curculionid beetle *Sitophilus* (Grinyer and Musgrave, 1966 ; Nardon, 1971) and other Calandrinae (Nardon et al, 1985) (Figure 1) ; in the glossine midgut (Hübner and Davey, 1974), the companion symbiotes in the leafhopper *Helochara communis* (Chang and Musgrave, 1972) and in *Cimex lectularius* ("RC" symbiote of Louis, 1980, and rod-shaped symbiote of Chang and Musgrave, 1973). Sometimes, the free symbiote appears surrounded by a halo as for the "type 3" of *Cimex* symbiotes. This is interpreted as a reaction of the host cell (Louis, 1980). In most insects, symbiotes are surrounded by a host membrane (Figure 1). The latter is either very close to the microorganism and only visible by electron microscopy as in cockroaches, glossines, leafhoppers, or it may be at some distance from it, leaving a vacuole around it as in some Aphids (Hinde, 1971b). In *Cimex lectularius* (Louis, 1980) the mode of inclusion seems symbiote-specific since in the same cell the three known possibilities coexist : "RC" is free, "type 3" with a halo and "type 4" is enclosed in a host membrane. However, in other insects it differs according to the host cell : in *Glossina* the symbiote is free in the midgut cell, and encapsulated in the ovary (Huebner and Davey, 1974). The "t" symbiotes of *E. plebejus* are free only in bacteriocytes (Louis et al, 1976), not in the embryo (Körner, 1976).

Usually only one symbiote, yeast (Riba et Chararas, 1976) or bacterium-like (Hinde, 1971b) is enclosed in a vacuole, but, occasionally, two or more symbiotes are seen in the same envelope, probably arising from recent divisions. Generally vacuole and symbiote divisions are correlated. Nevertheless, in some insects, like *Pediculus* or Pseudococcines, symbiotes always form microcolonies inside large chambers. In *Planococcus citri*, these "enclaves muqueuses" (Louis, 1980) are delimited by three membranes, the outer being a host membrane, the two inner ones being still of unknown origin. These "enclaves" are filled with a mucous and proteinaceous substance (Louis, 1967), and with ribosomes liberated by degenerating symbiotes. "Enclaves" and vacuoles behave as micro-ecological niches where symbiotes live, divide and exchange with the host cell. In *Cimex*, during transmission from bacteriocytes to oocytes, symbiotes are kept inside their vacuoles and are stable from one generation to the next. Such "enclaves" or simple vacuoles can be considered as protection devices for symbiotes in some circumstances, since they are destroyed when they escape from them as has been described in *Pseudococcus adonidum* (Louis, 1980) or in Aphids (Hinde, 1971a). We can also imagine that the host membrane is a reaction against the presence of symbiote. Certainly the system is interactive.

E) Specificity and cell recognition

Establishment of endocytobiosis supposes the existence of recognition mechanisms not only at the species level, but also at the cellular level. Concerning *Euscelis* symbiotes we have reported above that, during embryogenesis, they penetrate only in some defined cells. In the bedbug *Cimex* the companion symbiote "RC" and the symbiote "type 3" are found inside numerous tissues while the "type 4" symbiote is restricted to ovaries and bacteriomes (Louis, 1980). In the leafhopper *Nephotettix cinctipes* (Mitsuhashi and Kono, 1975) "a" and "t" symbiotes are also found only in bacteriomes and ovaries, in contrast to the third symbiote which infects other tissues and male germ cells, penetrating most frequently in cell nuclei. In Aphids (Hinde, 1971a) the symbiotes, despite their migration through the body cavity, never infect other tissues and those which are mislaid in hemolymph are inevitably phagocytised and broken down by haemocytes. The bacteriocytes protect the insect from the bactericidal response of the host. Several species have been reported to have lytic enzymes (lysozyme like) in their hemolymph. Outside the bacteriocyte, symbiotes may be attacked and then no more tolerated by host cells, as would be the case for damaged insect cells themselves. In the same cell the tolerance to symbiote penetration is often strictly limited in time, as in oocytes. From all these morphological observations and others not reported here (see Buchner, 1965), it is quite evident that mechanisms of cell recognition exist in insects despite the fact that their chemical nature is still unknown. Perhaps the symbiote has no intrinsic signal to infect one cell rather than another cell, and has to be pre-conditioned. This idea comes from the results of Brooks and Richards (1956) who failed to reinfect aposymbiotic cockroaches by implanting or injecting bacteroids obtained from symbiotic insects (see also below).

Few studies concern the species-specific recognition. In the bug *Coptosoma* (Buchner, 1965) symbiotes are externally transmitted with the eggs, and can be easily replaced by foreign bacteria, but none are able to play the role of the true symbiote : they are digested or they kill the insect. Reinfection is not possible with the true symbiote when previouly cultivated on agar where it probably loses its recognition signals. Similar attempts have been made in the Anobiid bettle *Stegobium paniceum* (Foeckler, 1961). Of the 15 foreign yeasts tested for reinfection, success was only obtained with dried *Torulopsis utilis* when added to the diet. Not only the caeca (as in normal symbiosis) but also the remainder of the midgut epithelium were infected by the yeasts. Therefore, a barrier impenetrable to the normal symbiotes is not set up against a foreign yeast. Furthermore, infected cells retain their brush-border and, during metamorphosis, the new midgut epithelium and the transmission organs remain empty, contrary to what is observed in normal symbiosis. From these two examples it can be concluded that *species-specificity is relatively strict between insects and their symbiotes*, bacteria or yeasts.

III - Mutual induction phenomena and physiological signals

From the morphological features reported above it is quite evident that the symbiote(s) and the insect mutually influence each other. The association is not additive, but interactive, and, in most cases (particularly with bacterium-like symbiotes) the partners have lost their independence : insects do not survive to symbiote elimination, and bacteria are not culturable *in vitro*. We know that the trophic relationship is essential for maintenance of the symbiosis (Koch, 1967 ; Brooks, 1970 ; Houk and Griffiths, 1980 ; Schwemmler, 1987), but in this final section, we want only to examine regulation and differentiation processes, and point out the principal features.

A) Control of symbiotes by the host

1. The host controls the location of its symbiotes perfectly. This conclusion emerges from cytological studies. The implied recognition mechanism is yet unknown, but we have no reason to imagine that it is different to other cell-to-cell chemical communication.

Certainly surface glycoproteins play a major role, and their presence or absence is under genetical control at cell level.

2. Concerning the pleomorphy of symbiotes according to their cell location, it is difficult to assume it is an intrinsic expression of the symbiote genome itself, or the result of the action of some effector from the host. Several "natural experiments" lead to the latter conclusion. In the beetle *Rhizopertha dominica* (Buchner, 1965), during embryogenesis, transmission forms sometimes penetrate into accessory cells, but they retain their dense structure and never divide. Therefore the host possibly controls cell division of its symbiotes, probably by inducing DNA replication. In *Cimex* (Louis, 1973), it has been claimed that the companion symbiote, non-motile inside the cell, acquires six flagella very rapidly when it is experimentally liberated from the cell, and becomes motile. Such an observation suggests that synthesis or assembly of flagellar components is inhibited in the host cell.

3. In the Aphid *Acyrthosiphon pisum*, in some very elegant experiments, genetical interactions between host and its symbiotes have been investigated (Ishikawa, 1982, 1987, Ishikawa et al, 1985, 1986) ; the results are very exciting and it is hoped they will be found in other insect species. It seems that most of the genes are not expressed when the symbiote is present intracellularly. Under *in vitro* conditions (does the possibility of artefacts exist ?), the endocytobiotes are able to synthesize at least a hundred protein species, while under *in vivo* conditions they synthesize only one protein, *symbionin* (63000 daltons), whose role is unknown. The effect of antibiotics suggests to the author that *symbionin is coded by the host genome but seems to be synthesized in the symbiotes.*

Morphogenes and regulation of cell differentiation are extensively studied in *Caulobacter* (Ely and Shapiro, 1984) and *E. coli* (Donachie et al, 1984). Similar, or more complicated sequences certainly exist in endocytobiotes and we can imagine that the host, in the course of evolution, has evolved signals initiating or modulating the gene expression at each step of the host-symbiote cycle.

4. Control of symbiote number

This phenomenon will be examined in more detail in a later paper in this book, so the feature will only be mentioned here. The question that arises is : what are the mechanisms implied in symbiote number regulation ? In this section we only discuss hormonal influence. Since in numerous insect species, symbiotes change their location during metamorphosis or even disappear, the question of an hormonal action has been considered. We have to keep in mind that molts are under the control of ecdysterone, and that juvenile hormone (JH), at a high level in larval stages falls during pupal and adult molts and reappears in young adult. Buchner (1965) has reported in several insects (*Fulgora, Haematoptinus* and others) the possible action of a substance diffusing from ovaries upon the bacteriocytes (induction of migratory forms). We can only remark that in several species ecdysterone is known to be synthesized in ovaries. Further studies would help to elucidate this point.

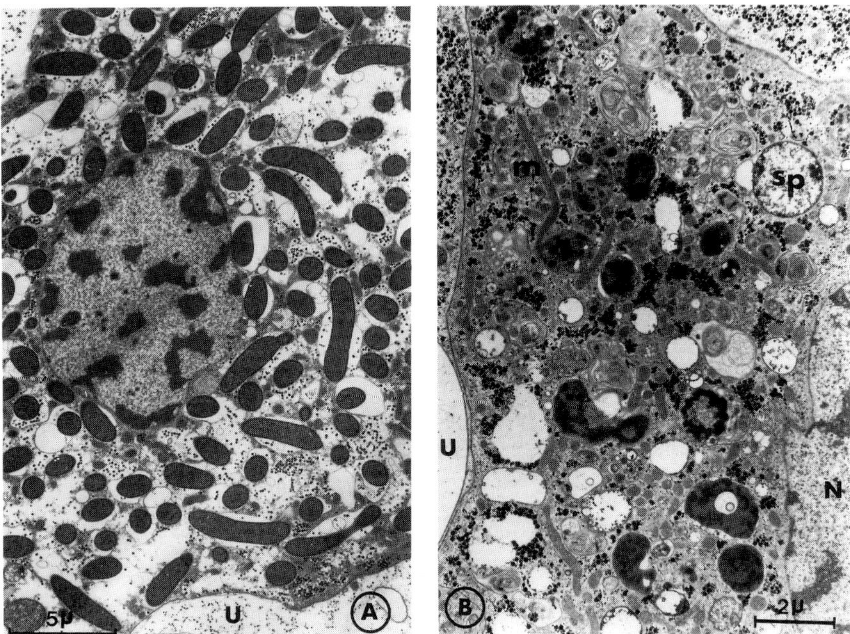

Figure 2 - *In vitro* culture of *Periplaneta americana* bacteriocytes in absence of hormones. A : bacteriocyte at day 0, with numerous bacteroids. B : after 15 days culture. Bacteria have disappeared or become dense bodies (arrow) or spheroplasts (sp). Mitochondria (m) increase in number and lenghten. U : urate cell. N : nucleus. From Courtesy of C. Philippe.

Tsang (1981) has investigated JH and ecdysterone influence upon *Blattella germanica* symbiotes. Normally, in adult females, fat body symbiotes fluctuate in synchrony with egg production. In isolated abdomens, deprived of hormones, development of terminal oocytes was prevented and growth of symbiotes was drastically curtailed. Injection of JH induced oocyte development but failed to have a growth promoting effect on symbiotes in either sex. On the contrary, injected ecdysterone failed to increase oocyte growth and ovarian bacteria, but it induced a population doubling of fat body symbiotes. In other cockroach species, Milburn (1966) described modification of symbiotes during metamorphosis, supposing them to be under the control of corpora cardiaca, but no experimental data were given. In *Periplaneta americana*, it has been shown that JH affects directly the distribution of protein particles on the convex face of the plasma membrane of symbiotes (Liu, 1974). The relocation of particles may indicate a shift in the functional state of symbiotes under JH influence. In her thesis (1986), C. Philippe investigated the fate of *Periplaneta* symbiotes when bacteriocytes are cultivated *in vitro* and pointed out three very interesting features. Firstly, adipocytes are necessary for the fat body bacteriocyte survival, and vice versa (reciprocal action). Secondly, in the absence of JH, the bacteroids are rapidly eliminated (2 weeks) from the bacteriocytes (Figure 2), and this does not happen when the latter are cocultivated with corpora allata (CA) (source of JH). Only 20 % of bacteriocytes do not respond to the presence of CA. In others, the symbiotes do not degenerate and are able to divide. Thirdly the multiplication and lengthening of mitochondria (Figure 2) in the absence of JH is correlated with the elimination of symbiotes. She suggests that juvenile hormone would act in protecting the peptidoglycan wall against lytic enzymes. A point which remains unexplained and requires further experiments is : why does ecdysterone have a stimulating effect upon symbiote growth *in vivo* (according to Tsang), and not *in vitro* (according to Philippe's findings) ?

In the human body louse *Pediculus humanus*, the migration of symbiotes attracted by oviducts, is initiated by an unidentified humoral factor associated with the adult molt (Eberle and Mc Lean, 1982). It is present in both sexes but the male bacteriome is unable to respond to it.

Besides humoral signals, symbiote growth is affected by nutritional factors, phagocytosis, lytic enzymes (Brooks, 1963 ; Nardon and Wicker, 1981) and probably by host signals preventing cell division (see further section).

B) Inductive action of symbiotes on the host

1. It is well established that endocytobiotes have a great metabolic activity inside host cells since they divide (as it has been described by all authors, see figure 1), synthesize ADN, ARN (Nardon, 1978 ; Louis, 1980 ; Ishikawa, 1982 a,b) and proteins (Louis et Kuhl, 1972 ; Ishikawa, 1984).

2. By their presence (surface glycoproteins) and/or their biological activity (synthesis of specific molecules), *symbiotes can modify the morphology of their host cells.* One of the best known examples is the disappearance of brush borders in infected gut cells or Malpighian cells (see in Buchner, 1965).

3. *Symbiotes can stimulate protein synthesis in their host.* Such a phenomenon is strongly suggested by some experiments carried out in cockroaches, as those of *in vitro* culture of adipocytes which need bacteriocyte presence for division (Philippe, 1986). According to Garthe and Elliott (1971), the bacteroids of *Blaberus craniifer* would be necessary for the synthesis of proteins in the fat body (this is in agreement with Philippe's findings), and a correlation between the number of symbiotes and the quantity of proteins is depicted.

4. Bacteriome and mycetome differentiation.

The problem of the role of symbiotes in the differentiation of symbiotic organs is still under discussion. In fact, no general philosophy can be traced and several cases occur. In the beetle *Oryzaephilus surinamensis* (Koch, 1967) bacteriomes differentiate (during 25 generations) in the absence of symbiotes, while in *Sitophilus oryzae* ovarian bacteriomes disappear in aposymbiotic strains (Nardon, 1973). In Cerambycidae the transmission organs exist also in non-symbiotic species (Buchner, 1965). In *Pediculus*, Aschner (1934) demonstrated that the embryo stomach disc and the ovarial ampullae (OA) in the female are formed even in the absence of bacteria. Nevertheless, this does not mean that the symbiotes have no inductive action at all since the structure of OA cells is greatly modified in aposymbiotic insects (Aschner and Ries, 1933). Furthermore, in embryos of aposymbiotic females, the stomacal disc differentiation is frequently disturbed ; follicular cells are also damaged and oviposited eggs soon die. So we can conclude that the bacteria are needed to maintain the normal structure of different kinds of cells, either harbouring them (OA) or not (follicular cells). The formation of a bacteriome (with some anomalies) in the absence of symbiotes can be interpreted in two ways : firstly we can consider that it was differentiated before the symbiotes were acquired and that it now represents a secondary adaptation ; secondly (the idea of Aschner, 1932) we can consider that the bacteriome was induced by the symbiotes in the course of evolution, and became a hereditary organ ("Sie sind also zu erblich fixierten Organen geworden"). Such a process would be plausible if some gene transfer occurred from bacteria to the louse. The bacteriocytes, in that case, would be interpreted as transformed cells. Their differentiation would result from cooperation between louse and bacteria. Such a feature closely resembles that found in the legume symbiosis.

Concerning the "transformed cells", we have to point out that bacteriocytes or mycetocytes are generally giant polyploid cells (see in Buchner, 1965). Moreover, in *Blattella germanica, the mitotic division of bacteriocytes seems to be controlled by the symbiotes* . In the absence of bacteroids the bacteriocytes remain small and do not divide (Brooks and Richards, 1955 ; Richards and Brooks, 1958).

The morphogenetic action of symbiotes has also been investigated in *Euscelis plebejus* where interaction between host and bacteria was clearly demonstrated (Sander, 1968 ; Körner, 1978). Nevertheless, these two authors do not agree with Schwemmler (1974) when he claims the necessity of symbiotes in promoting a normal development.

5. Action of symbiotes on reproduction

Finally symbiotes can interfere with reproduction in two ways. In a Coccid, *Stictococcus spoestedti*, according to Buchner (1965), the infected oocytes develop parthenogenetically to produce males. A similar mechanism is described in *Xyleborus ferrugineus* (Peleg and Norris, 1973) where the nuclear symbiote (a *Staphylococcus*) is able to reactivate oocytes as spermatozoïdes do.

The most surprising phenomenon is found in the crustacean Isopod *Armadillidium vulgare* where a symbiotic bacterium produces a factor which feminizes the genetic males (entirely or only partly) by modifying the functioning of the neurosecretory system controlling the synthesis and the release of the androgenous hormone (Juchault and Legrand, 1974).

In other circumstances, cytoplasmic symbiotes are responsible for reproductive cytoplasmic incompatibility between crosses of different geographic strains (Kellen et al, 1981 ; Wright and Barr, 1980 ; Legrand et al, 1985).

CONCLUSION

The types of endocytobiosis in insects are extremely diversified at morphological and physiological levels. The host and its symbiotes are integrated in a new organism. The signals implied in mutual control are far from being all identified. At present we know that symbiotes not only furnish the host with growth substances (not studied here) but have also a morphogenetic action. This has sometimes been contested since its importance varies according to the species considered, which probably reflects different coevolutionary relationships. At present, symbionin of *Acyrthosiphon pisum* is the only symbiote-specific chemical well characterized in insects, but its role is still unknown.

Concerning the host, its influence upon symbiotes is mediated by hormones (juvenile hormone, ecdysterone) or hormone-like substances, lytic enzymes and nutrients.

During coevolution between the host and its symbiotes, mutual genetical mechanisms of control have been established. Signals acting directly on DNA to modulate gene expression do exist and it would be very exciting to discover their chemical nature and their mode of action. Endocytobiosis reveals the most sophisticated characteristics of living matter : the tendency to association, which probably explains eucaryote cell genesis according to the endocytobiosis theory of origin of living creatures.

Acknowledgements : I wish to thank Dr D.C. Smith for correcting the English, C. Nardon for her technical assistance, C. Louis and C. Philippe for illustrations.

REFERENCES

Aschner M. - 1932 - Experimentelle Untersuchungen über die Symbiose der Kleiderlaus - *Naturwissen* , **20**: 501-505.

Aschner M. - 1934 - Studies on the symbiosis of the body-louse : 1 : Elimination of the symbionts by centrifugation of the eggs - *Parasitology* , **26**: 309-314.

Aschner M. & Ries E. - 1933 - Das Verhalten der Kleiderlaus bei Ausschaltung ihrer Symbionten. Eine experimentelle Symbiosestudie - *Z. Morphol. Ökol. Tiere* , **26**: 529-590.

Bismanis J.E. - 1976 - Endosymbionts of Sitodrepa panicea. - *Can. J. Microbiol.* , **22**, 10 : 1415-1424.

Brooks M.A. - 1963 - Symbiosis and aposymbiosis in Arthropods - *in "Symbiotic associations". University Press Cambridge* , : 200-230.

Brooks M.A. - 1970 - Comments on the classification of intracellular symbiotes of cockroaches, and a description of the species - *J. Invert. Pathol.* , **16**: 249-258.

Brooks M.A. & Kurtti T.J. - 1972 - Male rudimentary ovaries : a case of cellular symbiosis in Blatella germanica - *J. Insect Morph. Embryol.* , **1**, 2 : 169-180.

Brooks M.A. & Richards A.G. - 1955 - Intracellular symbiosis in cockroaches. II - Mitotic division of mycetocytes - *Science* , **122**: 242-.

Brooks M.A. & Richards A.G. - 1956 - Intracellular symbiosis in cockroaches. III - Re-infection of aposymbiotic cockroaches with symbiotes - *J. Exptl. Zool.* , **132**: 447-466.

Buchner P. - 1965 - *"Endosymbiosis of animals with plant microorganisms." WILEY J. and Sons (Ed.) Intersciences Publishers N.Y.* , : 1-909.

Carayon J. - 1952 - Les mécanismes de transmission héréditaire des Endosymbiotes chez les Insectes. - *Tijdschrift v. entomologie - Symposium AMSTERDAM 1951* , **95**: 111-142.

Chang K.P. & Musgrave A.J. - 1972 - Multiple symbiosis in a leafhopper Helochara communis Fitch (Cicadellidae, Hom.) envelopes, nucleoids and inclusions of the symbiotes - *J. Cell. Sci.* , **11**: 275-293.

Chang K.P. & Musgrave A.J. - 1973 - Morphology, histochemistry, and ultrastructure of mycetome and its rickettsial symbiotes in Cimex lectularius - *Can. J. Microbiol.* , **19**: 1075-1081.

Chang K.P. & Musgrave A.J. - 1975 - Endosymbiosis in a leafhopper Helochara communis Fitch. (Cicadellidae : Homoptera) : symbiote translocation and auxillary cells in the mycetome - *Can. J. Microbiol.* , **21**, 2 : 186-195.

Daniel R.S. & Brooks M.A. - 1967 - Chromatographic evidence for murein from the bacteroid symbiotes of Periplaneta americana (L.) - *Experientia* , **23**: 499-502.

Daniel R.S. & Brooks M.A. - 1972 - Intracellular bacteroids : Electron microscopy of Periplaneta americana injected with lysozyme - *Exp. Parasit.* , **31**: 232-246.

Dasch G.A., Weiss E. & Chang, K.P. - 1984 - - *"BERGEY'S manual of systematic Bacteriology",Krieg N.R. and Holt J.G. (Eds.), Williams and Wilkins, Baltimore / London* , : 811-836.

Donachie W.D., Begg K.J. & Sullivan N.F. - 1984 - Morphogenes of Escherichia coli - *In "Microbial Development" Losick R. and Shapiro L. (Eds.) Cold Spring Harbor Laboratory, New-York* , : 27-62.

Eberle M.W. & Mc Lean D.L. - 1982 - Initiation and orientation of the symbiote migration in the human body louse Pediculus humanus L. - *J. Insect Physiol.* , **28**, 5 : 417-422.

Eberle M.W. & Mc Lean D.L. - 1983 - Observation of symbiote migration in human body lice with scanning and transmission electron microscopy - *Can. J. Microbiol.* , **29**, 7 : 755-762.

Ely B. & Shapiro L. - 1984 - Regulation of cell differenciation in Caulobacter crescentus - *In "Microbial Development" Losick R. and Shapiro L. (Eds.) Cold Spring Harbor Laboratory, New-York* , : 1-26.

Foeckler F. - 1961 - Reinfektionsversuche steriler Larven von Stegobium paniceum L. mit Fremdhefen und die Beziehungen zwischen der Entwicklungsdauer der Larven und dem B Vitamingehalt des Futters und der Hefen. - *Z. Morph. Ökol. Tiere* , **50**: 119-162.

Garthe W.A. & Elliott M.W. - 1971 - Role of intracellular symbionts in the fat body of cockroaches - influence on haemolymph proteins - *Experientia* , **27**: 593-.

Gharagozlou I.D. - 1966 - Les bactéries symbiotiques du tissu adipeux des Blattes ; ultrastructure et mode de transmission - *Ann. Sci. nat. Zool. Biol. anim. Fr.* , **8**: 567-576.

Grinyer I. & Musgrave A.J. - 1966 - Ultrastructure and peripheral membranes of the mycetomal microorganisms of Sitophilus granarius - *J. Cell. Sci.* , **1**: 181-186.

Hamon C. - 1971 - Etude au microscope électronique des symbiontes à transmission héréditaire chez quelques insectes Homoptères auchénorhynques femelles - *Z. Zellforsch.* , **119**: 244-256.

Hinde R. - 1971a - The control of the mycetome symbiotes of the aphids Brevicoryne brassicae, Myzus persicae, and Macrosiphum rosae - *J. Insect Physiol.* , **17**: 1791-1800.

Hinde R. - 1971b - The fine structure of the mycetome symbiotes of the aphids B. brassicae, M. percicae and M. rosae - *J. Insect Physiol.* , **17**: 2035-2050.

Houk E.J. & Griffiths G.W. - 1980 - Intracellular symbiotes of the Homoptera - *Ann. Rev. Entomol.* , **25**: 161-187.

Houk E.J., Griffiths G.W., Hadjokas N.E. & Beck S.M. - 1977 - Peptidoglycan in the cell wall of primary intracellular symbiote of the pea aphid - *Science* , **198**, 4315 : 401-403.

Huebner E. & Davey K.G. - 1974 - Bacteroids in the ovaries of a tse-tse fly - *Nature* , **249**: 260-261.

Ishikawa H. - 1982a - Host-symbiont interactions in the protein synthesis in the pea aphid, Acyrthosiphon pisum - *Insect. Biochem.* , **12**: 613-622.

Ishikawa H. - 1982b - Isolation of the intracellular symbionts and partial characterizations of their RNA species of the elder aphid, Acyrthosiphon magnoliae - *Comp. Biochem. Physiol.* , **72B**, 2 : 239-247.

Ishikawa H. - 1984 - Characterization of the protein species synthesized in vivo and in vitro by an aphid endosymbiont - *Insect Biochem.* , **14**, 4 : 417-425.

Ishikawa H. - 1987 - Nucleotide composition and kinetic complexity of the genomic DNA of an intracellular symbiont in the pea aphid Acyrthosiphon pisum. - *J. Mol. Evol.* , **24**: 205-211.

Ishikawa H. & Yamaji M. - 1985 - Symbionin, an aphid endosymbiont-specific protein -I. Production of insects deficient in symbiont - *Insect Biochem.* , **15**, 2 : 155-163.

Ishikawa H., Hashimoto H. & Yamaji M. - 1986 - Symbionin, an aphid endosymbiont-specific protein -III. Symbionin present in the male, ovipara and fundatrix - *Insect Biochem.* , **16**, 2 : 299-306.

Ishikawa H., Yamaji M. & Hashimoto H. - 1985 - Symbionin, an aphid endosymbiont-specific protein -II. Diminution of symbionin during post-embryonic development of aposymbiotic insects - *Insect Biochem.* , **15**, 2 : 165-174.

Juchault P. & Legrand J.J. - 1974 - Nature et action interspécifique du facteur épigénétique féminisant responsable d'une perturbation totale ou partielle de l'équilibre endocrinien contrôlant le phénomène sexuel du crustacé Armadillidium vulgare (Isopode Oniscoides) - *Ann. Endocrin.* , **35**: 387-392.

Kellen W.R., Hoffmann D.F. & Kwock R.A. - 1981 - Wolbachia sp. (Rickettsiales : Rickettsiaceae) a symbiont of the almond moth, Ephestia cautella : Ultrastructure and influence on host fertility - *J. Invertebr. Pathol.* , **37**, 3 : 273-283.

Koch A. - 1967 - Insects and their endosymbionts . - *In "Symbiosis" HENRY (Ed.), Ac. Press New-York and London.* , **2**: 1-96.

Körner H.K. - 1972 - Elektronenmikroskopische Untersuchungen am embryonalen mycetom der Kleinzikade Euscelis plebejus Fall. (Homoptera. Cicadina). I Die Feinstruktur der a-symbionten - *Z. Parasitenkd* , **40**, 3 : 203-226.

Körner H.K. - 1974 - Elektronenmikroskopische Untersuchungen am embryonalen Mycetom der Kleinzikade Euscelis plebejus Fall. (Homopt. Cicadina). II - Die Feinstruktur der t-symbioten - *Z. Parasitenkd.* , **44**: 149-164.

Körner H.K. - 1976 - On the host-symbiont-cycle of a leafhopper (Euscelis plebejus) endosymbiosis - *Experientia* , **32**: 463-464.

Körner H.K. - 1978 - Intraovarially transmitted symbionts of leafhoppers - *Zool. Beiträge* , **24**: 59-68.

Legrand J.J., Juchault P. & Martin G. - 1985 - Inoculation chez la femelle du Crustacé Oniscoide Porcellio dilatatus dilatatus Brandt, d'une bactérie symbiote caractéristique de la sous-espèce P. d. petiti, et ses conséquences sur l'issue du croisement des deux sous-espèces - *C. R. Acad. Sci. D, Paris* , **300**, 4 : 147-150.

Liu T.P. - 1974 - The effect of corpora allata on the plasma membrane of the symbiotic bacteria of the oocyte surface of Periplaneta americana - *Gen. Comp. Endocrinol.* , **23**: 118-123.

Louis C. - 1967 - Cytologie et cytochimie du mycétome de Pseudococcus maritimus (Homoptera Coccidae) - *C.R. Acad. Sci. D, Paris* , **265**: 437-440.

Louis C. - 1980 - Recherches ultrastructurales sur les cycles de procaryotes intracytoplasmiques d'Arthropodes. - *Thèse de Doctorat ès Sciences, MONTPELLIER.* , : 1-191.

Louis C. & Kuhl G. - 1972 - Synthèse in situ de protéines chez les symbiotes et chez les mycétocytes de Pseudococcus obscurus en absence ou en présence d'inhibiteurs spécifiques - *C. R. Acad. Sci. D, Paris*, **274**: 715-718.

Louis C. & Laporte M. - 1969 - Caractères ultrastructuraux et différenciation de formes migratrices des symbiotes chez Euscelis plebejus (Hom. Jassidae) - *Ann. Soc. entomol. Fr.*, **5**, 4 : 799-809.

Louis C. & Nicolas G. - 1976 - Ultrastructure of the endocellular procaryotes of Arthropods as revealed by freeze-etching. I - A study of "a" type endocymbionts of the leafhopper Euscelis plebejus - *J. Microscopie*, **26**: 121-126.

Louis C., Laporte M., Carayon J. & Vago C. - 1973 - Mobilité ciliature et caractères ultrastructuraux des microorganismes symbiotiques endo et exocellulaires de Cimex lectularius L. (Hemiptera, Cimicidae) - *C. R. Acad. Sci. D., Paris*, **277**, 6 : 607-611.

Louis C., Nicolas G. & Pouphile M. - 1976 - Ultrastructure of the endocellular procaryotes of arthropods as revealed by freeze - etching. II - "t" type endosymbionts of the leafhopper Euscelis plebejus Fall. (Homoptera, Jassidae) - *J. Micr. Biol. Cell.*, **27**, 1 : 53-58.

Malke H. & Bartsch G. - 1966 - Elektronoptische Untersuchung zur intracellulären Bakterien symbiose von Nauphoeta cinerea (Olivier) (Blattariae) - *Z. allg. Mikrobiol..*, **6**: 163-176.

Milburn N.S. - 1966 - Fine structure of the pleomorphic bacteroids in the mycetocytes and ovaries of several cockroaches - *J. Insect Physiol.*, **12**: 1245-1254.

Mitsuhashi J. & Kono, Y. - 1975 - Intracellular microorganisms in the green rice leafhopper, Nephotettix cincticeps Uhler (Hemiptera : Deltocephalidae) - *Appl. Ent. Zool.*, **10**, 1 : 1-9.

Musgrave A.J.,Grinyer I. & Homan R. - 1962 - Some aspects of the fine structure of the mycetomes and mycetomal microorganisms in Sitophilus. - *Can. J. Microbiol.*, **8**: 747-751.

Nardon P. - 1971 - Contribution à l'étude des symbiotes ovariens de Sitophilus sasakii : localisation, histochimie et ultrastructure chez la femelle adulte. - *C. R. Acad. Sci.*, **272D**: 2975-2978.

Nardon P. - 1973 - Obtention d'une souche aposymbiotique chez le charançon Sitophilus sasakii Tak. : différentes méthodes et comparaison avec la souche symbiotique d'origine. - *C. R. Acad. Sci.*, **277D**: 981-984.

Nardon P. - 1978 - Etude des interactions physiologiques et génétiques entre l'hôte et les symbiotes chez le Coléoptère Curculionide Sitophilus sasakii (= S. oryzae). - *Thèse de doctorat INSA-Université Lyon I*, : 1-285.

Nardon P. & Wicker C. - 1981 - La symbiose chez le genre Sitophilus (Coléoptère Curculionide). Principaux aspects morphologiques, physiologiques et génétiques. - *Ann. Biol.*, **20**: 327-373.

Nardon P., Louis C., Nicolas G. & Kermarrec A. - 1985 - Mise en évidence et étude des bactéries symbiotiques chez deux charançons parasites du bananier : Cosmopolites sordidus (Germar) et Metamasius hemipterus (L.) (Col. Curculionidae). - *Ann. Soc. ent. Fr.*, **21**, 3 : 245-258.

Peleg B. & Norris D.M. - 1973 - Oocyte activation in Xyleborus ferrugineus by bacterial symbionts - *J. Insect Physiol.*, **19**: 137-145.

Philippe C. - 1986 - Développement d'un nouveau modèle in vitro pour la culture à long terme de tissus de Periplaneta americana : applications à l'étude de la symbiose intracellulaire et de la physiologie du corps gras et de l'ovaire - *Thèse de Doctorat d'état, Université Pierre et Marie Curie, Paris VI*, : 1-247.

Riba G. & Chararas C. - 1976 - Etude anatomo-histologique et ultrastructurale des cryptes à symbiontes des larves de Criocephalus rusticus (Coleoptère Cerambycidae). - *Bull. Soc. Zool. Fr.*, **101**, 4 : 597-602.

Richards A.G. & Brooks M.A. - 1958 - Internal symbiosis in Insects. - *Ann. Rev. Entomol.*, **3**: 37-56.

Sander K. - 1968 - Entwicklungsphysiologische Untersuchungen am embryonalen Mycetom von Euscelis plebejus F. (Homoptera, Cicadina). I- Ansschaltung und abnorme Kombination einzelner Komponenten des symbiontischen Systems - *Develop. Biol.* , **17**: 16-38.

Schwemmler W. - 1974 - Endosymbionts : factors of egg pattern formation - *J. Insect Physiol.* , **20**, 8 : 1467-1474.

Schwemmler W. - 1980 - Endocytobiosis : general principles. - *Biosystems* , **12**, 1-2 : 111-122.

Schwemmler W. - 1987 - *"Handbook of Insect Endocytobiosis : Morphology, Physiology, Genetics, Evolution" Schwemmler W. (Ed.), CRC Press Inc. (Boca Raton, Florida, USA)* , : -.

Tsang K.R. - 1981 - Regulation of growth and transmission of an intracellular symbiote by its host - *Diss. Abstr. int. Sect. B : Sci. Eng.* , **41**, 7 : 2474-.

Wright J.D. & Barr A.R. - 1980 - The ultrastructure and symbiotic relationships of Wolbachia of mosquitoes of the Aedes scutellaris group - *J. ultrastruct. Res.* , **72**: 52-64.

LUMINESCENT BACTERIA : SYMBIONTS OF NEMATODES AND PATHOGENS OF INSECTS

K. Nealson, T.M. Schmidt and B. Bleakley
Center for Great Lakes Studies, University of Wisconsin-Milwaukee,
Milwaukee, WI 53204, U.S.A.

Luminous bacteria are a well-known group of (primarily marine) bacteria that are ecologically quite diverse (Nealson and Hastings, 1979). They participate in a wide variety of symbioses, including loose associations as gut symbionts, species-specific associations as extracellular symbionts of light organs of marine fishes and squids, and intracellular associations as symbionts of luminous tunicates (Nealson et al., 1981). Some examples of these associations are shown in Table 1.

In the past decade, considerable progress has been made in the understanding of these symbioses. This progress has come in two areas : 1) structural studies of light organs, and 2) studies of the factors that regulate bioluminescence in the bacterial symbionts. The structural studies have revealed that the light organs are complex, highly structured organs, evolved for the culture and maintenance of the bacterial symbionts, and for the dissipation and control of the bacterial luminescence (Bassot, 1975 ; Kessel, 1978 ; Tebo et al., 1979 ; McFall-Ngai, 1983).

Bacterial studies have focused on the factors that control both growth and bioluminescence. These include a mechanism called autoinduction (Nealson, 1977), catabolite repression (Nealson et al., 1972), osmotic regulation (Dunlap, 1984, 1985), regulation by oxygen tension (Nealson and Hastings, 1977), and iron repression (Makemson and Hastings, 1982 ; Haygood and Nealson, 1984, 1985) (Table 2). As shown, the various species of luminous bacteria are markedly different with regard to these variables.

Based on these studies, and the determination of growth rates in light organs (Haygood et al., 1984), several models of symbiosis have been proposed that involve nutrient exchange, oxygen limitation, iron limitation and osmotic control of both growth and luminescence (McFall-Ngai and Dunlap, 1963 ; Tebo et al., 1979 ; Nealson, 1979).

NATO ASI Series, Vol. H17
Cell to Cell Signals in Plant, Animal and
Microbial Symbiosis. Edited by S. Scannerini et al.

Table 1 : Symbiotic associations of luminous bacteria

Hosts	Kind of symbiosis	Bacteria	Reference
Marine invertebrates	gut symbionts	all species	Nealson, Hastings, 1979 Nealson *et al.*, 1981
Marine fishes and squids	gut symbionts	all species	Nealson, Hastings, 1979 Nealson *et al.*, 1981
	extracellular	V.fischeri	Tebo
	extracellular light organ symbionts	*V.fischeri* *P.phosphoreum* *P.leiognathi*	Tebo *et al.*, 1979 Nealson *et al.*, 1981 Bassot, 1975 ; McFall-Ngai, 1983
	extracellular light organ symbionts	non-culturable	Nealson, Hastings, 1979 Nealson *et al.*, 1981 Kessel, 1978
Pyrosomes	intracellular light organ symbionts	non-culturable	Nealson *et al.*, 1981

Table 2 : Factors that regulate bioluminescence in luminous bacteria

Regulatory Factor	*V. fischeri*	*V. harveyi*	*P. leiognathi*	*P. phosphoreum*
Autoinduction	+	+	+	+
High oxygen	-	+	+	-
Low oxygen	+	-	-	+
High Salt	NE	-	-	+
Low Salt	NE	+	+	-
Catabolite Repression	+	+	-	-
Iron Repression	+	+	+	+

NE indicates no effect of the variable tested, + indicates that the property is present in the species tested, or that luminescence is stimulated by that condition. - indicates an absence of the property being tested, or that luminescence is inhibited by that condition.

Recently, another complex relationship involving luminous bacteria has been revealed, one that differs in many ways from the more well-known examples discussed above. This is the involvement of a terrestrial genus Xenorhabdus in a life cycle that includes symbiosis and pathogenesis (Figure 1). The bacteria are carried as symbionts in specialized organs of the gut tract of entomopathogenic nematodes in the genera Heterorhabditis and Neoaplectana (Poinar and Thomas, 1967 ; Bird and Akhurst, 1983). The non-feeding (foraging) stages of the nematodes, upon locating suitable insect prey, enter the insect via the mouth or spiracles, and then penetrate the gut wall into the hemocoel. Once in the hemocoel, the bacteria are released, and the nematode begins its reproductive cycle. Upon completion of the life cycle, non-feeding stages are again formed, and the nematodes emerge in search of new insect prey (Poinar, 1966).

Many different insects are susceptible to killing by this symbiotic pair (Table 2) (Gaugler, 1981 ; Morris, 1985), with the host range most likely being dictated by the nematode host. The successful pathogenesis is a function of both symbiotic partners : without the nematode, the bacteria are unable to penetrate the insects and are therefore not potential pathogens, but rather opportunists ; without the bacteria, the nematodes are much less efficient at killing the insects (Dunphy et al., 1985) and at completing their own life cycle (Poinar and Thomas, 1966).

The bacterial symbionts, which have been placed in the genus Xenorhabdus (Grimont et al., 1984), are reported to have several unusual properties when grown in pure culture. These include production of red pigment, antibiotics, intracellular cristals, and bioluminescence. We have examined these factors for one strain of Xenorhabdus luminescens in an effort to unravel some of the organismic properties that are important in this unusual association, and present a summary of these results here.

Primary and secondary forms

One of the curious properties of Xenorhabdus is its propensity to form spontaneous variants that are altered in many properties (Table 3). These variants, referred to as secondary (2°) forms, are commonly identified as red colonies not surrounded by a zone of clearing on NBTA plates (Akhurst, 1980). They are reported to be much less virulent in terms of killing the

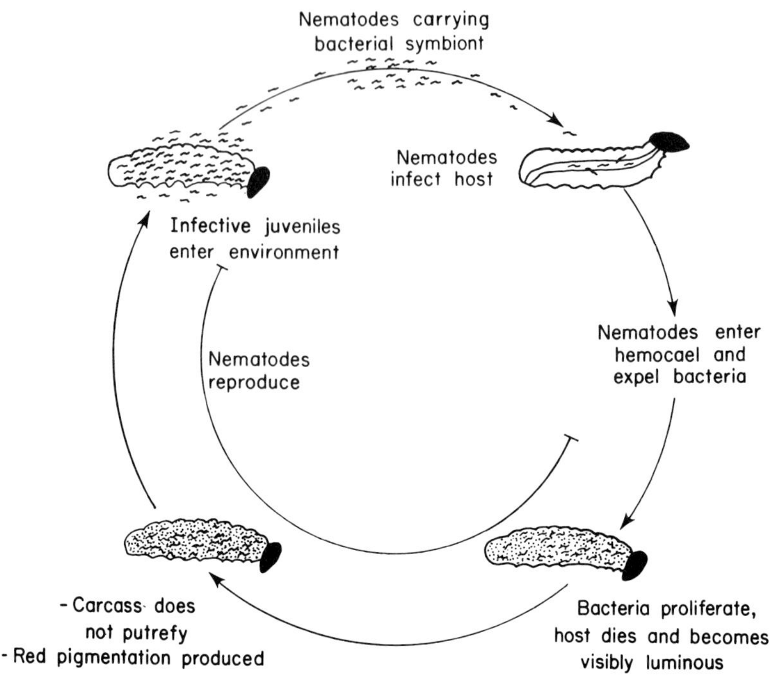

Bacterial/Nematode/Insect Life Cycle

Figure 1 : The life cycle of the bacterial/nematode/insect system. This diagram depicts the various interactions that occur between *Xenorhabdus* the bacterial symbionts, *Heterorhabditis*, the nematode symbionts, and *Galleria*, the insect prey. Adapted from Poinar (1966).

VIBRIO CHOLERAE

XENORHABDUS LUMINESCENS

VIBRIO HARVEYI

PHOTOBACTERIUM PHOSPHOREUM

VIBRIO FISCHERI

PHOTOBACTERIUM LEIOGNATHI

Figure 2 : Positive enzyme subunit complementation results obtained with subunits of bacterial luciferases from various species. The points of the arrows indicate the source of the beta subunit for positive results. Data summarized from Kopecky and Nealson, 1984 ; Meighen and Bartlett, 1980 ; Ruby and Hastings, 1980 .

insect prey, and are also much less effective in allowing the nematodes to complete their reproductive cycle. Thus it is important for successful infections that primary forms be maintained by the nematode. In terms of the overall symbiosis/pathogenesis, the transformation of 1° to 2° form would appear to be very important.

The mechanism whereby the 1° and 2° forms of *Xenorhabdus* are interconverted is not understood. In fact, there is disagreement in the literature as to whether the conversion occurs in both directions. In our experiments we have observed transformation in only one direction, from 1° to 2°. In virtually every case of spontaneous variant formation, all of the properties (pigment, antibiotic, bioluminescence and protease) are simultaneously altered. However, if luminescence is any indication, the structural genes are not lost, but strongly repressed. The mechanism is reminescent of the phase variation system seen in other enteric bacteria (Silverman and Simon, 1983), which is mediated by a transposable insertion element, but no detailed studies have yet been done on the *Xenorhabdus* system. Since bioluminescence is accurately and sensitively measured, it provides a good measure with which to follow the primary-secondary transitions.

Bioluminescence

The luminous system of *Xenorhabdus* has previously been shown to be similar in its biochemical requirements to that of other luminous bacteria, requiring reduced flavin ($FMNH_2$), a long chain aldehyde, and molecular oxygen (Poinar *et al*., 1980). Purification of the enzyme yielded luciferase subunits that were identical in size to those of other luciferases (Schmidt and Nealson, 1987). Furthermore, complementation of the subunits from *Xenorhabdus* strain Hb and those of the marine luminous bacterium *Vibrio harveyi* yielded active hybrids (Kopecky and Nealson, 1984), indicating that the active site of the enzyme is substantially conserved. These data are summarized in Figure 2, in which *X. luminescens* luciferase is shown to be related to luciferases from the other known species. These results have recently been substantiated in our laboratory by the cross reaction of cloned luciferase genes (*luxA*) from three species of luminous bacteria with DNA of *Xenorhabdus luminescens* (Nealson, unpublished). Even though this group is taxonomically and ecologically quite distinct from the other luminous bacteria, it appears that the enzyme system has come from the same genetic origin.

Bioluminescence in vivo showed a pattern of induction similar to that seen in other luminous bacteria (Nealson, 1977), in the sense that a lag in light emission occurred early in the growth cycle, followed by a period of rapid increase (Figure 3, left panel). The amount of luminescence produced in vivo was markedly stimulated by the addition of long chain aldehyde to the cells, suggesting that the cells are aldehyde limited. This aldehyde limitation is even more extreme during the lag phase, where the luminescence in vivo decreases by nearly a factor of 100. In vitro luciferase assays indicated luciferase and aldehyde biosynthetic genes are expressed concomitantly with the rise in in vivo luminescence. Figure 3, right panel, shows the luminescence of a 2° variant of Hm, which is markedly altered. The cells are very dim in vivo yet produce a substantial amount of luciferase as judged by enzyme assays. Furthermore, the cells appear to have lost the ability to induce luciferase synthesis (they are now constitutive) and are severely aldehyde limited. These data suggest that while the structural genes for both luciferase and aldehyde synthesis are still present, they are strongly repressed in the 2° form.

The function of luminescence in the symbiosis is not proven, although it has been hypothesized that the light emission might be of value in attracting other insects (potential prey) to the infected glowing insects, thus increasing the chance for the nematodes finding new prey (Poinar et al., 1980). Careful experiments with dark mutants of luminous bacteria are needed to test this hypothesis. For now, it should simply be noted that the regulation of the luminescence follows the same pattern as that of several other secondary metabolites, and it has value as an easily measured parameter to judge 1° versus 2° forms.

Production of Antibiotics

Antibiotics are commonly produced by Xenorhabdus isolates (Meighen and Bartlett, 1980 ; Ruby and Hastings, 1980), and it is thought that the ability to inhibit other invading bacteria is an important aspect of the symbiosis. Antibiotics from several different Xenorhabdus species have been purified and identified (Paul et al., 1981 ; Couche and Gregson, 1986 ;

Table 3 : Properties of primary and secondary forms of Xenorhabdus luminescens

Traits	Primary	Secondary
Pigmentation	+	-
Antibiotic Activity	+	-
Relative Bioluminescence	1.0	0.01
Intracellular Protein Crystals	+	-
Protease Activity	+	-
Lipase Activity	+	-

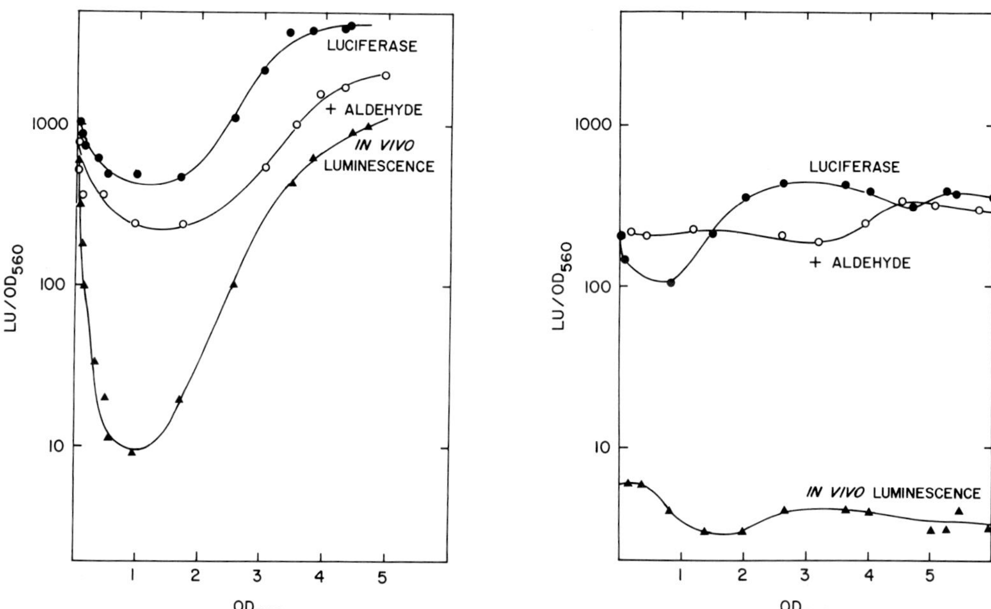

Figure 3 : Development of the luminous system in X. luminescens strain Hm. Luciferase, luminescence in vivo, and luminescence in vivo with added aldehyde are all plotted as a function of growth (OD_{560}). The 1° form (left) exhibits an inducible pattern of luminescence while the 2° form exhibits a constitutive pattern of synthesis with strong aldehyde limitation.

Richardson et al., 1987), and shown to fall into several different classes. We isolated and purified the antibiotic produced by Hk, and identified it as the hydroxystilbene compound shown in Figure 4. This compound has strong antibacterial activity against a variety of Gram (+) and Gram (-) types as determined by disk assay (Bleakley & Nealson, 1987). Interestingly, the 2° form of Hb produced no detectable antibiotic activity (Bleakley & Nealson, 1987).

The antibiotic is probably involved in the inhibition of other invading bacteria that might enter the hemolymph when the nematode enters. Since the completion of the life cycle requires the maintenance of the proper conditions, it seems likely that the antibacterial activity is an important, if not absolutely necessary component of the system.

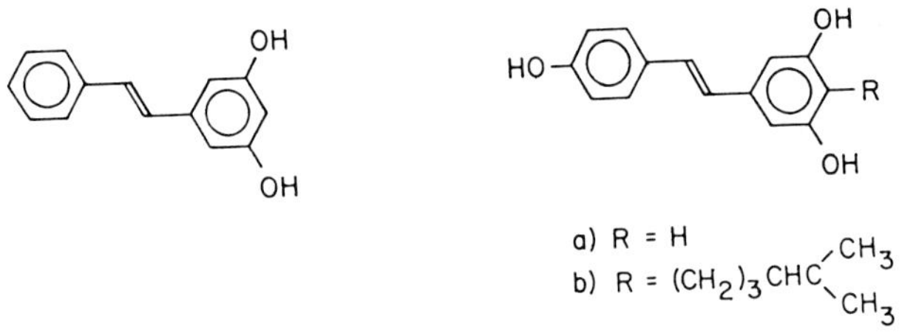

a) R = H
b) R = $(CH_2)_3CHC(CH_3)_2$

Figure 4 : Structure of the hydroxystilbene antibiotic produced by *X. luminescens* strain Hk.

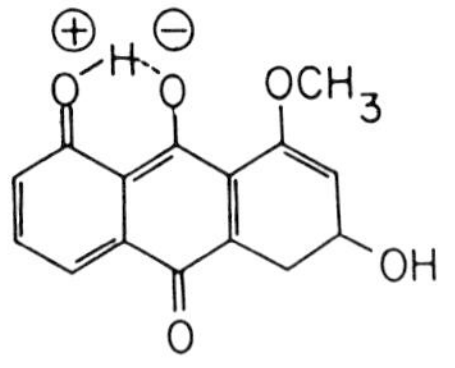

Figure 5 : Structure of the anthraquinone pigment produced by *X. luminescens* strain Hk.

Production of Pigment

In addition to producing antibiotic activity, the primary form of Hk is also strongly pigmented, producing a red color late in growth. This color probably accounts for the strong coloration of infected insects during the infection. We purified and identified the pigment as a pH sensitive anthraquinone compound (Figure 5 ; Richardson et al., 1987). This compound is yellow or cream colored at pH values below 9, and red at pH's above 9. In cultures grown on protein-containing media the pH is usually quite basic late in growth, and the colonies are accordingly red. On defined media, the color of the colonies is dependent on the substrates supplied (e.g. whether or not basic products can be produced). Preferred substrates for X. luminescens strains include K-malate ; as the malate is metabolized, the medium becomes alkaline, leading to strongly colored media either liquid or solid (Bleakley and Nealson, 1987). As with the antibiotic production, the secondary forms do not produce visible pigment on any media tested.

As with bioluminescence, the pigment could be involved with the attraction of new prey to the vicinity of the infection. During the day, colored carcasses might well attract other insects or higher level predators, resulting in distribution of the nematodes. Experiments with mutants of pigment-less mutants of luminous bacteria as symbionts are needed in order to carefully test this hypothesis.

Extracellular Enzyme Production

Xenorhabdus isolates are strongly proteolytic, as might be expected considering their role as pathogens in the hemocoel of the insect, where proteins are abundant. We have purified and partially characterized the protease from X. luminescens Hb (Schmidt et al., 1987) and found it to be an alkaline protease that can be inhibited by metal chelaters. The enzyme has a molecular weight of approximately 61,000, and a maximum activity near pH 8. As with the other properties considered above, the protease is not produced in detectable quantities by the secondary form of Hm.

Plate Assays with Tween 40 and Tween 80 revealed that the 1° forms were strongly lipolytic, while the 2° forms were either very weak or negative for this trait (Bleakley & Nealson,1987). No characterization of this activity has been reported.

The protease is undoubtedly involved with the breakdown of the insect protein material, and its absence in the 2° form is curious. The mode of growth in the absence of the protease would undoubtedly be quite different, with the emphasis being pointed toward carbohydrate and amino acid utilization. Protein degradation, with accompanying ammonia generation as amino acids are deaminated and consumed, would tend to make the primary forms drive the pH upward ; the effect of growth on pH of the 2° forms is less predictable. In terms of the nematode life cycle this might very well be quite important.

Intracellular Crystals and Buoyant Density of Cells

A distinguishing feature of the *Xenorhabdus* group is the presence of intracellular crystals (Boemare *et al.*, 1982). These crystals render the cells bright when viewed in the phase microscope. While we have not studied this property in detail, we have noted that there is a marked difference between 1° and 2° forms with regard to the crystals. The 1° forms contain phase bright crystalline inclusions, while the 2° forms lack them. This is consistent with the reports of Bowen and Ensign (1987) who have reported that 2° isolates of Hm do not contain intracellular crystals. In addition, the 1° and 2° forms differ markedly with respect to buoyant density. Percoll gradients can be used to separate the two forms from mixed cultures (Schmidt, unpublished results). Whether the presence of crystals is associated with this change in density is not known.

Both Bowen and Ensign (1987), working with *X. luminescens*, and Couche and Gregson (1987), working with *X. nematophilus*, have reported that the crystals are composed of proteins, and have reported molecular weights for the subunits of the crystals (10K for the former, and 22K and 26K for the latter). No functions have been reported, or even postulated for the crystals at this time, other than the suggestion of Boemare *et al.* (1982) that they might be involved with the luminescent reaction.

The symbiosis we have discussed here is intimately connected with a pathogenesis, and as such the relationships that occur between the bacteria and the eucaryotes may be quite different from those seen in other symbioses. The ultimate goal of this symbiosis is presumably to achieve a system that is compatible with successful reproduction of the nematode/bacterial pair. To this end, we suspect that the bacterial products produced during growth must be intimately involved, and our work has thus focused on the products of the bacteria. Once the identity and the patterns of production of the secondary metabolites are known, we may begin to appreciate the role that chemical communication plays in this unusual symbiotic system.

As a final point, it should be noted that the mechanism whereby the insects are killed by this symbiosis is not known. While the bacteria may be involved either directly or indirectly, it seems equally likely that the role of the bacteria is in preparing and maintaining conditions for successful reproduction and distribution of the nematodes. These issues can be resolved with careful experiments using bacterial-nematode systems in the future.

REFERENCES

Akhurst, R.J. 1980. Morphological and functional dimorphism in Xenorhabdus spp., bacteria symbiotically associated with the insect pathogenic nematodes Neoplectana and Heterorhabditis. J. Gen. Microbiol. 121: 303-309.

Akhurst, R.J. 1982. Antibiotic activity of Xenorhabdus spp., bacteria symbiotically associated with insect pathogenic nematodes of the families Heterorhabditidae and Steinernematidae. J. Gen. Microbiol. 128: 3061-3065.

Bassot, J.M. 1975. Les organes lumineux a bacteries symbiotiques de quelques teleosteens Leiognathides. Arch. Zool. Exp. Gen. 116: 359-373.

Bird, A.F., Akhurst, R.J. 1983. The nature of the intestinal vesicle in nematodes of the family Steinernematidae. Int. J. Parasit. 13: 599-606.

Bleakly, B., Nealson, K.H. 1987. Characterization of primary and secondary forms of Xenorhabdus luminescens: Growth, luminescence, and secondary metabolite production in defined media. J. Bacteriol. (in press).

Boemare, N., Louis, C., Kuhl, G. 1982. Etude ultrastructurale des cristaux chez Xenorhabdus spp., bacteries infeodees aux nematodes entomophages Steinernematidae et Heterorhabditidae. C.R. Soc. Biol. 177: 107-115.

Bowen, D., Ensign, J.C. 1987. Intracellular protein crystal of the insect pathogen Xenorhabdus luminescens. Proc. Amer. Soc. Microbiol. p. 183 (abstract).

Couche, G.A., Gregson, R.P. 1986. Metabolites produced during in vitro growth of Xenorhabdus sp. Proc. Int. Conf. Insect. Parasitol. Netherlands (abstract).

Couche, G.A., Gregson, R.P. 1987. Protein inclusions produced by the entomopathogenic bacterium, Xenorhabdus nematophilus subsp. nematophilus.

J. Bacteriol. (in press).

Dunlap, P.V. 1984. The ecology and physiology of the light organ symbiosis between Photobacterium leiognathi and ponyfishes. PhD Thesis, Univ. Cal. Los Angeles, CA.

Dunlap, P.V. 1985. Osmotic control of luminescence and growth in Photobacterium leiognathi from ponyfish light organs. Arch. Microbiol. 141: 44-50.

Dunphy, G.B., Rutherford, T.A., Webster, J.M. 1985. Growth and virulence of Steinernema glaseri influenced by different subspecies of Xenorhabdus nematophilus. J. Nematol. 17: 476-482.

Gaugler, R. 1981. Biological control potential of neoaplectanid nematodes. J. Nematol. 13: 241-249.

Grimont, P.A., Steigerwalt, A.G., Boemare, N., Hickman-Brenner, F.W., Deval, C., Grimont, F., Brenner, D.J. 1984. Deoxyribonucleic acid relatedness and phenotypic study of the genus Xenorhabdus. Int. J. Syst. Bacteriol. 34: 378-388.

Haygood, M.G., Nealson, K.H. 1985. Mechanisms of iron regulation of luminescence in Vibrio fischeri. J. Bacteriol. 162: 209-216.

Haygood, M.G., Nealson, K.H. 1985. The effect of iron of the growth and luminescence of the symbiotic bacterium Vibrio fischeri. Symbiosis 1: 39-51.

Haygood, M.G., Tebo, B., Nealson, K.H. 1984. Luminous bacteria of a monocentrid fish (Monocentris japonicus) and two anomalopid fishes (Photoblepharon palpebratus and Kryptophanaron alfredi): Population sizes and growth within the light organs, and rates of release into the seawater. Mar. Biol. 75: 249-254.

Kessel, M. 1978. The ultrastructure of the relationship between the luminous organ of the teleost fish Photoblepharon palpebratus and its symbiotic bacteria. Cytobiologie Z. Exp. Zellforsch. 15: 145-158.

Kopecky, Nealson, K.H. 1984. Cross reaction of luciferase subunits from different species of bacteria. Proc. Amer. Soc. Microbiol. (abstract).

Makemson, J., Hastings, J.W. 1982. Iron represses bioluminescence and affects catabolite repression of luminescence in Vibrio harveyi. Curr. Microbiol. 7: 181-186.

McFall-Ngai, M.J. 1983. Adaptations for reflection of bioluminescent light in the gas bladder of Leiognathus equulus (Perciformes: Leiognathidae) J. Expt. Zool. 227: 23-33.

McFall-Ngai, M.J., Dunlap, P. 1963. Three new modes of luminescence in the leiognathid fish Gazza minuta: Discrete projected luminescence, ventral body flash, and buccal luminescence. Mar. Biol. 73: 227-237.

Meighen, E., Bartlet, I. 1980. Complementation of subunits from different bacterial luciferases. J. Biol. Chem. 255: 1181-1187.

Morris, O.N. 1985. Susceptibility of 31 species of agricultural insect pests to the entomagenous nematodes Steinernema feltiae and Heterorhabditis bacteriophora. Can. Ent. 117: 401-407.

Nealson, K.H. 1977. Autoinduction of bacterial luciferase: Occurrence, mechanism and significance. Arch. Microbiol. 112: 73-79.

Nealson, K.H. 1979. Alternative strategies of symbiosis of marine luminous fishes harboring light emitting bacteria. Trends Biochem. Sci. 4: 105-110.

Nealson, K.H., Hastings, J.W. 1977. Low oxygen is optimal for luciferase synthesis in some bacteria: Ecological implications. Arch. Microbiol. 112: 9-16.

Nealson, K.H., Hastings, J.W. 1979. Bacterial bioluminescence: Its control and ecological significance. Microbiol. Rev. 43: 496-518.

Nealson, K.H., Eberhard, A., Hastings, J.W. 1972. Catabolite repression of bacterial bioluminescence: Functional implications. Proc. Nat. Acad. Sci. (USA) 59: 1073-1076.

Nealson, K.H., Cohn, D., Leisman, G., Tebo, B. 1981. Coevolution of luminous bacteria and their eukaryotic hosts. N.Y. Acad. Sci. 361: 76-91.

Paul, V.J., Frautschy, S., Fenical, W., Nealson, K.H. 1981. Antibiotics in microbial ecology: Isolation and structure assignment of several new antibacterial compounds for the insect-symbiotic bacteria Xenorhabdus spp. J. Chem. Ecol. 7: 589-597.

Poinar, G.O. 1966. The presence of Achromobacter nematophilus in the infective stage of a Neoplectana sp. (Steinernematidae: Nematoda). Nemagologica 12: 105-108.

Poinar, G.O., Thomas, G.M. 1966. Significance of Achromobacter nematophilus Poinar & Thomas (Achromobacteriaceae: Eubacteriales) in the development of the nematode, DD136 (Neoplectana sp. Steinernematidae). Parasitol. 56: 385-390.

Poinar, G.O., Thomas, G.M. 1967.The nature of Achromobacter nematophilus as an insect pathogen. J. Invert. Pathol. 9: 510-514.

Poinar, G.O. Jr., Thomas, G., Haygood, M., Nealson, K.H. 1980. Growth and luminescence of the symbiotic bacteria associated with the terrestrial nematode, Heterorhabditis bacteriophora. Soil Biol. Biochem. 12: 5-10.

Richardson, W.H., Schmidt, T.M., Nealson, K.H. 1987. Secondary metabolite production by Xenorhabdus luminescens. J. Bacteriol. (in press).

Ruby, E.G., Hastings, J.W. 1980. Formation of hybrid luciferases from subunits of different species of Photobacterium. Biochem. 19: 4989-4993.

Schmidt, T.M., Nealson, K.H. 1987. Regulation of bioluminescence and properties of luciferase from Xenorhabdus luminescens. J. Bacteriol. (in press).

Schmidt, T.M., Bleakley, B., Nealson, K.H. 1987. Purification and characterization of an extracellular protease from the insect pathogen Xenorhabdus luminescens. J. Gen. Microbiol. (in press).

Silverman, M., Simon, M. 1983. Phase Variation and Related Systems (J.A. Shapiro, Ed.) Academic Press, N.Y.

Tebo, B.M., Linthicum, D.S., Nealson, K.H. 1979. Luminous bacteria and light emitting fish: Ultrastructure of the symbiosis. BioSystems 11: 169-180.

Cell-to-cell interactions during the establishment of the Hydra-*Chlorella* symbiosis.

P.J. McAuley
Department of Plant Sciences
Oxford University
Parks Road
Oxford OX1 3PF
United Kingdom

Introduction

This review paper will describe research into the problem of whether specific signals or properties of the algal symbionts of green hydra elicit a response from the host digestive cells which may constitute discriminatory "recognition". It will also address two paradoxes raised by attempting to compare reinfection of artificially produced aposymbiotic hydra in the laboratory with the life cycle and habitat of green hydra in nature. Firstly, although hydra have been successfully reinfected with a variety of *Chlorella* algae in the laboratory, including several which do not release maltose (Jolley and Smith 1980; Rahat and Reich 1985), hydra collected from nature invariably have been found to contain maltose-releasing algae. Secondly, there is the problem of the significance of discriminatory processes shown during reinfection of aposymbiotic hydra. Based on an original scheme given by Pardy and Muscatine (1973), McAuley and Smith (1982a) described four stages in the recolonization of digestive cells by symbiotic algae. These are: phagocytosis; sorting (avoidance of lysosomal attack); transport to base; and integration with the metabolic processes of the host cell. These stages are summarised in Figure 1. Maltose-releasing algae have been shown

NATO ASI Series, Vol. H17
Cell to Cell Signals in Plant, Animal and
Microbial Symbiosis. Edited by S. Scannerini et al.

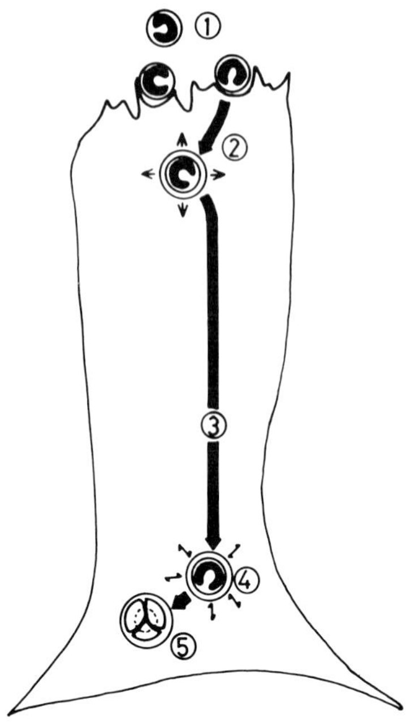

Figure 1. Stages of reinfection of aposymbiotic hydra by symbiotic algae: 1, phagocytosis; 2, avoidance of host lysosomes; 4, integration with host cell metabolic processes leading to; 5 division of algae and repopulation of the host cell. After McAuley and Smith (1982).

to be phagocytosed in larger numbers, and to avoid lysosomal attack and be transported to the base with greater frequency, than non-maltose releasing algae (McNeil et al. 1981; Hohman et al. 1982; McAuley and Smith 1982a). However, algal symbionts are normally transmitted directly to asexually produced offspring during cell division and, during sexual reproduction of the hydra, algae enter the egg before maturation and fertilisation of the oocyte (Muscatine and McAuley 1984). Why, therefore, do hydra digestive cells possess the ability to discriminate between maltose and non-maltose releasing algae? Is there evidence that any of the stages of the reinfection process constitutes specific recognition of potential symbionts?

The process of reinfection

1. Phagocytosis. After algae are injected into the coelenterons of aposymbiotic hydra, the digestive cells take them up

by phagocytosis. Freshly isolated symbionts (FIS) infect a higher proportion of digestive cells, and enter in higher numbers, than heat-treated FIS, cultured symbiotic or free-living algae, and latex spheres. It is well known that, during fertilisation or tissue differentiation in plants and animals, cell-to-cell recognition is mediated by interaction between specific macro-molecules on the cell surfaces. However, there is no evidence that higher infectivity of FIS is due to a host cell response to unique surface characteristics. Attempts to remove or mask possible recognition sites on FIS and host cell surfaces using proteases, lectins and antibodies (Pool 1979; Jolley and Smith 1980; Meints and Pardy 1980) failed to distinguish between specific and general inhibitory effects on host cell phagocytosis, and did not take into account contamination of FIS with host cell material which is not removed by conventional washing techniques (Douglas and Smith 1983; McAuley 1986a). Scanning electron microscope studies of the modes of phagocytotic entry into host cells of FIS, other algae, and a range of similarly-sized particles, using the Florida and European strains of hydra, showed that quantitative differences in uptake were not correlated with qualitative differences in phagocytotic modes (McNeil 1981a; McNeil and Smith 1982). Instead, higher uptake of FIS appears to be due to interaction of a high net negative surface charge of the algal cells and a net positive surface charge of the digestive cells (McNeil et al. 1981). Abolition of the negative charge by treatment with polycations reduces phagocytosis of FIS by digestive cells, and this inhibitory effect can be reversed by subsequent addition of polyanions, which also increases uptake of latex spheres (Table 1). Thus, although FIS enter host cells in greater numbers than other similarly sized particles, there is no discrimination specifically relevant to algal symbionts at the digestive cell surface. As will be discussed below, mode of entry into the digestive cells is not related to subsequent establishment of symbiosis.

2. Avoidance of host digestion. Once potential symbionts have

Table 1. Polyanion reversal of polycation inhibition of uptake of freshly isolated symbionts (FIS).

Treatment of FIS		No. of FIS/cell	% of control
Hour 1	Hour 2	($\bar{x}$ ± S.E.)	
M	M	13.5 ± 0.5	100
Poly-L-lysine	M	2.9 ± 0.3	21
Poly-L-lysine	BSA	13.3 ± 0.3	99
Poly-L-lysine	Poly-L-glutamic acid	17.0 ± 0.7	127
BSA	Poly-L-lysine	7.7 ± 0.4	53

FIS were incubated for 1h in the polyanion poly-L-lysine (2.5 mg/ml), the polycation poly-L-glutamic acid (2.5 mg/ml) or M solution (hydra culture medium). They were then washed 4 times and incubated for a second 1h in another of these 4 solutions. Finally, treated FIS were injected, without washing, into Florida aposymbionts. Results from McNeil et al. (1981) by kind permission of Dr. P.L. McNeil.

entered a host cell, they must avoid the digestive processes which normally follow phagocytosis. Although early work suggested that successful symbionts were able to resist host digestive enzymes, it was subsequently demonstrated that symbiotic algae are not exposed to these enzymes, since they are able to prevent fusion of host lysosomes with the phagosome in which they are contained (Hohman et al. 1982; O'Brien 1982; McNeil and McAuley 1984). Normally, algal symbionts of green hydra release a large proportion of photosynthetically fixed carbon in the form of maltose, which is rapidly utilised by the host (Mews 1980). The ability of algae to avoid lysosomal fusion is correlated with their maltose release capacity. Treatments which reduce or perturb maltose release increase the frequency of fusion of primary and secondary lysosomes with algal-bearing phagosomes, shown by appearance of acid phosphotase and ferritin or thorium markers. Maltose release is dependent upon external pH, and is maximal at about pH 4.0

(Mews 1980); therefore, avoidance of lysosomal fusion must be dependent upon rapid acidification of the phagosome. In amoebae and in mammalian cells, pinosomes and phagosomes have been shown to become acidic even before primary lysosomal fusion, within the first five minutes after entry into the cell (McNeil et al. 1983; Galloway et al. 1983). The proportion of phagocytosed FIS algae which escape digestion varies between hydra strains. In digestive cells of Florida hydra, almost all FIS escape digestion while, in European hydra, only 50% of FIS do so (Hohman et al. 1982; McAuley and Smith 1982a; McNeil and McAuley 1984). Digestion of viable, maltose-releasing FIS may be due to competition between rates of maltose-release and lysosomal fusion, or to variability in acidification of phagosomes.

The precise nature of the signal which prevents lysosomal fusion is unknown, although it must interfere in some way with the interaction between the outer membrane layers of the phagosome and the lysosomes. It is not known how the membrane of a perialgal vacuole (PAV) containing FIS or other maltose-releasing algae specifically differs from vacuoles containing freeliving algae or food particles, although freeze-fracture studies in symbiotic Paramecium bursaria as well as green hydra have shown that the distribution of membrane particles differs between perialgal and food vacuoles (Meier et al. 1980, 1984; McNeil 1984). The time-course of changes in membrane particle topography and the influence of maltose release during transformation of algal-bearing phagosomes to PAVs may provide insight into the way in which lysososmal fusion is prevented.

3. Transport to cell base. After avoiding lysosomal fusion, perialgal vacuoles containing symbiotic algae must be transported from the digestive cell apex to the base, where the algae normally reside. Movement of PAVs appears to be mediated by host microtubules. Tracts of microtubules running between apex and base have been visualised in digestive cells using electron microscopy (Hawes, pers. comm.), and treatment of newly reinfected hydra with inhibitors of microtubule poly-

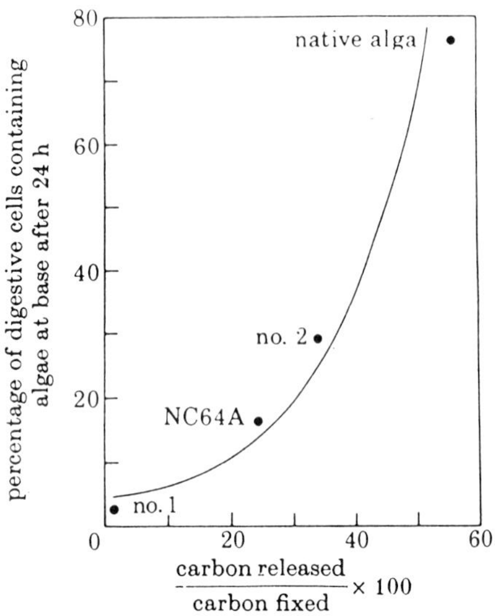

Figure 2. Correlation of photosynthate release of four strains of algae and their retention in digestive cells of European aposymbionts. Exponential curve fitted by regression and analysis: $y = 1.68^{1.1x}$, $r = 0.98$. From McAuley and Smith (1982a).

merization have been shown to delay movement of algae from apex to base (Cooper and Margulis 1977; Hohman et al. 1982; McAuley and Smith 1982b). These experiments also demonstrated that avoidance of lysosomal fusion and initiation of transport are separate events, as treatment with microtubule inhibitors increased the time PAVs spent at the cell apex but did not increase the number of algae digested (Hohman et al. 1982; McAuley and Smith 1982b). Prevention of transport away from the cell apex does not appear to cause digestion of FIS. The mechanism by which symbiotic algae stimulate transport of their enveloping PAVs is not known but, like avoidance of lysosomal fusion, transport appears to depend upon the capacity for maltose release by the algae. Further, Figure 2 shows that the proportion of phagocytosed algae retained by the digestive cells is related to the amount of maltose released by the algae.

4. Integration. Once at the base of the digestive cell, symbiotic algae must integrate with host cell metabolic processes, to allow exchange of metabolites in both directions, and so that subsequent growth and division of the algal population is regulated to that of the host cell. In reinfected Florida hydra, the number of algae retained is similar to that of the normal population level, and no further algal division is needed to repopulate the digestive cells (Pool 1979; Neckelmann and Muscatine 1983). However, in European hydra, the average number of algae retained per cell is lower than the normal population level, and algae must temporarily divide faster than host cells to reach that level, a process that can take up to three weeks (Jolley and Smith 1980; McAuley and Smith 1982a). During the first 5-7 days of this reinfection period, movement of some algae from the base of the digestive cell and subsequent digestion and ejection has been observed, a process known as "resorting". The significance of this is unclear. It may be due to failure of some algae to integrate properly with host cell metabolism, or because a function or functions lacking in aposymbiotic digestive cells, necessary for resumption of symbiosis, is recovered only slowly.

Differences in function and metabolism between aposymbiotic and algal-bearing digestive cells suggest that the host cells actively respond to the presence of symbiotic algae. Experiments with European and Florida hydra have shown that establishment of maltose-releasing symbiotic algae within digestive cells affects the host cell so that algae which are subsequently phagocytosed are not transported to the base and are unable to avoid lysosomal fusion and digestion (McAuley and Smith 1982b; Neckelmann and Muscatine 1983). This occurs even in digestive cells which do not contain a full complement of algae, and both high and low-maltose releasing strains of *Paramecium* algae are able to "protect" their habitat, even from the native strain of symbiotic algae. This suggests that the signal which, in effect, causes a shut-down of "recognition" processes in host cells, is not quantitatively associated with maltose release. Recent research has shown that the presence of maltose-releasing algae increases the level of host glutamine synthetase, thereby either preventing or controlling access by the algae to ammonium produced by host catabolic processes (Rees 1986). The level of host glutamine synthetase appears to be related to the amount of maltose re-leased by the algae. Hydra symbiotic with a low maltose-releasing strain possessed levels of the enzyme similar to that in aposymbionts.

Reinfection - specific or non-specific?

The process of reinfection leading to establishment of a stable symbiosis is not activated by any single unique character of symbiotic *Chlorella*. Instead, potential symbionts must pass through a serial process of selection based upon interaction between a variety of algal and host cell characters. At each stage in the reinfection process, an algal cell has a finite chance of passing onto the next even if it lacks the properties which lead to symbiotic algae being selectively retained at digestive cell bases. This is very different from cell-to-cell recognition processes during fertilisation or tissue differentiation, where interaction between macromolecules on the cell surfaces invoke an all-or-nothing response. During reinfection of hydra, FIS and other symbiotic algae which possess a high net negative surface charge and the capacity to release maltose are more readily phagocytosed and retained by digestive cells than symbiotic and freeliving algae which do not possess these characters. However, on occasion, small numbers of freeliving algae able to avoid or to resist host digestive processes are subsequently able to colonise cells of aposymbiotic hydra. A small proportion of non-maltose releasing strains of *Chlorella* may be retained at the bases of European hydra digestive cells (McAuley and Smith 1982a), and a variety of freeliving *Chlorella* may form temporary or more permanent symbioses with hydra in the laboratory (Jolley and Smith 1980; Rahat and Reich 1984).

Since reinfection of hydra digestive cells takes place by a number of defined stages, it has been suggested that reinfection is analogous to colonization of an empty ecological niche rather than cell-to-cell recognition. Rahat and Reich (1987) suggested that infection is determined not by specific host-symbiont interactions but by characteristics of algae which enable them to survive within the host vacuole. This model of reinfection suggests that the host cell is essentially the passive partner in the symbiosis, and does not respond to the presence of symbionts. Unfortunately, while the model of the host cell as an ecological niche is an attractively simple idea, critical examination reveals a number of flaws.

Firstly, the "ecological niche" concept regards the host cell as a passive, steady-state system whose conditions the algae are able to exploit readily. However, the presence of a host vacuolar membrane surrounding each algal cell presents a barrier to unregulated uptake of nutrients by the algae. A range of nutrients, from phosphate to amino acids, require active transport systems to traverse the membrane. Not only must these systems be derived from the host cell, but they must also be powered by host ATP. Nor can the host cell be considered analogous to a free pathway down which nutrients diffuse to the PAVs. Host cell utilisation of nutrients and sequestration into pools unavailable to the algae would present formidable barriers against free access. Indeed, recent research has shown that nitrogen supply is strictly controlled by the host. Rather than being unchanged, digestive cells containing maltose-releasing algae possess higher levels of glutamine synthetase than aposymbiotic cells or cells containing algae which only release small amounts of maltose (Rees 1986). It has been suggested that high levels of this enzyme in the host cell prevents or controls access by the algae to ammonium produced by host catabolic processes (Rees 1986), and symbiotic algae isolated from green hydra exhibit properties of amino acid uptake and metabolism similar to those possessed by nitrogen deficient algae (McAuley 1986b; in press).

A second problem is that, while symbioses between hydra and free-living algae can be produced and maintained in the laboratory, hydra isolated from nature invariably contain maltose-releasing algae. Also, artificial symbioses between hydra and free-living algae exhibit a number of characteristics not found in the normal symbiosis. Free-living algae are frequently found in vacuoles containing a number of cells rather than in individual vacuoles; they are distributed throughout the length of the digestive cell rather than being sequestered at the base; and they appear to be continuously ejected from the digestive cells, unlike native symbionts (Rahat and Reich 1984, 1985). Furthermore, when double infections of various types of free-living algae and the native symbiont were made, it was found that in every case the native symbiont displaced the free-living algae (Rahat 1985). As described above, maltose-releasing algae established at the bases of digestive cells prevent further infection, since the mechanism whereby newly-phagocytosed symbiotic algae interact with the cells to avoid digestive processes is switched off (McAuley and Smith 1986b). Free-living algal strains appear unable to prevent subsequent infection with other types of algae. Thus, while they may form opportunistic infections in hydra digestive cells which do not contain maltose-releasing algae, subsequent infection with symbiotic algae displaces any free-living algae. Rather than behaving as a passive "ecological niche", the host cell responds to the presence of maltose-releasing algae so that further, possibly deleterious, infection is prevented.

How often does infection occur in nature?

In the laboratory, the reinfection process has been studied by introducing algae to artificially produced aposymbiotic hydra which have never been found in nature. Usually, hydra reproduce asexually and, at mitosis, algae appear to be invariably transmitted to the daughter cells. However, in nature, hydra undergo at least one round of sexual reproduction a year (Ellard 1986), and the gametes are derived from ectodermal cells which do not contain algae. Therefore, is reinfection necessary when hydra sexually reproduce?

Electron microscopical studies have shown that, in at least

two strains of hydra, Frome and Florida, algae are present in the eggs, entering by cytoplasmic extensions from the endodermal cells which may also transport nutrients to the developing ooplasm (Muscatine and McAuley 1984). This study also showed that the algae lie outside the developing embryo. It is not known how and at what stage of embryo development the algae enter the digestive cells, but it is not inconceivable that they do so by a process closely analagous to the reinfection processes described in aposymbiotic hydra. In this case, digestive cells need not distinguish between different types of algae, since only native symbionts would be present in the egg. However, since the algae must enter digestive cells by the same or by similar phagocytotic modes as food particles, a signal (maltose release) would be necessary to enable digestive cells to distinguish between food and symbionts. This may explain why "recognition" occurs after initial contact, and why it is dependent upon the non-specific character of maltose release.

The number of algae found in hydra eggs varies considerably, between as few as one and as many as 2,000 (Muscatine and McAuley 1984). Furthermore, about 40% of Florida and 20% of Frome eggs appeared to contain no algae at all, although the possibility that a single infecting algal cell was over-looked cannot be ruled out. In a study of hydra hatching from eggs produced by wild-type hydra it was found that, although four out of seven individuals hatched as aposymbionts, all later greened (Table 2). Since artificially produced aposymbionts never re-green in the conditions in which hatched hydra were cultured it is likely that freshly-hatched hydra which are apparently aposymbiotic are infected at a very low level. During the period of greening, individuals with low numbers of symbionts may be vulnerable to infection with other types of algae. Algae in the gut of a captured zooplankter may be taken up by the digestive cells (cf infection method of Rahat and Reich, 1984, 1985), while protozoa symbiotic with *Chlorella* may be a vector for infection if taken into the hydra's coelenteron (M. Christopher, unpublished). However, experiments in which hydra were infected with combinations of different algal strains showed that the native algae always

Table 2. Description of hydra hatched from eggs produced between 3-11 June, 1985 by wild-type hydra collected from Duck Decoy Pond, Wytham, Oxford.

Date hatched	Number of tentacles	Comments	Interval before 1st bud (days)
1. 17/7/85	6	Normal green hydra.	5
2. 19/7/85	6	Normal aposymbiotic hydra. Patchy greening after 10 days.	4
3. 23/7/85	5	Normal aposymbiotic hydra. Patchy greening after 10 days.	5
4. 30/7/85	8	Normal green hydra.	6
5. 15/8/85	0	Headless aposymbiotic hydra containing a green pellet. After 3 days developed 10 tentacles and had greened.	11
6. 2/9/85	0	Headless aposymbiotic hydra which greened but produced no tentacles or buds.	-
7. 3/9/85	4	Normal green hydra.	10

displaced free-living algae (Rahat 1985), and the establishment of maltose-releasing algae within digestive cells prevents further infection (McAuley and Smith 1982b). Thus, while transitory infections of freeliving algae may occur in newly hatched hydra which initially possess low numbers of symbionts, preferential response of host cells to maltose-releasing symbionts would ensure that freeliving algae were eventually lost.

Conclusions

Maltose release by symbiotic algae elicits responses from host digestive cells of green hydra both during reinfection of aposymbionts and in the established symbiosis. During reinfection, the presence of maltose-releasing algae within phagosomal vacuoles prevents fusion of lysosomes and causes the vacuoles to be transported from the digestive cell apex to the base (Hohman et al. 1982; McAuley and Smith 1982a,b; O'Brien 1982). In the established symbiosis, maltose-releasing algae prevent further infection by inducing shut-down of the host cell mechanisms involved in selection of suitable symbionts (McAuley and Smith 1982b); and maltose release also causes elevation of host glutamine synthetase levels, involved in regulating algal cell growth and division via the nitrogen supply (Rees 1986). The former reaction is elicited by both high and low-maltose releasing strains of algae; the latter by only high-maltose releasing strains.

As Smith (1980) pointed out, "the arrival of an alga at the distal end of a digestive cell is not a consequence of a positive and specific act of recognition". Rather than a signal specifically produced by the symbiont to elicit a single specific reaction by the host cell, in green hydra the general property of maltose release by the algae serves to elicit a variety of responses from the host cell. Green hydra may not have developed specific recognition mechanisms because of direct transmission of algae into the egg and to asexually produced offspring. The symbiotic algae need only to avoid host digestive processes upon entering the digestive cells of the developing embryo, and are not in competition with other potential symbionts. Thus, although artificially produced aposymbionts demonstrate a degree of non-specificity regarding the algae with which they form symbioses, in nature the symbiosis is highly specific and its stability is dependent upon a continuous flow of signals between the partners.

References

Cooper, G. & Margulis, L. (1977). Delay in migration of symbiotic algae in Hydra viridis by inhibitors of microtubule polymerization. Cytobios 19, 7-19.

Douglas, A.E. & Smith, D.C. (1983). The cost of symbionts to their hosts in green hydra. In Endocytobiology II (eds. W. Schwemmler and H.E.A. Schenk) pp 631-648, Walter de Gruyter, Berlin.

Ellard, F.M. (1986). Growth in symbiotic hydra. Ph.D. thesis, Oxford University.

Galloway, C.J., Dean, G.E., Marsh, M., Rudnick, G. & Mellman, I. (1983). Acidification of macrophage and fibroblast endocytic vesicles in vitro. Proc. Natl. Acad. Sci. U.S.A. 80, 3334-3338.

Hohman, T.C., McNeil, P.L. & Muscatine, L. (1982). Phagosome-lysosome fusion inhibited by algal symbionts of Hydra viridis. J. Cell Biol. 94, 56-63.

Jolley, E. & Smith, D.C. (1980). The green hydra symbiosis. II. The biology of the re-establishment of the symbiosis. Proc. R. Soc. Lond. B. 207, 311-333.

McAuley, P.J. (1986a). Isolation of viable uncontaminated Chlorella from green hydra. Linnol. Oceanogr. 31, 222-224.

McAuley, P.J. (1986b). Uptake of amino acids by cultured and freshly isolated symbiotic Chlorella. New Phytol. 104, 415-427.

McAuley, P.J. Nitrogen limitation and amino acid metabolism of Chlorella symbiotic with green hydra. Planta, in the press.

McAuley, P.J. & Smith, D.C. (1982a). The green hydra symbiosis. V. Stages in the intracellular recognition of algal symbionts by digestive cells. Proc. R. Soc. B. 216, 7-23.

McAuley, P.J. & Smith, D.C. (1982b). The green hydra symbiosis. VII. Conservation of the host cell habitat by the symbiotic algae. Proc. R. Soc. Lond. B. 216, 415-426.

McNeil, P.L. (1981). Mechanisms of nutritive endocytosis. I. Phagocytic versatility and cellular recognition in Chlorohydra digestive cells, a scanning electron microscopical study. J. Cell Sci. 49, 311-339.

McNeil, P.L. (1984). Mechanisms of nutritive endocytosis. III. A freeze-fracture study of phagocytosis by digestive cells of Chlorohydra. Tissue and Cell 16, 519-533.

McNeil, P.L., Hohman, T.C. & Muscatine, L. (1982). Mechanisms of nutritive endocytosis. II. The effect of charged agents on phagocytic recognition by digestive cells. J. Cell Sci. 52, 243-269.

McNeil, P.L. & Smith, D.C. (1982). The green hydra symbiosis. IV. Entry of symbionts into digestive cells. Proc. R. Soc. Lond. B. 216, 1-6.

McNeil, P.L., Tanasugarn, L., Meigs, J.B. & Taylor, D.L. (1983). Acidification of phagosomes is detected before lysosomal enzyme activity is detected. J. Cell. Biol. 97, 692-702.

McNeil, P.L. & McAuley, P.J. (1984). Lysosomes fuse with one half of alga-bearing phagosomes during the re-establishment of the European green hydra symbiosis. J. Exp. Zool. 230, 377-385.

Meier, R., Reisser, W., Wiessner, W. & Le Fort-Tran, M. (1980). Freeze-fracture evidence of differences between membranes of perialgal and digestive vacuoles in Paramecium bursaria. Z. Naturf. 35c, 1107-1110.

Meier, R., Le Fort-Tran, M., Pouphile, M., Reisser, W. & Wiessner, W. (1984). Comparative freeze-fracture study of perialgal and digestive vacuoles in Paramecium bursaria. J. Cell Sci. 71, 121-140.

Meints, R.H. & Pardy, R.L. (1980). Quantitative demonstration of cell surface involvement in a plant-animal symbiosis: lectin inhibition of reassociation. J. Cell Sci. 43, 239-251.

Mews, L.K. (1980). The green hydra symbiosis. III. The biotrophic transport of carbohydrate from alga to animal. Proc. R. Soc. Lond. B. 209: 377-401.

Muscatine, L. & McAuley, P.J. (1982). Transmission of symbiotic algae to eggs of green hydra. Cytobios. 33, 111-124.

Neckelmann, N. & Muscatine, L. (1983). Regulatory mechanisms maintaining the Hydra-Chlorella symbiosis. Proc. R. Soc. Lond. B. 219, 193-210.

O'Brien, T.L. (1982). Inhibition of vacuolar membrane fusion by intracellular symbiotic algae in Hydra viridis (Florida strain). J. Exp. Zool. 223, 211-218.

Pardy, R.L. & Muscatine, L. (1973). Recognition of symbiotic algae by Hydra viridis. A quantitative study of the uptake of living algae by aposymbiotic H. viridis. Biol. Bull. 145, 565-579.

Pool, R.R. Jr. (1979). The role of antigenic determinants in the recognition of potential algal symbionts by cells of Chlorohydra. J. Cell Sci. 35, 367-379.

Rahat, M. (1985). Competition between Chlorellae in chimeric infections of Hydra viridis: the evolution of a stable symbiosis. J. Cell Sci. 77, 87-92

Rahat, M. & Reich, V. (1984). Intracellular infection of aposymbiotic Hydra viridis by a foreign freeliving Chlorella sp.: Initiation of a stable symbiosis. J. Cell Sci. 65, 265-277.

Rahat, M. & Reich, V. (1985). Correlations between characteristics of some free-living Chlorella sp. and their ability to form stable symbioses with Hydra viridis. J. Cell Sci., 74, 257-266.

Rahat, M. & Reich, V. (1987). A re-evaluation of alleged requirements for the establishment of algal/hydra symbiosis and host/symbiont specificity: pre-adaption and vacuolar ecology. Endocyt. C. Res. 4, 13-23.

Rees, T.A.V. (1986). The green hydra symbiosis and ammonium. I. The role of the host in ammonium assimilation and its possible regulatory significance. Proc. R. Soc. Lond. B. 229, 299-314.

Smith, D.C. (1980). The role of nutrition in the establishment of the green hydra symbiosis. In Nutrition in the lower Metazoa (Smith, D.C. & Tiffon, Y., eds.) pp 129-139. Pergamon Press, Oxford.

SPECIFICITY IN THE *CONVOLUTA ROSCOFFENSIS/TETRASELMIS* SYMBIOSIS

A.E. Douglas,
John Innes Institute,
Colney Lane,
NORWICH, NR4 7UH.
U.K.

1. Introduction

Convoluta roscoffensis is an intertidal acoel turbellarian ('flatworm') that is locally abundant on sandy beaches of the Brittany coast of France and nearby Channel Islands. It is a conspicuous member of the beach fauna, for two reasons: first, the animals are green, due to the presence of algal symbionts of the genus *Tetraselmis* (=*Platymonas*, *Prasinocladus*) in their tissues (Parke & Manton, 1967); and, secondly, they are gregarious, forming large dark-green patches on the surface of the beach at low tide (Holligan & Gooday, 1975). The details of the life cycle of *C. roscoffensis* was established by Keeble & Gamble (1907). The adult animals reproduce exclusively sexually and the fertilized eggs are deposited, within a common gelatinous capsule, in the sand. The algal symbionts are not transmitted directly into the eggs and the association is re-established at each generation of *C. roscoffensis* when recently-hatched juveniles feed on free-living algae in the interstitial water. The symbiosis appears to be obligate for *C. roscoffensis*; all adult animals are green in the field and, in the laboratory, aposymbiotic juveniles die within a few weeks of hatching. The adult animals do not feed holozoically. They are believed to derive a substantial proportion of their energy and nutritional requirements from the algal symbionts, although the extent of nutrient translocation and identity of mobile products remain uncertain (Boyle & Smith, 1975; Meyer *et al*, 1979; Douglas, 1983).

C. roscoffensis exhibits a high specificity for its algal symbionts; each adult worm contains only one species of *Tetraselmis*, usually *T. convolutae* (= *Platymonas convolutae*)

NATO ASI Series, Vol. H17
Cell to Cell Signals in Plant, Animal and
Microbial Symbiosis. Edited by S. Scannerini et al.

(Parke & Manton, 1967; McFarlane, 1982a). However, a wide variety of unicellular algae, including Tetraselmis species, flourish on many of the beaches with C. roscoffensis (Parke & Manton, 1967). One might therefore anticipate specific interactions between the algal and animal cells, leading to the selection of one algal taxon as symbiont and discrimination against alternative algae during the development of the symbiosis. Such interactions have been termed recognition by Smith (1981).

The symbiosis in C. roscoffensis is very amenable to the study of alga-animal interactions in the developing association. T. convolutae and related species are available in axenic culture and large numbers of aposymbiotic juvenile animals are produced by laboratory cultures of adult C. roscoffensis. Further, the symbiosis is established simply by allowing juveniles to feed on a dilute suspension of algae. The course of the establishment of the association is illustrated in Figure 1.

2. Laboratory studies of specificity

A useful approach to study recognition underlying the specificity of the symbiosis is to compare the development of the association with T. convolutae firstly, to algae which do not naturally form a symbiosis (heterologous algae) and, secondly, to experimentally-manipulated T. convolutae. Evidence for discrimination between 'acceptable' and 'unacceptable' algae (and hence recognition) has been obtained at three stages in the development of the association (Provasoli et al, 1968; Douglas, 1983b).

a) On initial contact (stage 1 in Figure 1). Juvenile C. roscoffensis ingest cells of all Tetraselmis species to a similar extent, but do not take up other algae. This suggests that phagocytosis is triggered by a common surface feature of Tetraselmis species. The nature of this putative determinant is unknown, but the failure of animals to phagocytose athecate T. convolutae cells indicates that the determinant is borne on the theca.

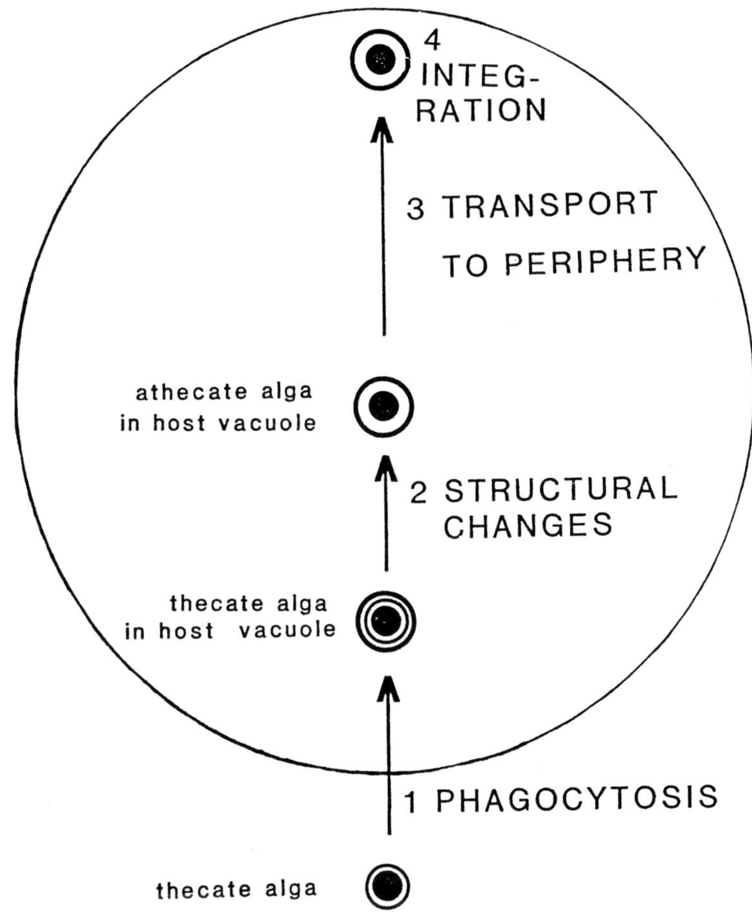

Figure 1 The development of the symbiosis between juvenile *C. roscoffensis* and *T. convolutae* can be divided into four stages: 1) contact & phagocytosis: the free-living algal cells are bounded by a theca and may be flagellate. They are brought to the mouth by the action of cilia and phagocytosed into the syncytial parenchyma. (Like other acoel flatworms, *C. roscoffensis* lacks a gut.) 2) structural changes in algal cells: soon after ingestion, the algal cells in the central parenchyma lose their theca, flagella and eyespot. 3) transport to the periphery: the athecate and non-flagellate algal cells are transported to the periphery of the parenchyma, where they take up a final position near the body wall musculature. 4) integration: several features of integration are apparent. (a) the athecate cells are oriented with the posterior end facing the surface and projections of the cell extend into the epidermal layer of the animal; (b) nutritional interactions are established, resulting in growth of the animal from 12-14 days after infection; (c) a stable alga:host biomass ratio (of 1:20) develops and is maintained by control of algal growth and division rate.

b) In the central parenchyma (stage 2 in Figure 1). The structural changes and persistence of *T. convolutae* cells depend on their capacity to photosynthesize or a related characteristic. Thus, when juveniles, recently infected with *T. convolutae*, are incubated in darkness or the photosynthetic inhibitor, DCMU, thecate cells in the central parenchyma are lost. No information on the underlying mechanism(s) is available, but the discarded algal cells are expelled in an intact condition. By contrast, phagocytosis (stage 1) and persistence of algal cells at the periphery in both juvenile animals and the intact symbiosis in adults are independent of photosynthesis.

c) At the periphery of the animal (stage 4 in Figure 1). All *Tetralsemis* species so far tested can form a viable association with *C. roscoffensis*, as indicated by the animals' sustained growth and development to sexual maturity and the maintenance of a stable biomass ratio of algal cells to animal (Nozawa *et al*, 1971). However, under the culture conditions adopted by Provasoli *et al* (1968), juveniles containing *T. convolutae* are longer than those with alternative algae. Further, in multiple infections, the algal species which promotes greatest growth of the animal is selected and alternative algae are expelled once they have reached the periphery, probably via the mucous glands. This selection only occurs in the developing symbiosis; by 25-30 days after infection, a symbiosis with heterologous algae is sustained when the animal is exposed to free-living *T. convolutae*.

These laboratory studies indicate that selection of algal symbionts, and hence recognition, occurs at several different stages in the development of the symbiosis and on the basis of different properties of the algae; to date, surface characteristics, capacity to photosynthesize and promotion of animal growth have been identified. These results can account for the observed specificity of the symbiosis in the field, where each animal contains one species of *Tetraselmis*, usually *T.convolutae*.

3. Specificity in the field

In general, it is very difficult to study the specificity of alga - invertebrate symbioses in the natural situation. This is either because no variation in the identity of the algal symbionts is apparent or because the variants are difficult to identify, requiring isolation into long-term culture and detailed ultrastructural or cultural investigations. The symbiosis in C. roscoffensis at the only known site on mainland Britain, Aberthaw in South Glamorgan (Mettam, 1979), represents an exciting exception to this generality. Several morphologically distinct algae of the genus Tetraselmis, none of which is T. convolutae, have been isolated from C. roscoffensis at Aberthaw (McFarlane, 1982a). The algae belong to two subgenera, Tetraselmis and Prasinocladia, but their specific identity has not yet been established. The two subgenera can be distinguished readily by their pyrenoid characteristics, both at the ultrastructural level and by phase contrast microscopy of the lightly-compressed living animal (Figure 2). Each animal contains algal cells of one pyrenoid type.

Study of the association at Aberthaw is at a preliminary stage, limited to the initial characterization of McFarlane (1982a & b) and two subsequent collections in January, 1984 (Douglas, 1985) and January, 1987 (Douglas, unpublished). The C. roscoffensis are limited to a narrow strip of sand at high water mark. Animals containing Prasinocladia are more common than those with Tetraselmis (McFarlane, 1982a; Douglas, 1985). The animals containing algae of subgenus Tetraselmis do not differ markedly in gross morphology or ultrastructural characteristics from those containing Prasinocladia and there is no reason to believe the animals are taxonomically distinct (McFarlane, 1982a; A.E. Dorey, pers. comm.). However, the structural relationship between C. roscoffensis and Prasinocladia differs markedly from that with Tetraselmis at Aberthaw (see table) and T. convolutae (also of subgenus Tetraselmis) on the Channel Islands (see Doonan et al, 1980). In the latter associations, virtually all the algal cells are in the peripheral region of the parenchyma and most are oriented with the posterior end towards the surface and are of irregular shape (see also legend to Figure 1, stage 4). By contrast, Prasinocladia cells are invariably ovate in outline.

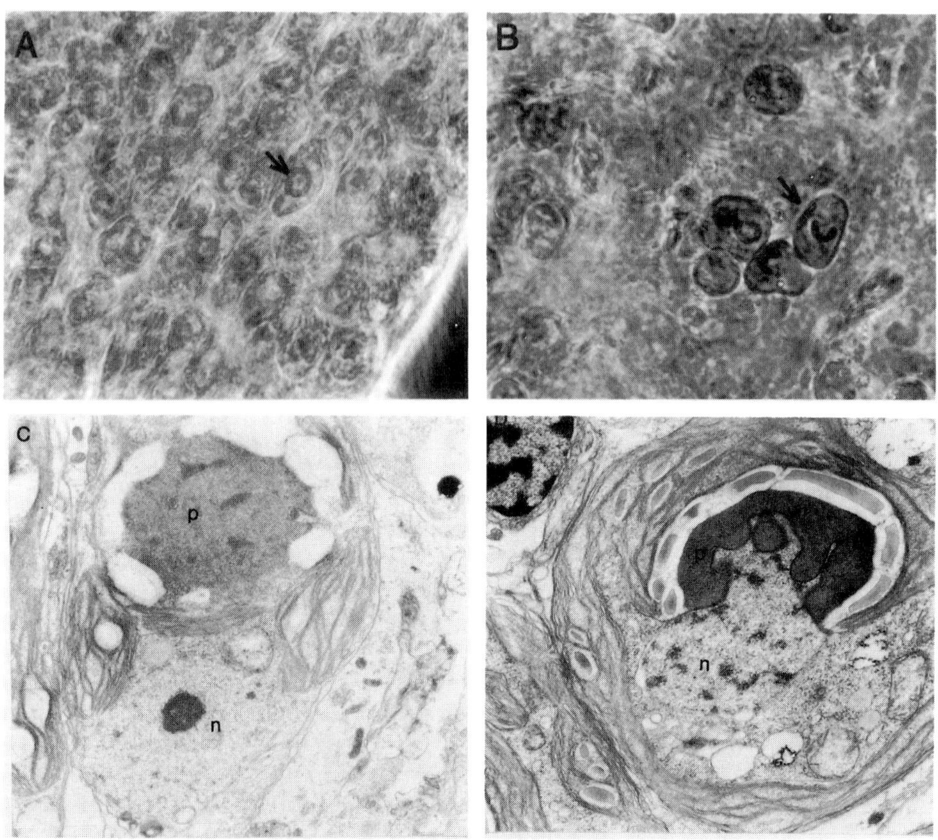

Figure 2. The structural characteristics of algal symbionts of C. roscoffensis at Aberthaw at the light microscope (A, B) and ultrastructural (C, D) level. The pyrenoid of algal cells of the subgenus Tetraselmis is spherical and interrupted by many narrow channels of cytoplasm whereas that of Prasinocladia is cup-shaped, with a single large extension of the nucleus projecting into it. n - algal nucleus, p - pyrenoid.

Many of the cells are in the central parenchyma and the orientation of approximately 40% of those at the periphery is different from that of Tetraselmis cells (see Figure 3a). The absence of interdigitation between *Prasinocladia* cells and the animal tissue has also been observed in the experimental association between *C. roscoffensis* and *T. marinus* (also of the subgenus *Prasinocladia*), but in the latter more than 90% of the algal cells are at the periphery and of normal orientation (Nozawa *et al*, 1972; Douglas, unpub.). The significance of the position and orientation of algal cells is unknown, but the irregularities of the association with *Prasinocladia* suggest that algae of this subgenus are less well integrated into the symbiosis than *Tetraselmis*.

Table Structural features of the symbiosis between *C. roscoffensis* and algae of subgenera *Tetraselmis* and *Prasinocladia*. Animals collected in January 1987 were prepared for electron microscopy by a method modified from Parke & Manton (1967). Sections from each of three positions along the body of 5 animals containing each subgenus of symbiont were scored. No consistent differences were recorded between sections containing one type of symbiont. Therefore the data were amalgamated and mean values are shown.

	Tetraselmis	*Prasinocladia*
Position of algal cells (%)		
dorsal	68.8	60.0
ventral	30.3	29.4
central	0.8	15.6
total number of cells scored	4701	3821
Orientation of algal cells (%)[1]		
'normal'	99.8	59.2
inverted	0	18.7
intermediate	0.2	22.1
total number of cells scored	1408	1070
Dividing algal cells (%)[2]	0.4	8.4
total number of cells scored	4791	3821

[1]Scored for cells at dorsal surface only. A cell of 'normal' orientation has its posterior end facing the epidermis of *C. roscoffensis*.

[2]Cells undergoing cytokinesis and 'doublets' (two closely apposed daughter cells of inverted orientation relative to each other).

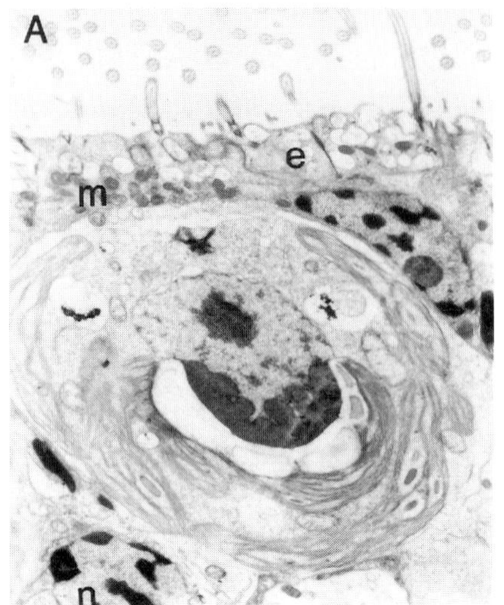

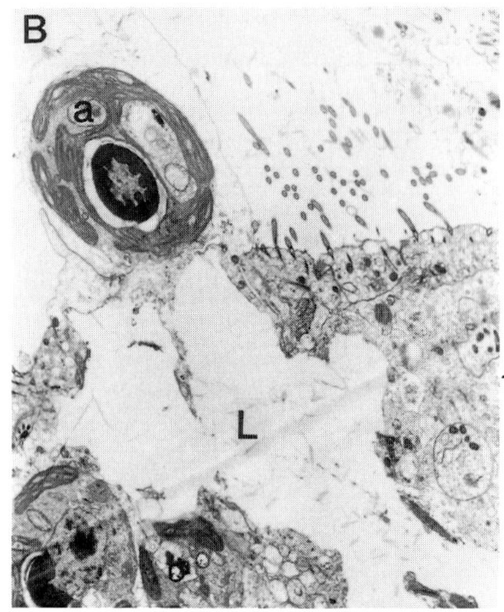

Figure 3. Unusual features in the structural relationship between C. roscoffensis and algae of the subgenus Prasinocladia. A. Prasinocladia cell with anterior end facing surface of animal ('inverted cell'); B. apparent expulsion of algal cell from mucous gland. (a - algal cell and enclosing animal material, apparently mid-expulsion, c - epidermal cell, l - lumen of mucous gland, m - mitochondria in epidermal cell, n - animal cell nucleus).

Consistent with these data, the performance (or fitness) of C. roscoffensis containing Tetraselmis is superior to that of animals containing Prasinocladia by several criteria, including the size and fecundity of adults and growth rate of juveniles (Douglas, 1985). Further, infection of juveniles with algal cells of both subgenera leads to a stable symbiosis with Tetraselmis and exclusion of Prasinocladia (McFarlane, 1982b).

These data cannot easily be reconciled with the high frequency of C. roscoffensis containing Prasinocladia in the field. One possible explanation is that free-living Prasinocladia cells are more abundant than Tetraselmis at the zone of C. roscoffensis (mean position of high water) and many juveniles fail to encounter Tetraselmis. McFarlane (1982b) has suggested that Prasinocladia can tolerate the desiccating conditions in this position more readily than Tetraselmis. The relationship between the percentage of C. roscoffensis containing Tetraselmis and position of C. roscoffensis on the beach is consistent with this view. The frequency of animals with Tetraselmis is highest at western end of the C.roscoffensis zone. This is close to the saltmarsh, the drainage waters of which irrigate the zone of C. roscoffensis. The sand in this area probably has a higher moisture content than at other parts of the beach. Further, at sites which are definitely intertidal (e.g. Channel Islands, coast of France), T. convolutae (subgenus Tetraselmis) is the dominant symbiont. However, the capacity of isolates from C. roscoffensis at Aberthaw to withstand desiccation has not been investigated.

These data raise the possibility that specificity in the field is determined not only by recognition mechanisms during the establishment of the symbiosis (see Section 2), but also by the availability of algal symbionts in the free-living condition. The latter may depend on properties of the symbiont of no apparent relevance to the symbiosis (e.g. tolerance of desiccation). These considerations highlight the potential hazards or disadvantages to the host of a dependence on free-living forms for symbionts. Further study of the symbiosis with C. roscoffensis at Aberthaw may contribute to current discussion on the relative advantages to the host of direct transmission of symbionts and infection with free-living symbionts at each host generation (e.g., Smith & Douglas, 1987).

The physiological basis of the superiority of algae of subgenus Tetraselmis as symbionts is unknown, but one may speculate that they release nutrients, such as photosynthate, to the animal tissues at a higher rate than algae of subgenus Prasinocladia. Structural studies of the association provide two strands of potentially relevant information. First, a substantially higher percentage of Prasinocladia cells (8%) is in a dividing condition than Tetraselmis (0.4%, see Table) or T. convolutae in animals from the Channel Islands (less than 5%, Doonan et al, 1980)). Secondly, in virtually all sections of animals from Aberthaw with Prasinocladia, mucous glands contained structurally intact algal cells, but no cell of Tetraselmis was observed in the lumen of a mucous gland. These Prasinocladia cells were probably destined for expulsion. A cell, with surrounding animal material, apparently in the

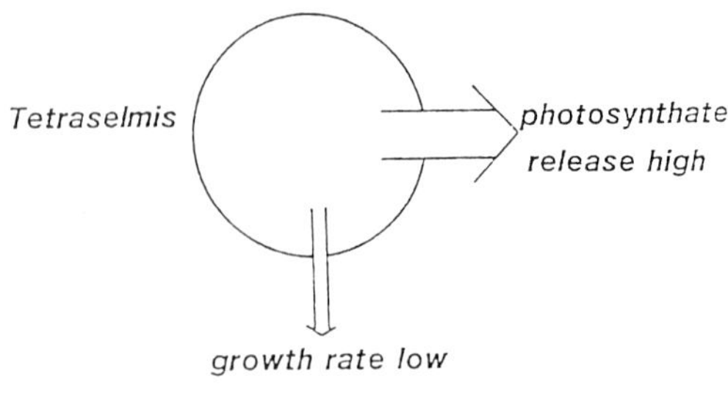

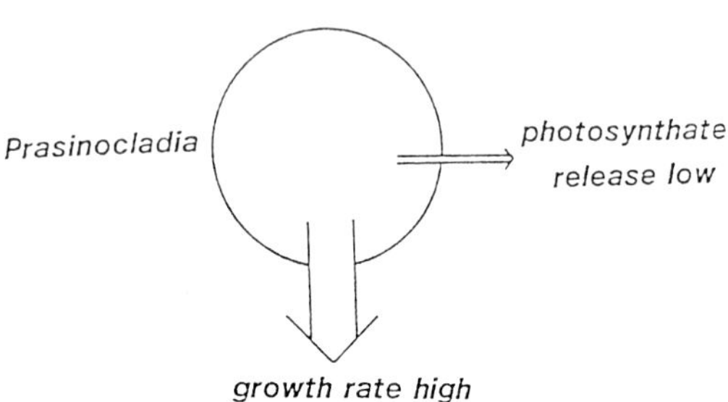

Figure 4. Proposed relationship between photosynthate release (and hence performance of C. roscoffensis) and growth rate of algal cells (and hence expulsion) in C. roscoffensis containing Tetraselmis and Prasinocladia. See text for details.

---

process of expulsion is shown in Figure 3b. These two characteristics of Prasinocladia may be linked. Specifically, Prasinocladia cells may retain sufficient photosynthetic carbon (as a result of low levels of release, see above) to fuel growth and division at a higher rate than that of the animal. The constant relative biomass of algae to animal is maintained by expulsion of excess algal cells (see Figure 4). The characteristics of the natural symbiosis of C. roscoffensis with Prasinocladia are closely similar to the experimental association between green hydra and Chlorella vulgaris strain NC64A, an alga which releases very little photosynthate (see Douglas, this volume). Nutrient release by the symbionts of C. roscoffensis at Aberthaw remains to be investigated directly. However, the evidence to date suggests that the mechanisms of regulation identified in the green hydra association may not be unique to that symbiosis, but possibly are of general occurrence in alga-invertebrate symbioses.

Acknowledgements

I thank Dr. A.G. & Mrs. S. Searle for their assistance with the collection of *C. roscoffensis* in January 1987 and Mr. B. Wells and Dr. K. van den Bosch for their advice on the preparation and staining of semi-thin sections. The structural study of *C. roscoffensis* at Aberthaw was financed by a grant from the Royal Society of London.

References

Boyle, J.E. & Smith, D.C. (1975). Biochemical interactions between the symbionts of *Convoluta roscoffensis*. *Proc. R. Soc. Lond. B*, 189, 121-135.

Doonan, S.A., Douglas, A.E. & Gooday, G.W. (1980). Acquisition of algae by *Convoluta roscoffensis*. In: *Endocytobiology Endosymbiosis & Cell Biology*, Volume 1 (W. Schwemmler & H.E.A. Schenck, editors) pp. 293-304 Walter de Gruyter, Berlin.

Douglas, A.E. (1983a). Uric acid utilization in *Platymonas convolutae* and symbiotic *Convoluta roscoffensis*. *J. mar. biol. Ass. UK*, 50, 199-208.

Douglas, A.E. (1983b). Establishment of the symbiosis in *Convoluta roscoffensis*. *J. mar. biol. Ass. UK*, 63, 419-434.

Douglas, A.E. (1985). Growth and reproduction of *Convoluta roscoffensis* containing different naturally occurring algal symbionts. *J. mar. biol. Ass. UK*, 65, 871-879.

Holligan, P.M. & Gooday, G.W. (1975). Symbiosis in *Convoluta roscoffensis*. *Symp. Soc. exp. Biol*. 29, 205-227.

Keeble, F. & Gamble, F.W. (1907). The origin and nature of the green cells of *Convoluta roscoffensis*. *Quart. J. microscop. Sci*. 51, 167-219.

McFarlane, A.E. (1982a). Two species of algal symbiont in naturally occurring populations of *Convoluta roscoffensis*. *J. mar. biol. Ass. UK*, 62, 235.

McFarlane, A.E. (1982b). *Ultrastructural & Immunological Studies of the Symbiosis between Convoluta roscoffensis and Prasinophycean Algae*. Ph.D. Thesis, University of Bristol.

Mettam, C. (1979). A northern outpost of *Convoluta roscoffensis* in South Wales. *J. mar. biol. Ass. UK*, 59, 251-252.

Meyer, H., Provasoli, L. & Meyer, F. (1979). Lipid biosynthesis in the marine flatworm *Convoluta roscoffensis* and its algal symbiont *Platymonas convolutae*. *Biochim. Biophys. Acta*, 573, 464-480.

Nozawa, K., Taylor, D.L. & Provasoli, L. (1972). Respiration and photosynthesis in *Convoluta roscoffensis*, infected with various symbionts. *Biol. Bull*., 143, 420-430.

Parke, M. & Manton, I. (1967). The specific identity of the algal symbiont of *Convoluta roscoffensis*. *J. mar. biol. Ass. UK*, 47, 445-464.

Provasoli, L., Yamasu, T. & Manton, I. (1968). Experiments on the resynthesis of the symbiosis in *Convoluta roscoffensis* with different flagellate cultures. *J. mar. biol. Ass. UK*, 48, 465-479.

Smith, D.C. (1981). The role of nutrient exchange in recognition between symbionts. *Ber. Deutsche Bot. Ges.*, 94, Supplement 517-528.
Smith, D.C. & Douglas, A.E. (1987). *The Biology of Symbiosis* Edward Arnold.

THE CELL STRUCTURES OF PLANT, ANIMAL AND MICROBIAL SYMBIONTS, THEIR DIFFERENCES AND SIMILARITIES.

Silvano Scannerini
Dipartimento di Biologia Vegetale dell'Università di Torino.
Viale P.A. Mattioli, 25. I-10125 Torino (Italia).

INTRODUCTION.

Organisms from different kingdoms and with different cell biologies living together in symbiosis have to solve the common problems of mutual recognition and of establishing an interchange of substances with each other. However, the different symbioses are often considered as disconnected phenomena, so that the common problems are not dealt with from a similar point of view. This leads to further problems of standardization of terminology, for example the definition of "symbiont", "parasite", as well as their cell-to-cell relationships in geometrical terms.

D.C. Smith (1979) has proposed a return to the original definition of "symbiosis" (in the wide sense), given by De Bary, which distinguishes between "mutualistic symbiosis" (mutualism) and "parasitic symbiosis" (parasitism). This is frequently not accepted and "symbiosis" is often considered as a synonym of "mutualism" (e.g. Cook *et al*. 1980, Lugtenberg, 1986).

The distinction between parasitism and mutualism would be simple and effective were it not for the wide range of interactions which cannot be so rigidly classified. For example there are complexes which were originally parasitic, but in which the original pathogen evolves into a mutualistic organism, for example the bacterial symbionts of *Amoeba proteus* (Jeon, 1983). There are other situations where a phagocytized organism has such a long survival period as to suggest a transition towards mutualism, as in *Leishmania*-macrophage complex (Chang, 1983). Moreover

NATO ASI Series, Vol. H17
Cell to Cell Signals in Plant, Animal and Microbial Symbiosis. Edited by S. Scannerini et al.

when new associations are discovered, it is usually some time before the metabolic relationships are elucidated. The same is true when hosts acquire non-native symbionts (Bonfante-Fasolo et al.-1984- for Pezizella ericae-Legume root associations; Rahat & Reich, 1986 for Chlorella-Hydra).

The choice of Smith's definition (1979), apart from being semantically more correct, has the advantage of allowing the term "symbiosis" to be used for those stable associations in which the interrelationships of the partners are not clear.

The problem of the terminology used to distinguish the localization of symbionts with respect to space has been discussed by Smith (1979) with particular reference to the meaning of "intracellular" and "intercellular". For example he illustrates that a sequence of stages from inter- to intra cellularity exists for algal symbionts. This scheme may be easily extended to fungal partners of mycorrhizae according to Bonfante-Fasolo & Scannerini (1983). If the cell-wall is considered as the boundary of the plant, fungal and procaryotic cells then incorporation into this scheme would be possible for the "periplastic parasite" Bdellovibrio in Gram negative bacteria (Stolp, 1979) as well as lichens.

The problem is not only topological but also semantic. For example, "endosymbiont" and "ectosymbiont" have been used with ambiguous meaning by Koch (1967), Hartzell (1967) and more recently by Whatley and Whatley (1984). Comparing the definition of Smith (1979) with those of Taylor (1979), Schwemmler (1983) and Nardon (in this book), the acceptance of "endosymbiosis" as a term meaning the simple inclusion of one organism within another, and of "endocytobiosis" for an organism, or a stage in its cycle, enclosed within a host cell may be valid starting points in standardize the terminology.

If one takes into consideration the alga-fungus interactions in lichens described by Honneger (in this book), which include different degrees of fungal penetration into the algal cells, the lichens may also be tentatively inserted in the classification by considering the fungus as an endosymbiont of the alga, at least in those cases in which fungal penetration structures exist. In some cases, these terms may not accurately describe the cell-to-cell relationships, as in lichen thalli with Coccomyxa symbionts (wall-to-wall appositions) although by way of a compromise, it may be possible to use the term "endosymbiontic partners". Thus, the term "endosymbiosis" could include all those direct cell-to-cell relationships which do not imply the presence of the symbiont inside the host cell, while "endocytobiosis" would refer to those in which one organism is included in the cell or cells of another, and "ectosymbiosis" for those associations where the most important portions of the organisms do not have a direct cell-to-cell contact.

The following set of terms could therefore be proposed: "mutualistic symbiosis" and "parasitic symbiosis"; "ectosymbiosis", "endosymbiosis" and "endocytobiosis". This would unify the topological and semantic terms of our vocabulary. Thus, the title of this paper might be changed to "The cell structures of plant, animal and microbial mutualistic symbionts etc." and I would suggest that it forms the basis for a wider discussion of the standardization of the significance of "ectosymbiont", "endosymbiont" and "endocytobiont".

UNIFIED MODELS FOR EUCARYOTIC AND PROCARYOTIC CELLS.

The basic reasons for the lack of communication between specialists in the different types of mutualistic symbiosis do not depend on semantic and topological problems so much as on the more prominent facts of cell biology. For too long have the differences between plant and animal cells been stressed. In whatever way the classification of organisms is considered, plant and animal cells have been considered as different. Cell-wall, vacuoles and plastids were originally considered peculiar to plant cells and not homologous to any animal cell structure (Frey-Wyssling and Mühlethaler, 1965). More recently, however it is believed that there is a nearly complete homology between their components. In fact the correspondence is complete, if we exclude the obvious exception of plastids that may be found in, or artificially introduced into animal cells, but which cannot be considered as organelles because of their transient nature (e.g. Hinde and Smith, 1974, Laval-Peito & Febvre, 1986).

These are the reasons why L. Margulis was able to build up this nice unified scheme in 1981.

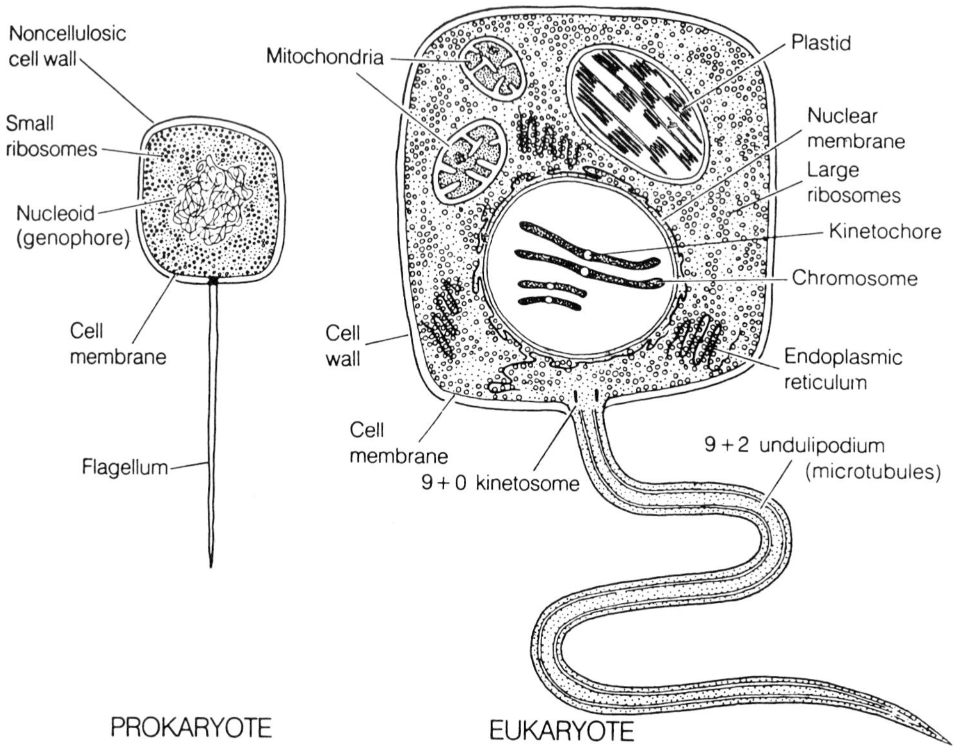

(from Margulis, 1981).

More precisely:

1) The plant cell-wall and the animal cell glycocalix (cell-coat) (and parallel structures of algae, fungi and protozoan glycocalyx) can be seen as structures with significant similarities, in agreement with Bennett (1969). Heslop-Harrison & Linskens objections to this (1984) were based on the heterogeneity of wall components and on the fact that the plant cell wall can act as a very effective barrier isolating the plasmalemma from

direct impact of external factors. This is not, in my opinion, particularly important because the cell walls of fungi and plants can form part of their means of communication (as demonstrated by Dazzo et al.; Callow; Bonfante-Fasolo in this book) exactly as the glycocalyces of the animal cells. Moreover, advances in knowledge of primary wall is now emphasizing the similarities between the carbohydrate and protein investment of plant and animal cells; for example the similarities between extensin, arbinogalactan-protein (AGPs) (Lamport & Caty, 1981) and Hp rich proteins of collagen (Eyre, 1980).

Obviously these considerations do not contradict the extreme biochemical and structural variability of algal, fungal and plant cell walls as Heslop-Harrison & Linskens correctly stress. However this mainly concerns the secondary walls which often isolate the cells which possesses them,but are not any greater than that which is present in the extracellular material of animal cells of connective tissue.

2) In the same way, vacuoles have been considered as absolutely peculiar to plants, fungi and algae. Today, however, there is a sufficient evidence to consider them as homologous to lysosomal systems, (Matile, 1978) because their content of plant hydrolases (Boller, 1982) because the vacuolation sequences in the plant cell (Marty, Branton, Leigh, 1980) and because of their capability to carry out autophagic activities in higher plant (Marty, 1978) and in fungi (Scannerini et al. 1975).

3) The similarity between plant and animal cells is further confirmed by the membrane flow model which connects the plasmalemma and the plant cells endomembrane system. If we compare the membrane flow diagram (Whatley & Whatley, 1984) with those known for animal cells (e.g. Alberts et al. 1985) it is difficult to find significant differences.

4) The microtubule systems which constitute the cytoskeleton represents the most "conservative" structure, from the biochemical point of view, of the whole cell.Although we are relatively ignorant of the cytoskeletal structures of higher plant cells, there is no reason to suppose significant differences from animal cell. In fact, as actin microfilaments, once

considered to be exclusively animal, are now known to be present in higher plant (Jackson, 1982); the absence of flagellar structures may be due to secondary loss.

5) The same is true for the so-called semiautonomous organelles. The mitochondria have a common type of organization in plant and animal cells. Furthermore the symbiotic origin of mitochondria, which is also extended to peroxisomes (Cavalier-Smith, 1987), is widely accepted (see Margulis and Bermudes in this book for references) thus reinforcing the general homology of cell components.

6) Without going into details, it may be said that the site, localization, and expression of nucleic acids are similar in plant and animal cells (e.g. Gull, 1981; Brown, 1981) and this is confirmed by molecular biological data over and above the classical structural aspects.

On the other hand, present knowledge of cytology distinguishes clearly between eucaryotic and procaryotic cells and Cavalier-Smith (1981) lists 22 differences between them.

Procaryotic cells are characterized (Margulis, 1981) by the almost complete absence of endomembranes, by the exclusive presence of S_{70} ribosomes, by the inability of its plasma membrane to carry out phagocytosis, by the different DNA organization and by the presence of cell walls containing muramic acids (excepts _Archaebacteria_ and _Mycoplasma_).

Thus unified models for eucaryotic and procaryotic cells can be built which may be used for a comparative approach to symbiotic interactions.

COMMON CYTOLOGICAL FEATURES OF MUTUALISTIC SYMBIONTS.

The picture resulting by the comparison of the present knowledge of the functional morphology in different types of mutualistic symbiosis (see for references Smith & Douglas, 1987 and papers in the first chapter of this book) can be summarized as follows:

1) Absence of cytopathological symptoms in the partners during the active phase of mutualism; 2) presence of complex interfaces between cells of the partners with a predominant type of perisymbiotic membrane surrounding

intracellular symbionts and occasional presence of endocytobionts free in the host cytoplasm; 3) presence of various types of phagocytic-like structures during the establishment of symbiosis and during "harvesting" phase of symbiont population control by the host.

1) Absence of cytopathological symptoms.

a) On the part of the host. At the cellular level the host of a parasite shows a varied set of cytological symptoms of "defence reaction" well known both in animal cells (see for example Palade and Farquar, 1981) and in plant cells (see for example Ouchi, 1983). The "defence reaction" in animal cells is initially antibody-mediated while plants have not antibody--mediated mechanism. Defence barriers in plants are related to the induction of cytological modifications (papillae, callose layers, lignified cells walls, production of phytoalexins or of pathogen related proteins - e.g. chitinase -). The cytopathological symptoms in parasitic interactions are of two types (Keen, 1986), the "Thug" and the "Confidence man"; in which the defence reaction is always preceded by a period of "accessibility" more or less characterized by "susceptibility" (= "compatibility") symptoms. Susceptibility is therefore an indicator of accessibility: "the state of the cell in which pathogens or non pathogens are permitted to coexist in harmony" (Ouchi, 1983). The same may more or less be said for animals (Palade and Farquar, 1981). So quite paradoxically the first stage of a biotrophic parasitic interaction is cellular compatibility between partners indistinguishable from that involved in mutualistic symbioses. It may therefore be said that the two associations essentially differ in the period of accessibility (compatibility). Throughout this period symptoms of biological damage do not appear in the endomembrane systems, the semiautonomous organelles, in the nucleus or in the glycocalyx.

b) On the part of the symbiont. The endosymbionts and endocytobionts also do not show symptoms of pathological degeneration during accessibility (active phase of mutualism). On the other hand, the glycocalix and wall structures are subject, more generally, to progressive ultrastructural modifications and simplifications, but without functional alteration of unicellular and multicellular populations. While these modifications serve to reduce the apoplastic barrier, they also imply deep morphogenetic effects both on procaryotes and on eucaryotes (see for references the papers in the chapter first of this book).

2) Interfaces between partners during compatibility.

For a long time, the concept of interface between pathogenics symbionts (Bracker & Littlefield, 1973) has been applied to symbiotic organisms. The applications of this criterion to mutualistic symbionts results in a relatively simple scheme (Table 1).
None of these interfaces, in agreement with what has been said about symbiotic structures, may be considered by themselves, as an indicator of cell compatibility. For this purpose two problems remain to be solved: the nature of the interface material in the different kind of mutualism and the composition of perisymbiotic membranes. As far as the interface material is concerned, it is known to consist of polysaccharides and proteins in almost all systems so far studied. As far as perisymbiotic membrane is concerned it may be derived from the host cell plasmalemma, or from the membrane of a vacuole in which does not contain hydrolases, or from membrane of a vacuole containing hydrolases and therefore a lysosome. These alternatives are valid regardless the host cell type but in practice the information available is scanty.

3) Presence of various phagocytic-like structures.

As we have just seen, during the compatibility phase the structures contained within the perisymbiontic membrane escape digestion mainly by inhibition of lysosomal fusion (see e.g. Nardon, McAuley, Reisser in this book), resistence to lysosomal enzyme (Chang, 1983) or escape from the vacuole systems (Nardon in this book).
During the final phases of mutualistic symbiosis the endocytobionts may be lysed - regulation by farming - or be expelled - ecological regulation - or may outlive their host cell (ericoid mycorrhizae, perhaps some cases of nitrogen fixing bacteria). Therefore, both in the compatibility and in the population control phases, the phagocytic sequences are discovered widely

INTERFACES BETWEEN MUTUALISTIC SYMBIONTS

INTERFACES	ASSOCIATIONS	REFERENCES
WALL-WALL (and/or INTERFACE MATERIAL)	Lichens, Plant-Cyanobacteria, Plant-Dinitrogen fixing bacteria, Ectomycorrhizae, Endomycorrhizae, Nematode-luminescent bacteria	Honneger*, Grilli-Caiola & Albertano*, Torrey*, Piché & Peterson*, Gianinazzi-Pearson & Gianinazzi*, Nealson et al.*
PERISYMBIOTIC MEMBRANE - INTERFACE MATERIAL - WALL	Lichens, Plant-Dinitrogen fixing bacteria, Endomycorrhizae	Honneger*, Torrey*, Gianinazzi-Pearson & Gianinazzi, Bonfante-Fasolo*
PERISYMBIOTIC MEMBRANE - WALL	Plant-Dinitrogen fixing bacteroids, Endomycorrhizae, animal-algae, protozoa-algae, protozoa-bacteria, insect-bacteria, insect-yeasts	Torrey*, Bonfante-Fasolo*, McAuley*, Reisser*, Hinde*, Jeon (1983), Nardon*
PERISYMBIOTIC MEMBRANE - PLASMA MEMBRANE	Animal-Algae	Douglas*, Hinde*, Trench*
PERISYMBIOTIC MEMBRANE - SYMBIONT REMNANTS	"Harvesting phase" of intracellular associations	e.g. Reisser*, Scannerini & Bonfante Fasolo (1983)
CYTOPLASM - WALL (and/or PLASMA MEMBRANE)	Protozoa-xenosomes, Insect-bacteroids VA Fungi - bacteria-like organisms (BLOs)	Soldo (1983), Nardon*, Macdonald & Chandler (1981)

TAB. 1 - References marked with * refer to papers in this book. For more general informations see also Smith & Douglas (1987).

in different types of symbioses.

A TENTATIVE COMMON MODEL FOR CELL-TO-CELL INTERACTIONS BETWEEN MUTUALISTIC SYMBIONTS.

We have seen that the level of knowledge of the functional morphology of cell-to-cell interactions in mutualistic symbioses differs from system to system, but they are sufficient to permit proposal of a common model based on "frustrated phagocytosis". The time of blocking would correspond to the "compatibility" phase, with a final control by "cell harvesting" or "cell extrusion" in unicellular endocytobionts or by external digestion in the case of endosymbionts or multicellular endocytobionts.

This model is basically in agreement with the passage from inter- to intracellularity in Smith's classical model (1979). The various mutualistic symbiotic associations may thus be fitted into various stages of the heterophagocytic sequences outlined according to Glauman et al. (1981) and to Hoffstein's classical model (1980).

I) recognition and entrapment on the cell surface of the particle or organism to be phagocytized using surface receptors;

II) invagination of the plasmalemma, stimulation of peroxidase activity and formation of the open phagocytic vacuole;

III) closure of the vacuole with intervention on the part of microtubules and microfilaments;

IV) migration of the vacuole to the centre of the cell and its differentiation into a heterophagocytic vacuole, activation of the lysosomal system through membranes and lytic enzymes with the intervention of the Golgi apparatus and the granular endoplasmic reticulum (this phase requires the presence of Ca^{++} as activators); and

V) lysis of the segregated material by enzymes.

The sum of the first phase (entrapment and adhesion) together with the second and third (engulfment) make up the "accessibility" phase and it is valid both for mutualistic symbionts and for parasitic biotrophs, and corresponds to the active phase of the interaction involving the classical two-way interchange. In the case of plant symbionts the first stages of recognition and adhesion may be mediated by the cell wall and in the case of wall-wall interface symbionts there is not further development (ectomycorrhizas, almost all lichens, Azolla-Anabaena). In the case of an endocytobiotic mutualistic phase the niche is derived by the second and third stages, by means of the differentiation of a closed vacuole (stage III) in the case of animals. In the case of plants the niche may be stage II as the plasmamembrane invagination become possible because of the localized lysis of the cell wall and a decrease of the apoplastic mechanical barriers. If the endocytobiont is single celled the vacuole may be easily closed (nitrogen fixing-bacteria), but if it is multicellular (VA endomycorrhizas) this is not possible purely for geometrical reasons and the endocytobiont remains in a membrane sac. As already shown in the previous phagocytic model, nothing prevents us from believing that a higher plant cell is capable of eterophagy in the presence of the appropriate signals (Scannerini, 1985; Boller and Wiemken, 1986).

The differentiation of the vacuole in heterophagic vacuoles induces the harvesting stage for the control of the endocytobiont. The latter may escape stage IV and V in various ways (these are completely documented in the animal-algae, protozoan,bacterial, nitrogen-fixing bacteria and insect bacteroid symbioses) and therefore staying the third stage (perisymbiontic membrane delimited endocytobionts) or come out from the vacuole to be included into the cytoplasm (cytoplasmic endocytobionts, BLOs sensu McDonald and Chandler - 1981 -). The latter possibility may progress from endocytobiont to organelle according to Taylor (1983) and permits the survival of xenosomes and organelles (e.g. chloroplasts in invertebrates) naturally occurring or experimentally enclosed in other cells.

The possibility of stopping in the fourth stage and choosing it as an ecological niche by acquiring structures resistant to hydrolases is rarer in the case so far studied (Phagosomes-Leishmania).

If the control of the endocytobiont does not occur through harvesting or though exocytosis of supernumerary organisms, the endocytobiont can overcome the perisymbiotic barrier and survive after the death of the host cell (ericoid mycorrhizas, Pezizella-susceptible non-host plants, parasites).

Classical functional morphology, therefore leads us to a unifying model in which the modulation between parasitic and mutualistic symbionts is due to switches which, by increasing or decreasing the length of the "accessibility" (compatibility) phase, shift the equilibrium from mutualistic to parasitic symbiosis and vice-versa. This is possible by: A) modification and modulation of cell-wall and glycocalix ultrastructure and components; B) control of the membrane flow between plasmalemma invagination, lysosomes and Golgi resulting in exocytic or endocytic pathways with a possibility of escape from the perisymbiotic barrier.

In this tentative cytological model the following topics are clearly to be elucidated:

1) The signals for adhesion and engulfment;
2) the features of the perisymbiotic membranes;
3) the signals which determine the regulation of the membrane flow which guarantees harvesting of symbionts, the exocytic responses, the possibility of overcoming the perisymbiontic membranes so as to become completely integrated into the cell or, on the other hand, behaving as pathogen.

Therefore only an accurate study of the molecular structure of the surface elements (cell walls, glycocalices) of partner cells, and of the perisymbiotic membrane, carried out in parallel with a precise investigation of development with time, may allow us to assess the real differences, if they exist, between different mutualistic strategies.

REFERENCES.

Alberts, B., Bray, D., Lewis, J., Raff, M., Roberts, K., Watson, J.D. 1985. Molecular Biology of the cell. Garland Publishing Inc. New York and London.

Bennett, H. 1969. The Cell Surface: Components and Configurations. Handbook of Molecular Cytology. Edited by A. Lima de Faria. 1251-1293. North Holland Amsterdam-London.

Boller, T. 1982. Enzymatic equipment of Plant vacuoles. Physiol. Veg. 20: 247-257.

Boller, T., Wiemken, A. 1986. Dynamics of vacuolar compartmentation. Ann. Rev. Plant Physiol. 37: 137-164.

Bonfante-Fasolo, P., Scannerini, S. 1983. Mycorrhizae as a sequence running from extracellular to intracellular relations between symbionts. 3° International Mycological Congress Tokyo, Japan, 393.

Bonfante-Fasolo, P., Gianinazzi-Pearson, V., Martinengo, L. 1984. Ultrastructural aspects of endomycorrhiza in the Ericaceae. IV. Comparison of infection by Pezizella ericae in host and non host plants. New Phytologist 98: 329-333.

Bracker, C.E., Littlefield, L.J. 1973. Structural concepts of host-Pathogen interfaces. In: Fungal Pathogenecity and the plant's response. Edited by R.J.W. Byde and C.V. Cutting 159-318. Academic Press London.

Brown, D.D. 1981. Gene expression in eukaryotes. Science 211: 667-674.

Cavalier-Smith, T. 1981. The origin and early evolution of eukaryotic cell. Symp. Soc. Gen. Microbiol. 32: 33-84.

Cavalier-Smith, T. 1987. Eukaryotes with no mithocondria. Nature 326: 332-333.

Chang, K.P. 1983. Cellular and Molecular mechanisms in intracellular symbiosis in Leishmaniasis. International Review of Cytology, suppl. 14 Intracellular Symbiosis. Edited by K.W. Jeon 267-305.

Cook, C.B., Pappas, P.W., Rudolph, E.D. 1980. Preface in Cellular Interactions in Symbiosis and Parasitism. Edited by C.B. Cook, P.W. Pappas and E.D. Rudolph Ohio State University Press. Columbus.

Eyre, D.R. 1980. Collagen Molecular diversity in the body's proteins scaffold. Science 207: 1315-1322.

Frey-Wyssling, A., Mühlethaler, K. 1965. Ultrastructural Plant Cytology. Elsevier. Amsterdam.

Glaumann, H., Ericsson, J.L.E., Marzella, L. 1981. Mechanisms of intralysosomal degradation with special reference in autophagocytosis and heterophagocytosis of cell organelles. International Review of Cytology 73: 149-182.

Gull, J.G. 1981. Chromosome structure and the C- value paradox. J. Cell Biology 91: 3-14.

Hartzell, A. 1967. Insect ectosymbiosis. In: Symbiosis. Edited by S. Mark--Henry vol. 2: 107-140. Academic Press. New York and London.

Heslop-Harrison, A., Linskens, H.F. 1984. Cellular interaction; a brief conspectus. In: Encyclopedia of Plant Physiology 17. Cellular interactions. Edited by H.F. Linskens and J. Heslop-Harrison 2-17. Springer-Verlag. Heidelberg.

Hinde, R., Smith, D.C. 1974. "Chloroplast symbiosis" and the extent to which it occurs in Sacoglossa (Gastropoda Mollusca). Biological Journal of the Linneau Society 7: 161-171.

Hoffstein, S.T. 1980. Intra- and extracellular secretion from plymorphonuclear leukocytes. In: The cell biology of inflammation 387-480. Elsevier -North Holland. Amsterdam.

Jackson, W.T. 1982. Actomyosin. In: The cytoskeleton in plant growth and development. Edited by C.W. Lloyd 3-29. Academic Press London-New York.

Jeon, K.W. 1983. Integration of bacterial Endosymbionts in Amoebae. In: International Review of Cytology. Suppl. 14 Intracellular Symbiosis. Edited by K.W. Jeon 29-47.

Keen, N.T. 1986. Pathogenic strategies in Fungi. In: Recognition in Microbe -Plant Symbiotic and pathogenic interactions. Edited by B. Lugtenberg. Series H NATO ASI Series. 171-188. Springer-Verlag. Heidelberg.

Koch, A. 1967. Insects and their Endosymbiont. In: Symbiosis. Edited by S. Mark-Henry vol. 2: 1-106. Academic Press New York and London.

Lamport, D.T.A., Caty, J.W. 1981. Glycoproteins and enzymes of Cell Wall. Encyclopedia of Plant Physiology. 13 B. Extracellular Carbohydrates. Edited by W. Tanner and F.A. Loewus 132-165. Springer-Verlag, Heidelberg.

Laval-Peito, M., Febvre, M. 1986. On plastid symbiosis in Tontonia appendiculariformis (Ciliophora, Oligotrichina). Biosystems 19: 137-158.

Lugtenberg, B. 1986. Preface V-VI in Recognition in Microbe-Plant Symbiotic and Pathogenic interactions. Edited by B. Lugtenberg. Series H NATO ASI Series. 171-188. Springer-Verlag. Heidelberg.

Macdonald, R.M., Chandler, M.R. 1981. Bacterium-like organelles in the vesicular-arbuscular mycorrhizal fungus Glomus caledonius. New Phytol. 89: 241-246.

Margulis, L. 1981. Symbiosis in Cell Evolution. Freeman & Co. San Francisco.

Marty, F. 1978. Cytochemical studies on GERL, provacuoles and vacuoles in root meristematic cells of Euphorbia. Proc. Nat. Acad. Sci. USA 75: 852-856.

Marty, F., Branton, D., Leigh, R.A. 1980. Plant vacuoles. In: The Biochemistry of Plants: A comprehensive treatise. Edited by N.I. Tolbert 1: 625-658. Academic Press. New York.

Matile, Ph. 1978. Biochemistry and function of vacuoles. Ann. Rev. Plant Physiol. 28: 193-213.

Ouchi, S. 1983. Induction of resistance or susceptibility. Ann. Rev. Phytopathol. 21: 289-315.

Palade, G.E., Farquar, M.G. 1981. Cell biology in Pathophysiology. In: The Biological principles of disease. Edited by L.H. Smith, S.O. Thier 1-56. Sanders. Philadelphia.

Rahat, M., Reich, V. 1986. Algal endosymbiosis in brown hydra: host/symbiont specificity. J. Cell Sci. 86: 273-286.

Scannerini, S. 1985. Mycorrhizal symbiosis. 2. The process. Riv. Biol. 78: 546-553.

Scannerini, S., Bonfante-Fasolo, P. 1983. Comparative ultrastructural analysis of mycorrhizal associations. Can. J. Bot. 61: 917-943.

Scannerini, S., Giunta, C., Panzica, G.C. 1975. Lysosomes in cultivated mushroom (Psalliota bispora Quél.). Giorn. Batt. Virol. Immunol. 68: 17-31.

Schwemmler, W. 1983. Analysis of possible gene transfer between an Insect Host and its bacteria-like endocytobionts. International Review of Cytology. Suppl. 14: 247-263.

Smith, D.C. 1979. From extracellular to intracellular: the establishment of a symbiosis. Proc. R. Soc. London, B 204: 115-130.

Smith, D.C., Douglas, A.E. 1987. The biology of Symbiosis. Edward Arnold, London.

Soldo, A.T. 1983. The biology of xenosome, an intracellular symbiont; International Review of Cytology. Suppl. 14. Edited by K.W. Jeon 79-109.

Stolp, H. 1979. Interactions between Bdellovibrio and its host cell. Proc. R. Soc. London B 204: 211-217.

Taylor, F.J.R. 1979. Symbioticism revisited: a discussion of the evolutionary impact of intracellular symbioses. Proc. R. Soc. London B 204: 267--286.

Wathley, J.M., Whatley, F.R. 1984. Evolutionary aspects of eukaryotic Cell and its organelles. In: Encyclopedia of Plant Physiology. Cellular interactions. Edited by H.F. Linskens and J. Heslop-Harrison 17: 18-58. Springer-Verlag. Heidelberg.

SYMBIOSIS AND EVOLUTION: A BRIEF GUIDE TO RECENT LITERATURE

Lynn Margulis and David Bermudes
Boston University
Biological Science Center
2 Cummington Street
Boston, MA 02215

I. Speciation and the Origin of Higher Taxa

We have recently explored the hypothesis that, in certain taxa, hereditary symbiosis is the major mechanism of origin of that taxon. The reader is referred to Taylor, 1983; Margulis and Bermudes, 1985; and Bermudes and Margulis, 1987.

Three fundamental assertions about the evolutionary importance of symbiosis were made by Ivan Wallin in 1927. Paraphrased they are:

1. Bounded organelles capable of development and reproduction by division originated by symbionticism (which is defined by Wallin, 1927, p. 8 as "the establishment of intimate microsymbiotic complexes"). These organelles (mitochondria and plastids) of cells of animals, plants, fungi and protoctists are descendants of intracellular bacteria. In particular, mitochondria found in all animal and plant cells originated by symbionticism from oxygen-respiring bacteria.

2. "Symbionticism" is the major mechanism for generation of evolutionary innovation, as such symbionticism is the major mechanism of the origin of species.

3. Histogenesis (i.e., the generation of tissue types) both phylogenetically and ontogenetically is a consequence of interactions of populations of (primarily microbial) symbionts.

Because it was ignored or disdained by active biologists of his day, Wallin's work had no discernible effect on the subsequent development of 20th century biological literature (Mehos, 1983). Even though Wallin himself cannot be credited with the contemporary development of these ideas, we note that his assertion (1) the origin of plastids and mitochondria by hereditary cell symbiosis has been proven and accepted by most practicing biologists (Gray, 1983). Furthermore, we defend here his assertion (2) by referencing recent books and articles that argue the importance of symbiosis as a mechanism

NATO ASI Series, Vol. H17
Cell to Cell Signals in Plant, Animal and
Microbial Symbiosis. Edited by S. Scannerini et al.

of evolution. Finally, we defer Wallin's assertion (3) for discussion at a later date; nevertheless we suggest that the reader, through perusal of this workshop volume, keep in mind the effects of symbiosis on development and tissue formation. Contributions concerning lichen tissue morphology (Honneger), endomycorrhizal structure (Gianninazzi), insect öogenesis and histogenesis (Nardon), Azolla-Anabaena associations (Grilli Caiola and Albertano), bacterial and actinobacterial root associations (Torrey; Dazzo and his co-workers) all support Wallin's suggestion that symbiont-related hypertrophy is an essential factor in tissue differentiation.

NeoDarwinists have traditionally rejected "macromutation" as a mechanism of the origin of fundamentally new metabolic pathways, morphologies and behaviors. Furthermore, it is generally assumed in modern texts on evolution that the origin of higher taxa occurs by the same mechanism as the origin of species (viz., gradual accumulation of randomly acquired small genetic mutations). The term "macromutation", popularized primarily by Goldschmidt (1940), has been severely criticized and, in recent years, the concept generally has been rejected. Nevertheless, based on new findings of micro- cell- and molecular biology, we suggest serious reconsideration of the idea of "macromutation" in its literal sense of "large heritable change". We acknowledge evidence for evolutionary discontinuity and the necessity to explain it. Such discontinuity is evident in species and higher taxa of both living organisms and those known only from the fossil record. At least some of these extant and fossil discontinuities can be accounted for by the "inheritance of acquired characteristics" (Taylor, 1983) where the "characteristics" are entire genomes and "inheritance" refers to acquisition of symbionts that became behaviorally, metabolically and genetically integrated such that the associations evolved into permanently heritable associations.

Alternatives to the standard neodarwinian "gradualist" scenario have not been adequately explored. The appearance of semes, structural, functional or behavioral features of an organism (i.e., traits that are determined by at least several genes), can be directly correlated with the origin of certain taxa. New semes (i.e., neosemes) in many documented cases are directly attributable to hereditary symbiont acquisition. For a further discussion of semes and their evolutionary importance see Hanson, 1977. Furthermore, a survey of phyla from all four eukaryotic kingdoms (for example, apicomplexan protoctists, basidiomycote fungi, cycadophyte plants and vestiminiferan animals) indicates that the acquisition of heritable symbionts can be correlated with the origin of many higher taxa.

II. Cell Structures: Origin by Symbiosis

The best documented example of the importance of symbiont acquisition for evolutionary innovation is the role hereditary microbial symbiosis played as the major evolutionary mechanism of eukaryotic cells. Reviewed here are the main classes of organelles with known or possible symbiotic origin.

A. Nucleocytoplasm

In a series of papers Searcy and his colleagues (Searcy *et al*., 1981; Searcy, 1986, 1987) have argued that the nucleocytoplasm of eukaryotes evolved from an archaebacterial lineage that, through time, became associated with eubacterial ancestors to plastids and mitochondria.

B. Undulipodia

The amount of circumstantial evidence for the symbiotic origin of undulipodia (i.e., cilia, eukaryotic "flagella") has increased although the hypothesis cannot be considered proven. The evidence depends on (1) analysis of cortical genetic systems in ciliates, (2) comparison of undulipodia with other classes of cell organelles, their genetic and developmental behavior (3) application of Darwin's concept of the historical explanation of biological "peculiarities" (i.e., S.J. Gould's Panda Principle discussed in Bermudes *et al*., 1987a; Szathmary, in press) (4) identification of possible bacterial co-descendants of undulipodia (e.g., spirochetes) that tend to form motility symbioses, Bermudes, *et al*., 1987b. Ultrastructural, immunocytological and protein chemical studies seeking tubulin-like proteins in *Spirochaeta bajacaliforniensis* have been performed. *Spirochaeta bajacaliforniensis*, an anaerobic free-living bacterium, possesses at least two proteins (S1 and S2) which have been copurified by a temperature-dependent cycling method used for the isolation of brain tubulin. Warm aggregates of these proteins contain fibrous material as observed by light and electron microscopy. Anti-tubulin (bovine brain) antiserum is reactive against S1 protein and the fibrous aggregate (Bermudes, 1987). Sequencing studies to definitively establish the extent of homology between the spirochete proteins and tubulins are underway (Tzertzinis, 1987 personal communication). The consequences of the spirochete symbiosis concept for protist evolution, the origin of mitosis and meiotic sex are discussed in Margulis and Sagan (1986).

C. Mitochondria

The polyphyletic origin of mitochondria from two or more lineages of respiring gram-negative eubacteria is highly likely. See Margulis and Bermudes, 1985, and citations therein for review.

D. Plastids

Plastids (e.g., chloroplasts, rhodoplasts, chrysoplasts, etc.) evolved from various lineages of phototrophic bacteria - hence their polyphyletic origins can be considered established. For discussion and references see Margulis and Bermudes, 1985.

E. Xenosomes

A number of intracellular membrane-bounded organelles (e.g., "xenosomes", trichocysts, hydrogenosomes, scintillons) may have originated by intracellular symbiosis. The biology of xenosomes has been reviewed by Soldo (1983). The generality, significance and even the suggestion that the field dealing with symbiosis and cell organelles (endocytobiology) be renamed "xenosomology" has been discussed in Corliss (1987) and references cited therein. A vast reorientation of cell biology as well as systematics and taxonomy of eukaryotic microorganisms is required because of symbiotic concepts. This is outlined and documented in Corliss (1986a, 1987). The serial endosymbiotic theory (SET) has profound implications especially for the emerging field of protoctistology (protistology) (see Taylor, 1983; Corliss, 1986a,b; 1987 for reviews).

Major discoveries and recent literature on symbiosis and evolution derived from many different fields of biology (e.g., parasitology, plant pathology, tropical marine biology, etc.)

are described in two general texts (Ahmadjian and Paracer, 1986; Smith and Douglas, 1987). Symbiosis as a major mechanism of organellar evolution is detailed in book form in Margulis (1981) and Schwemmler (1984). Collections of recent papers on symbiosis all of which include at least some mention of evolutionary implications include Fredrick, ed. 1981; Jeon, ed. 1983; Schwemmler and Schenk, eds. 1980, 1983; Dyer and Obar, eds. 1985; Lee and Fredrick, eds. 1987.

Acknowledgments

We gratefully acknowledge Shirley Manditch for typing this manuscript. This work was supported by a National Science Foundation Graduate Fellowship (to DB), the Lounsbery Foundation (to LM), NASA Grant NGR-004-025 (to LM), and the Boston University Graduate School.

References

Ahmadjian, V. and S. Paracer, 1986. Symbiosis: An Introduction to Biological Associations. University Press of New England, Hanover, NH.

Bermudes, D., 1987. Distribution and Immunocytochemical Localization of Tubulin-like Proteins in Spirochetes. Ph.D. Thesis, Boston University Graduate School, Boston, MA.

Bermudes, D. and L. Margulis, 1987. Symbiont acquisition as neoseme: origin of species and higher taxa. Symbiosis (in press).

Bermudes, D., L. Margulis, and G. Tzertzinis, 1987a. Prokaryotic origin of undulipodia: Application of the Panda Principle to the centriole enigma. Annals of the New York Academy of Sciences 503:187-197.

Bermudes, D., S.P. Fracek Jr., R.A. Laursen, L. Margulis, R. Obar, and G. Tzertzinis, 1987b. Tubulin-like protein from Spirochaeta bajacaliforniensis. Annals of the New York Academy of Sciences 503:515-527.

Corliss, J.O., 1986a. Progress in protistology during the first decade following reemergence of the field as a respectable interdisciplinary area in modern biological research. Progress in Protistology 1:11-63.

Corliss, J.O., 1986b. Advances in studies on phylogeny and evolution of protists. Insect Science Applications 7:305-312.

Corliss, J.O., 1987. Protistan phylogeny and eukaryogenesis. International Review of Cytology 100:319-370.

Dyer, B.D. and R. Obar (eds.) 1985. The Origin of Eukaryotic Cells. Van Nostrand Reinhold Co., New York.
Fredrick, J.F. (ed.) 1981. Origins and Evolution of Eukaryotic Intracellular Organelles. Annals of the New York Academy of Sciences. Vol. 361.
Gray, M.W., 1983. The bacterial ancestry of plastids and mitochondria. BioScience 33:693-699.
Goldschmidt, R., 1940. The Material Basis of Evolution, Yale University Press, New Haven, pp. 390-393.
Hanson, E.D., 1977. The Origin and Early Evolution of Animals, Wesleyan University Press, Wesleyan, CT.
Jeon, K.W. (ed.) 1983. Intracellular Symbiosis, International Review of Cytology, Supplement 14. Academic Press, New York.
Lee, J.J. and J.F. Fredrick (eds.) 1987. Endocytobiology III. Annals of the New York Academy of Sciences, Vol. 503.
Margulis, L., 1981. Symbiosis in Cell Evolution. W.H. Freeman and Co., San Francisco.
Margulis, L. and D. Bermudes, 1985. Symbiosis as a mechanism of evolution: Status of cell symbiosis theory. Symbiosis 1:101-124.
Margulis, L. and D. Sagan, 1986. The Origins of Sex: Three billion years of Genetic Recombination, Yale University Press, New Haven, CT.
Mehos, D.C., 1983. Symbionticism as a Biological Principle: Ivan E. Wallin's Theory of Organic Evolution. Master of Arts Thesis, Boston University Graduate School., Boston, MA.
Schwemmler, W. and H.E.A. Schenk (eds.) 1980. Endocytobiology: Endosymbiosis and Cell Biology a Synthesis of Recent Research. Walter de Gruyter, New York.
Schwemmler, W. and H.E.A. Schenk (eds.) 1983. Endocytobiology II. Intracellular Space as Oligogenetic Ecosystems. Walter de Gruyter, New York.
Schwemmler, W., 1984. Reconstruction of Cell Evolution: A Periodic System. CRC Press, Boca Raton, FL.
Searcy, D.G., D.B. Stein, and K.B. Searcy, 1981. A mycoplasm-like archaebacterium possibly related to the nucleus and cytoplasm of eukaryotic cells. Annals of the New York Academy of Sciences 361:312-324.
Searcy, D.G., 1986. Some features of thermoacidophilic archaebacteria preadaptive for the evolution of eukaryotic cells. Systemic and Applied Microbiology 7:198-201.
Searcy, D.G., 1987. Phylogenetic and phenotypic relationships between the eukaryotic nucleo-cytoplasm and thermophilic archaebacteria. Annals of the New York Academy of Sciences 503: (in press).
Smith, D.C. and A.E. Douglas, 1987. The Biology of Symbiosis. Edward Arnold, London.
Soldo, A.T., 1983. The biology of the xenosome, an intracellular symbiont. International Review of Cytology, Supplement 14:79-109.
Szathmary, E., 1987. Early evolution of microtubules and undulipodia. BioSystems 20: (in press).
Taylor, F.J.R., 1983. Some eco-evolutionary aspects of intracellular symbiosis. International Review of Cytology, Supplement 14:1-28.

To, L.P., 1987. Are centrioles semiautonomous? Annals of the New York Academy of Sciences 503-83-91.
Wallin, I.E., 1927. Symbionticism and the Origin of Species. Williams and Wilkins Co., Baltimore, MD.

Molecular Signals in Plant Cell Recognition

J.A.Callow, T. Ray, T.M. Estrada-Garcia, J.R. Green

Dept. of Plant Biology
University of Birmingham
PO Box 363
Birmingham B15 2TT
UK

Introduction

Molecular recognition and associated signalling mechanisms in plants are fundamental to a wide range of biological processes including fertilisation, cell division, ordered cell growth and development, host-pathogen interactions, symbiosis and stress response. The general presumption (borne out in some cases) is that such recognition events involve surface-localised molecules associating in some complementary fashion, and that the surface events are linked to a cellular response through some form of signalling mechanism involving transmission and transduction of the stimulus. In this presentation our current understanding of the nature of some molecular signals and their receptors involved in recognition of 'self' and 'non-self' will be considered in the context of recognition between higher plants and parasitic microorganisms. Particular attention will be devoted to the application of newer molecular technologies.

Mutual molecular interactions between cell surfaces of host plants and of pathogens are presumed to play a fundamental role in the control of pathogenesis and disease resistance. On the one hand, the host plant has to be able to detect or recognize a potential pathogen as foreign or 'non-self' and to use this initial act of recognition to trigger induced resistance mechanisms. How does the plant detect the vast majority of potential pathogens in its environment as 'non-self' and how does the successful pathogen escape detection to cause disease? What are the signals generated directly or indirectly by pathogens, what are the host receptors for the signals and where are they located at the cellular level? How does reception of the signal become transduced into the biochemical mechanism of resistance?

On the other hand, a potential pathogen requires an ability to recognize those features of a plant that signal its suitability for parasitism. On a leaf surface, for example, spores of both saprophytic and parasitic fungi can be found in abundance but it is only those of the parasitic species that can penetrate and cause disease, the parasitic species not only possess the necessary 'factors' for pathogenesis, but through various 'cues' they may detect the presence of the host

NATO ASI Series, Vol. H17
Cell to Cell Signals in Plant, Animal and
Microbial Symbiosis. Edited by S. Scannerini et al.

plant and use these recognition events to modulate or control the expression of the mechanisms of pathogenesis.In the interaction between rust fungus and host leaf tissue for example, one can recognize a number of stages in the infection process, including spore germination, directional and adhesive growth of germ tubes to stomata, appressorial differentiation, vesicle expansion, haustorial initiation and penetration, haustorial maturation . All of these developmental stages are likely to involve responses of the pathogen to specific recognition cues from the host surface.

A. Recognition of pathogen signals in induced defence

Although much of resistance to natural disease in plants lies in static, constitutive or preformed defensive barriers, it is now clear that plants also possess effective inducible resistance mechanisms requiring *de novo* gene expression, synthesis of new proteins, and switching of metabolic pathways (e.g. Ryder *et al.* 1986; Hadwiger *et al.* 1986; Dixon *et al.*, 1987). Most current models (e.g. Bailey, 1987; Callow, 1987) invoke two classes of 'resistance gene' viz. 'response' genes, involved in the expression of the resistance pathway *per se* and 'recognition' genes which code for receptors, presumably located at the cell surface, and which are therefore involved in the signalling mechanisms leading to cascade expression of the response genes, in turn leading to local defence reactions including hypersensitivity, phytoalexin accumulation, lignification and other wall modifications, accumulation of hypro-rich glycoproteins, release of hydrolytic enzymes, and the possibility of systemic responses notably the induced synthesis of protease inhibitors (Fig. 1).

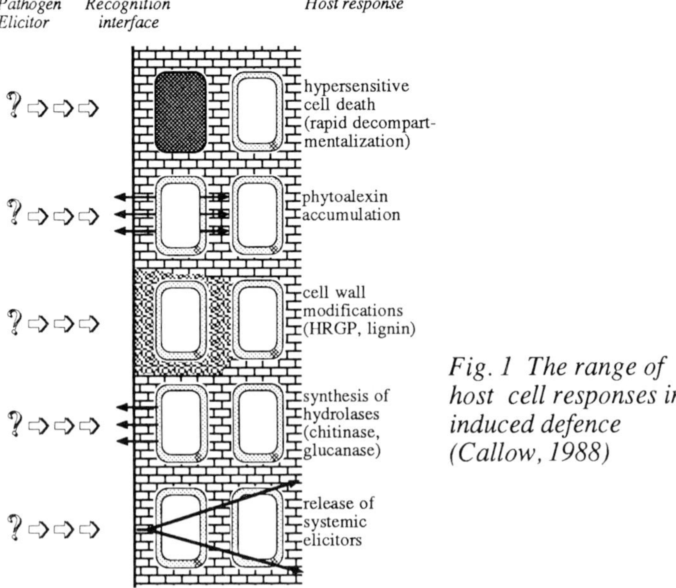

Fig. 1 The range of host cell responses in induced defence (Callow, 1988)

Specificity in the association is therefore held to be at the level of recognition rather than in the mechanism of induced resistance *per se* , and it has been argued by several authorities (Ellingboe, 1982; Bushnell & Rowell, 1981; Callow, 1984,1987) , often using elaborate models invoking evolutionary mechanisms, that in race/cultivar-specific resistance, controlled by oligogenic gene-for-gene interactions involving pathogen genes for avirulence and host genes for 'resistance' , that the R genes conferring host specificity, are in effect, 'recognition' genes. The search for factors involved in host-pathogen recognition has therefore become identified with the search for the primary gene products of pathogen genes controlling avirulence and host genes controlling resistance respectively.

On *a priori* grounds, there are those who favour a central role for polysaccharides and glycoconjugates in this initial recognition (e.g. Callow, 1987). What lines of evidence suggest that plants can use carbohydrate-based molecules to signal the presence of potential pathogens? A number of reviews (e.g. Darvill & Albersheim, 1984; De Wit, 1986), have considered how plants may use a variety of polysaccharide or glycoprotein molecules commonly found in fungal cell walls or secretions, as signals for the elicitation of host responses such as phytoalexin synthesis. The list of active components includes peptides , glycoproteins, and oligosaccharides derived from fungal wall polymers such as β-glucan , chitin and chitosan . Lipid fractions also have elicitor activity. None of these elicitors is race and cultivar-specific and the general conclusion is that they serve as recognition cues at the level of general resistance, i.e. they may be used by plants to signal the presence of most pathogens.

In only one of these cases is there a comprehensive molecular description of the signal. The cell walls of many fungi contain an alkali-insoluble β1,3-linked glucan with β1,6-branches. Glucan oligosaccharides of variable molecular weight are released from fungi or may be extracted by a variety of procedures. The oligosaccharides from *Phytophthora megasperma* f.sp. *glycinea* have been subject to partial acid and enzyme hydrolysis to yield fragments from which a highly active heptaglucoside has been isolated (Sharp *et al.* 1984a,b). The heptaglucoside is the smallest fragment with elicitor activity and is highly active, 1 pmole being sufficient to elicit phytoalexins in soybean. What is more remarkable, however, is that activity is highly stereospecific, being dependent on a backbone of five 1,6-linked glucosyl residues, with two nonreducing terminal glycosyl residues linked to C3 of the second and fourth residues. Seven closely related variants of the heptaglucoside have been structurally characterized and shown to be inactive biologically, and the structure of the active isomer has been confirmed by chemical synthesis (Ossowski *et al.*, 1983). It is not yet clear whether or not the active glucan isomer is released from larger glucan oligosaccharides following partial hydrolysis by plant enzymes.

Thus it seems that plants possess the ability to recognize 'foreign' carbohydrate-containing molecules with both a high degree of sensitivity and with stereochemical specificity, both of which characteristics encourage the view that such recognition is biologically relevant. However, against this it must be said that over the years many investigators have isolated glycoprotein and polysaccharide molecules from pathogen surfaces which elicit responses chararacteristic of resistance when applied to uninfected host plants. By and large, definitive evidence is lacking that such "elicitors" exert their effect through some special surface recognition event located at the host cell surface involving host receptors, and signal transmission through secondary messengers.

A key aspect of our model implicating carbohydrate signals is the predicted presence of receptors on plant cell surfaces. The plasma membrane offers a logical site for the receipt and transduction of stimuli originating from pathogens, and there are a number of lines of circumstantial evidence which support the existence of surface membrane receptors for the glucan type of elicitor. Thus, plant protoplasts will bind β-glucosyl Yariv antigens (Larkin 1977), cell wall-derived β-glucans will alter protoplast membrane potentials (Kota & Stelzig 1977) and agglutinate them (Peters *et al.*, 1978). More direct evidence has been obtained by Yoshikawa *et al.*, (1983) who used radiolabelled mycolaminaran, a β1,3-linked storage glucan with some 6-linked branches, as an elicitor analogue, The analogue bound to soybean membrane fragments, binding was specific based on the results of competition experiments with unlabelled ligand and other carbohydrates, saturation kinetics were rapid, and binding was temperature and protease-sensitive. However, many of these characteristics would also be true if the binding was due to glucanase or glucan synthetase activities at the cell surface (Ayers *et al.* 1985), and until such receptors are isolated and fully characterized, particularly with respect to the binding of pure, synthetic elicitors of known structure, then the existence of elicitor receptors at the cell surface must still be considered as hypothetical. In one case, that of the chitosan elicitor (Hadwiger *et al.* 1987), it seems that surface receptors and secondary signalling mechanisms need not be invoked, the elicitor may not only enter cells directly by some unknown route, but appears to target to nuclear DNA. In fact there is little direct evidence that elicitors are involved in host-pathogen interactions *in vivo*. For example, applied elicitors of the glucan type fail to protect plants from infection by virulent fungal races (Ward 1983). An exception to this appears to be the race-specific necrosis-inducing peptide elicitor isolated from apoplastic fluids of tomato leaves infected by compatible races of *Fulvia fulva* (De Wit & Spikman, 1982; De Wit *et al.* 1985).

Growing emphasis is being placed upon somewhat more indirect interpretations of the course of events involved in resistance triggering (Bailey, 1987; Ward, 1986). Cell injury involving plasmalemma dysfunction and eventually cell death i.e. hypersensitivity, is a very common reaction during resistance to pathogens of all types and to various fungal elicitors, and may therefore be an early consequence of pathogen recognition by the host. This cell injury or stress may then form a critical trigger for cascade gene expression leading to, e.g. phytoalexin synthesis without the need to invoke subtle secondary messenger mechanisms. Indeed, the action of β-glucan elicitors is potentially explained by membrane damage through sterol binding and solubilisation (Pillai & Weet, 1975). There is also evidence (Bailey, 1982) which indicates that the interaction of a primary elicitor, or pathogen with plant cells may result in the release, probably from the plant cell wall, of 'secondary' or 'constitutive' elicitors which may then diffuse to surrounding healthy cells promoting defence responses. Coincident with this, related work has shown that some pathogens may induce the release of pectic fragments from the plant cell wall which then serve as secondary elicitors (Hahn *et al.* 1981; West, 1981) and subsequent work (Albersheim *et al.* 1983) has shown that a whole range of polysaccharide fragments with diverse biological activity (the 'oligosaccharins') may be obtained from plant walls (Table 1).

TABLE 1. Oligosaccharins: oligosaccharides with regulatory properties

Type	Origin	Active on	Evoking	Concentration	Reference
βglucan heptasaccharide	acid hydrolysis of fungal walls	soybean	glyceollin phytoalexin	0.1ng/cotyledon	Sharp *et al.* 1984
α1,4 galact-uronan dodeca-saccharide	acid hydrolysis of soybean walls & citrus pectin	soybean	glyceollin phytoalexin	1μg/cotyledon	Nothnagel *et al.* 1983
α1,4 deca-galacturonide	PGA-lyase on soybean walls,PGA & pectin	soybean	glyceollin phytoalexin	2μg/cotyledon	Darvill *et al.* 1985
α1,4 trideca-galacturonide	PG hydrolysis of castor bean walls	castor bean	casbene phytoalexin		Bruce & West 1982
α1,4 oligo-galacturonan	extracts from wounded tomato leaves	tomato	inducer of a proteinase inhibitor (PIIF)	10μg/cotyledon	Bishop *et al.* 1982
xyloglucan nonasaccharide	endoglucanase hydro-lysis of *Acer* walls	pea epicotyls	inhibits 2,4-D enhanced growth	0.01-0.1 μg/ml	York *et al.* 1984
pectic oligo-saccharides	hydrolysis of *Acer* walls with acid or endopolygalacturonase	1)Acer cells	1)cell death and inhibition of protein synthesis	1-10 μg/ml	Yamazaki *et al* 1983
		2)*Lemna*	2)inhibition of flowering		Gollin *et al.* 1984
		3)tobacco cell layers	3)morphogenetic effects		Tran Thanh Van *et al.* 1985

Whilst these molecular signals are of tremendous interest and potential importance, their significance has yet to be properly evaluated in terms of *in vivo* relevance, existence of receptors etc. Interesting results have been recently reported by Schell's group (Schell, 1988). The gene for tuber-specific proteinase inhibitor II of potato is silent in potato leaves but is systemically induced throughout the potato plant as the result of wounding or by treatment of detached leaves with oligosaccharins. The tobacco genome does not carry genes homologous to the potato proteinase inhibitor gene II but when the gene was introduced into tobacco, high levels of mRNA transcript were detected following wounding or oligosaccharide treatment, and this was also evident as a systemic effect in non-wounded tissue. The importance of this demonstration is that the foreign potato gene is responsive to tobacco PIIF (proteinase inhibitor-inducing factor) signals, presumably an oligosaccharin, and such studies will eventually lead to detection and comparison of oligosaccharin-induced gene regulatory sequences in different plants.

B. Detection of host recognition cues by the pathogen

If we consider the evidence for plant recognition cues (Table 2) then most information at the molecular level is available for associations between plants and bacteria where the appropriate bacterial genes controlling early stages of infection are only turned on in the presence of the appropriate host. In the case of crown gall disease, virulence (*vir*) gene expression takes place in response to phenolics released in host exudates (Stachel *et al.* 1985). In *Rhizobium leguminosarum*, *R. meliloti* and *R. trifolii,* the induction of the *nod* (nodulation) gene cluster (up to 8 genes) is controlled by the constitutive product of one of these genes, the *nodD* gene, in the presence of low molecular weight exudates from legumes (Kondorosi & Kondorosi 1986; Rossen et al. 1986). The active host components appear to be flavonoids of the flavone and flavone glucoside type, such as luteolin, and in the case of *R. leguminosarum*, other flavonoids and isoflavonoids antagonize *nod* gene expression. The control of nodule number on legume roots may thus depend on the extent to which *nod* genes are expressed, as determined by the relative concentrations of inducer and antagonizer molecules in the plant and rhizosphere.

In fungi, during the infection of leaf tissue by, for example, a rust fungus, a number of developmental stages in the infection process can be recognized, all of which involve responses of the pathogen to specific recognition cues from the host surface. At present our understanding of such signals is very poor. There is some evidence for the existence of chemical cues for spore germination but otherwise, cues on the outer surface of the host determining developmental responses of the fungus appear to be primarily specific contact stimuli produced through the periodicity of leaf surface particles. We have very little information as to how such physical stimuli may be translated into a cellular response but there is some evidence that extracellular proteins of wheat rust are involved in the transmission of this stimulus (Epstein *et al.* 1985). Appressoria develop over stomata and this can be mimicked by scratches on glass or plastic. During this process, differentiation-specific proteins appear as a result of gene transcription and other proteins appear to be down-regulated (Staples *et al.* 1986). Mendgen *et al.* (1985) have shown that there are qualitative differences in the surface carbohydrates of the different infection structures of rust fungi (germ tubes, appressoria, sub-stomatal vesicle, infection hyphae) that may be related to host recognition.

TABLE 2 Host recognition "cues" used by plant pathogens & parasites

species	host signal	process triggered	reference
Bacteria			
Agrobacterium tumefaciens	acetosyringone	induction of *vir* genes	Stachel *et al.* 1985
Rhizobium meliloti	flavones & flavone glycosides	induction of *nod* genes	Mulligan & Long 1985
R. leguminosarum			Firmin *et al.* 1986
R. leguminosarum	isoflavonoids & flavonols	antagonism of *nod* genes	Firmin *et al.* 1986
Fungi			
Many species	non-specific metabolites	spore germination on leaf surface	e.g. Allen 1976
Ustilago violacea	α-tocopherol	formation of pathogenic dikaryotic mycelium	Castle & Day 1984
rust fungi	contact	directional growth of germ tubes	Dickinson 1970
	contact/extracellular proteins	appressorial differentiation	Staples *et al.* 1986
zoosporic Oomycetes	non-specific metabolites	chemotaxis	e.g. Mitchell 1976
Phytophthora cinnamomi	fucose-rich ligands of root mucilage	binding to root	Hinch & Clarke 1980
Phytophthora cinnamomi	pectin, root mucilage	zoospore encystment	Irving & Grant 1984
Pythium aphanidermatum	fucose-rich ligands of root mucilage	zoospore binding	Longman & Callow, 1987
		zoospore encystment	Ray *et al.* (in prepn.)

Considering other situations requiring fungi to respond to plant recognition "cues", in the heterothallic smut fungi, two growth forms may be exhibited. In *Ustilago* spp., haploid sporidia, the saprophytic yeast-like phase, fuse in opposite mating pairs to form a parasitic mycelial dikaryon. In some cases e.g. *U. maydis* this will only happen efficiently in the presence of the host, the presumption being that some host factor is required. Castle & Day (1984) were recently able to show that α-tocopherol (vitamin E) was the active component of plant extracts inducing this change in *U. violacea* . Since extracts of both host and non-host plants of *U. violacea* contained α-tocopherol it seems unlikely that this particular recognition cue operates at the level of host specificity.

Such examples of host molecular signals recognised by pathogenic fungi and converted into a physiological or developmental response are very few indeed. One reason for the slow progress is that most systems involving filamentous fungi are too complex for facile biochemical analysis and some attention has been recently devoted to simpler systems involving zoosporic Oomycetes. The largely non-specific chemoattraction of zoospores of Oomycete pathogens to plant roots is only a prelude to their binding to the root surface, consolidating adhesion, encystment, germination and production of penetrating hyphae. Such pathogens provide useful model systems for analysing recognition mechanisms since interacting surfaces of both plant and pathogen are freely available for manipulation and analysis, and recognition can be quantified in terms of root binding and encystment. On binding to the plant root surface, particularly in the elongation zone, the zoospore is triggered to release a glycoprotein contained within spheroidal cytoplasmic vesicles onto the zoospore surface (Hemmes & Hohl, 1971; Sing & Bartnicki-Garcia, 1975; Grove & Bracker, 1978). This glycoprotein constitutes a 'glue' promoting the firm adhesion of the 'pre-cyst' stage to the root surface. *De novo* synthesis of the cellulosic cyst wall then takes place possibly mediated by enzymes contained within another type of cytoplasmic vesicle ('flattened' vesicles) located just beneath the plasma membrane (Bartnicki-Garcia & Wang, 1983).

Host plant signals for promoting binding and subsequent triggering of these developmental processes reside within component(s) of the root surface mucilage polysaccharide. The binding of zoospores of *Phytophthora cinnamomi* to maize roots (Hinch & Clarke, 1980) and *Pythium aphanidermatum* to roots of cress (*Lepidium sativum*) (Longman & Callow, 1987) is inhibited if the root surface mucilage polysaccharide is oxidised *in situ* with periodate, if there is a selective *in situ* hydrolysis of terminal fucose residues, or if fucose residues are blocked by the pretreatment of the root with fucose-binding lectins. Isolated root mucilage is also an effective competitive inhibitor of *Pythium* binding to cress roots but mucilage that had been hydrolysed with exofucosidase or which had been oxidised with periodate was much less effective

(Longman & Callow, 1987). Terminal fucose residues can be shown to be exposed on maize and cress root surfaces through the use of FITC-labelled fucose-binding lectins.

One conclusion from such studies is that the recognition cue for zoospore binding to roots, as opposed to, say binding to other carbohydrate-containing materials in the soil, may lie in the specific recognition of complex oligosaccharide signals containing terminal fucose residues forming part of the root surface mucilage. This maximises the probability of an individual zoospore infecting a susceptible plant.

Root mucilage is in fact a mixture of relatively poorly characterised polymers. In cress, both neutral and charged acidic components are present and there is a small percentage of protein (ratio of three components is 4:2:1 respectively, Ray *et al.* in preparation). More than one protein band is evident on SDS gels, and amino acid analysis reveals that there is very little hypro-rich protein of the type common within plant cell walls. Analysis of neutral sugars reveals galactose, glucose, and arabinose as major components with rhamnose, fucose, xylose and mannose also present. In cress, the amount of fucose is somewhat less than the very fucose-rich maize root mucilage. The component neutral sugars and quantity of uronic acid present shows that the cress mucilage may contain pectin-type polysaccharides and fractionation of whole mucilage by ion-exchange on DEAE-Sepharose 6B separates a neutral, peak I component (20-30% of total) that does not bind to the exchanger and which contains very little uronic acid (only 5% of total sugars), and a more uronic-acid rich (75-80%) peak II fraction which is eluted by salt from the ion-exchanger in a broad band. The proportions of the neutral sugars in the two peaks are quite different although fucose is present in both. Fractionation of peaks I and II on Sephadex G100 reveals size heterogeneity but in both cases, the majority of material is of mol. wt. >100,000.

What about the biological activity of this mucilage? We have already seen that modifications to surface mucilage *in situ* reduces zoospore binding to the root. Also, native mucilage will apparently act as a competitive inhibitor of binding. In both cases terminal fucose residues were implicated. Notwithstanding this, independent studies on *Phytophthora cinnamomi* have highlighted another biological activity of mucilage in that **encystment** of zoospores can be triggered, in the absence of the plant root, by citrus pectin, polygalacturonic acid, and root mucilage (Irving and Grant, 1984). We have also found with *Pythium*, that encystment can be triggered both by whole root mucilage, and with polygalacturonic acid. This activity means that reduced binding of zoospores to roots in the presence of exogenous mucilage, may simply reflect a reduction in the availability of motile zoospores for binding to roots rather than a competitive effect. It also raises the possibility that we are observing a multi-signal form of zoospore-root recognition. It may be that initial binding is mediated through fucose-containing oligosaccharide ligands forming one component of mucilage, whilst the uronic-acid rich fraction triggers the separate processes of encystment and adhesion. Other signals may regulate germination and directional growth of germ tubes.

Some evidence for separate signals regulating the different processes was obtained by Byrt, Irving & Grant (1982) who showed that fucose did not induce encystment of *P. cinnamomi* zoospores even though fucose was previously implicated in root binding. Elucidation of such a potentially complex set of recognition processes requires careful studies on the biological activity of sub-fractions of mucilage which is currently in progress for cress. Preliminary results on the cress/*Pythium* interaction (Ray *et al.* in preparation) suggest that the two sub-fractions of cress mucilage (peaks I and II), both induce zoospore encystment. Since the two sub-fractions are chemically very different, this result may suggest that encystment triggering signals presented by mucilage are rather non-specific compared with the apparently more specific root binding although this requires further careful experimentation on root binding and mucilage adsorption characteristics.

Turning now to examine the surface of the zoospore, it is likely that the zoospore receptors for root mucilage oligosaccharides are proteinaceous since digestion of the zoospore surface of *Pythium* with low concentrations of trypsin (5 μg/ml) abolished root binding without affecting motility (Longman & Callow, 1987). Monoclonal antibodies (MAbs) have been raised against the zoospore surface, with the intention of using these in various assays and screens to locate and identify antigens concerned with receptor function (Estrada-Garcia, Green & Callow, in preparation). The value of this approach has been discussed by Callow *et al.* (1988) and is similar to that adopted for *Phytophthora cinnamomi* by Hardham *et al.* (1985, 1986).

The several MAbs raised against whole zoospores so far are domain-specific and can be grouped into five categories on the basis of immunofluorescence. Group 1 MAbs bind to the whole surface of zoospores including flagella. Although tested on a very small number of other Oomycetes, this group appears to have some species-specificity, but more particularly, binding of the Group 1 MAb induces zoospore encystment. It is interesting that Hardham *et al.* (1986) found that a flagellum-specific MAb induced encystment in *P. cinnamomi*. Group 2 MAbs are specific for the anterior flagellum. Group 3 MAbs bind to the protoplast of some zoospores and not to the protoplasts of others in the same preparation. The binding has an uneven 'spotty' distribution. Other zoospores in the same preparation do not bind it. This Group will also bind to pre-cysts but not mature cysts. Immunogold labelling shows that this MAb binds very specifically to spheroidal peripheral vesicles, and probably, therefore, to the adhesive glycoprotein contained within them. Group 4 antibodies bind to antigens on the whole zoospore surface including flagella, but differ markedly from Group 1 MAbs in that they also bind to the outer coat of the mature cyst. Finally, Group 5 antibodies bind to mature cyst walls only.

Whilst these studies with zoospore MAbs are in a very early stage, our proposition is that they provide highly specific tools with which to study the recognition and differentiation processes involved in root-zoospore binding. As yet we have not investigated their use in this way, except for the encystment triggering mentioned above.

C. Use of novel molecular technologies

It will be evident from the above discussion that in biochemical terms, we have a relatively poor understanding of the nature of molecular signals and their receptors in plant-pathogen interactions, and one must look forward to those areas where new developments are likely to be achieved. One of the problems in the more conventional approaches adopted to date is that certain assumptions of the nature of the interacting moieties tend to be embodied within experimental protocols, and this may have limited the chances of detecting factors controlling recognition.

1. An analysis of the complex interaction between host and pathogen in molecular terms calls for approaches directed to the precise interface between host and pathogen rather than gross studies on cultured pathogens and infected tissues, and far more subtle and precise tools are required than hitherto used. It is very relevant that the recent progress in isolating the race-specific peptide elicitor of *Cladosporium fulvum* was achieved by analyses, not of culture fluids, or fungus in culture, but of components secreted by the pathogen only *in planta* (De Wit and Spikman, 1982). In addition, use of MAbs in analysing zoospore recognition at the root surface has already been referred to. MAbs provide very specific tools for probing delicate interfaces at the cellular level such as the haustorial interface where intact host plasma membrane is in relatively stable, and functional contact with the fungal cell (Callow *et al.* 1988) This is the interface at which determination of 'self' and 'non-self' is, in part, achieved.

2. An alternative approach to which increasing attention is being given, is the use of recombinant DNA technologies to isolate and study those genes and their products determining recognition. DNA transformation systems for pathogenic fungi such as *Colletotrichum, Cochliobolus* and *Cladosporium* are rapidly being developed for the isolation of genes that can complement mutations for functions such as virulence, pathogenicity and host-specificity. Hopefully we may see this approach extended to the isolation of genes involved in early stages of host-pathogen recognition such as appressorial differentiation and germ tube adhesion,

although the limitation to progress here is likely to be the absence of a facile screen to detect mutants such as those defective in appresorial adhesion and their complemented transformants.

In the case of recognition and specificity in interactions between pathogen races and differential host varieties, genetic analysis has shown that such interactions are controlled by gene-for-gene systems involving R genes for resistance in the host and complementary A genes for avirulence in the pathogen, i.e. host R genes "recognise" pathogen A genes to initiate active resistance and models have been proposed (e.g. Ellingboe, 1982; Keen, 1982; Day, 1984; Callow, 1984,1987) invoking molecular complementary in the primary gene products. The ultimate test of the validity of such models will be provided by the isolation and cloning of genes encoding specificity determinants in the pathogen and recognition and response genes in the host. Progress in this has been made in bacterial-host systems. Staskawicz *et al* ., (1984) reported the successful cloning of a race-specificity gene from *Pseudomonas syringae* pv. *glycinea*. Cosmid clones of race 6 were transferred into three other races and one recombinant was detected which changed the specificity of these races from virulent to avirulent on appropriate host cultivars. The cosmid clone was mapped and two EcoR1 fragments (0.95 kb and 0.56 kb) were identified by *Tn* 5 mutagenesis as being important in determining race specificity. Southern blots failed to reveal the same race 6 fragments in other races. Apart from being of seminal importance in the isolation of parasite genes controlling race specificity, the work of Staskawicz *et al.*, (1984) has an immediate bearing in some aspects of our model, since as predicted from the gene-for-gene hypothesis, avirulence gene functions were shown to be dominant over virulence. Gabriel *et al.* (1986) have cloned ten different avirulence genes from *Xanthomonas campestris* pv. *malvacearum*, i.e. each cloned gene has a specificity for a different host R gene.

Eventual isolation of the protein products of cloned avirulence genes will allow direct testing of the recognition model. A recent preliminary report suggests that transfer of the cosmid clone specifying the race 6 phenotype of *P. syringae* pv. *glycinea* into race 4 resulted in the appearance of a new surface antigen as detected by MAbs (Wingate *et al.*, 1984). It remains to be seen whether or not this change in the surface architecture of the bacterium is directly related to changes in the recognition of the bacterium by the plant but it seems that specificity in the structure of the O-chain carbohydrates of LPS can be ruled out (Barton-Willis *et al.* 1987).

On the host side the important goal of cloning R genes, the products of which form the other component of the complementary recognition system, is currently some way off since R genes are known only for their phenotypic effect, their primary products being totally unknown. In such cases the use of transposable elements to 'tag' R genes, thus permitting their physical isolation, is one of the ways forward. Other approaches to clone R genes may eventually utilise molecular (RFLP) maps of plant genomes coupled with chromosome "walking" techniques and plant transformation. Over the next few years there wil clearly be a greater concentration of effort on the molecular and cell biology of plant-pathogen interactions in which the complementary use of tools provided by MAbs and recombinant DNA technologies will be critical.

References

Albersheim, P. Darvill, A.G., McNeil, M., Valent, B.S., Sharp, J.K., Nothnagel, E.A., Davis K.R., Yamazaki, N., Gollin, D.J., York, W.S., Dudman, W.F., Darvil, J.E., Dell, A. 1983. Oligosaccharins: naturally occurring carbohydrates with biological regulatory functions. In: Structure and Function of the Plant Genome, eds. O. Ciferri and L. Dure III, NATO ASI, pp 293-312

Allen, P.J. 1976. Control of spore germination and infection structure formation in the fungi. In: Encyclopedia of Plant Physiology, N.S., Physiological Plant Pathology, eds. R. Heitefuss and P.H. Williams, pp 51-85, Springer-Verlag, Berlin, Heidelberg, New York.

Ayers A.R., Goodell, J.J. & De Angelis, P. 1985 Plant detection by pathogens. In: Biochemical Interactions of Plants with Other Organisms. eds E.E. Conn, G. Cooper-Driver & T. Swain, Recent Advances in Phytochemistry. **19**, 1-20.

Bailey, J.A. 1987. Phytoalexins: a genetic view of their significance. In Genetics and Plant Pathogenesis. pp 233-244, eds P.R. Day and G.J. Jellis, Blackwells.

Bailey, J.A. 1982 Physiological and biochemical events associated with the expression of resistance to disease. In., Active Defense Mechanisms in Plants, ed. R.K.S. Wood, pp 39-65. Plenum Pres, New York & London.

Bartnicki-Garcia, S., and Wang, M.C. 1983. Biochemical aspects of morphogenesis in *Phytophthora*. In Phytophthora: Its Biology, Taxonomy, Ecology and Pathology. eds D.C. Erwin, S. Bartnicki-Garcia, and P.H. Tsao. pp 121-137, American Phytopathological Society, St. Paul, Minnesota.

Barton-Willis, P.A., Wang, M.C., Staskawicz, B., & Keen, N.T. 1987 Structural studies on the O-chain polysaccharides of lipopolysaccharides from *Pseudomonas syringae* pv. *glycinea*. Physiological & Molecular Plant Pathology. **30**, 187-197.

Bishop, P.D., Makus, D.J., Pearce, G., & Ryan, C.A. 1981 Proteinase inhibitor-inducing factor activity in tomato leaves resides in oligosaccharides enzymically released from cell walls. Proceedings of the National Academy of Sciences, USA, **78**, 3536-3540.

Bushnell, W.R. and Rowell, J.B., 1981. Suppressors of defense reactions: a model for roles in specificity. Phytopathology, **71**, 1012-1014.

Bruce, R.J., & West, C.A. 1982 Elicitation of casbene synthetase activity in castor bean. The role of pectic fragments of the plant cell wall in elicitation by a fungal polygalacturonase. Plant Physiology, **69**, 1181-1188.

Byrt P.N., Irving H.R. & Grant, B.R. 1982 The effect of organic compounds onm the encystment, viability and germination of zoospores of *Phytophthora cinnamomi*. Journal of General Microbiology. **128**, 2343-2351.

Callow J.A. 1984. Cellular and molecular recognition between higher plants and fungi. Encyclopedia of Plant Physiology N.S., **17**, 212-237.

Callow, J.A.,1987. Models for host-pathogen interaction. In Genetics and Plant Pathogenesis. pp 283-295, eds P.R. Day and G.J. Jellis, Blackwells.

Callow, J.A., Estrada-Garcia, M.T. & Green, J.R. 1988. Recognition of non-self: the causation and avoidance of disease. In: New Perspectives in Plant Science, Annals of Botany Centenary Symposium (in press).

Castle A.J. and Day A.W.,1984. Isolation and identification of α-tocopherol as an inducer of the parasitic phase of *Ustilago violacea*. Phytopathology, **74**, 1194-1200.

Darvill A.G.& Albersheim P. 1984. Phytoalexins and their elicitors. A defense reaction against microbial infection in plants. Annual Review of Phytopathology **35**, 243-275.

Darvill, A.G., Albershiem, P., McNeil, M., Lau, J.M., York, W.S., Stevenson, T.T., Thomas, J., Doares, S., Gollin, D.J., Chelf, P., & Davis, K. 1985. Structure and function of plant cell wall polysaccharides. In, The Cell Surface in Plant Growth and Development, eds. K. Roberts, A.W.B. Johnston, C.W. Lloyd, P. Shaw & H.W. Woolhouse. Journal of Cell Science (Supplement), pp 203-217.

Day P.R.1984. Genetics of recognition systems in host-parasite interaction. In: Cellular Interactions (Ed.by H.F. Linskens & J. Heslop-Harrison). Encyclopedia of Plant Physiology 17, 134-147

De Wit, P.J.G.M. 1986. Elicitation of active resistance mechanisms. In: In Biology and Molecular Biology of Plant-Pathogen Interactions. 149-170, ed. J.A. Bailey, Springer-Verlag.

De Wit, P.J.G.M., Hofmann, J.E., Velthuis, G.C.M. & Kuc, J. 1985. Isolation and characterisation of an elicitor of necrosis isolated from intercellular fluids of compatible interactions of *Cladosporium fulvum* (syn,. *Fulvia fulva*) and tomato. Plant Physiology, **77**, 642-647.

De Wit P.J.G.M.& Spikman G. 1982., Evidence for the occurrence of race and cultivar-specific elicitors of necrosis in intracellular fluids of compatible interactions of *Cladosporium fulvum* and tomato. Physiological Plant Pathology **21**, 1-11.

Dickinson, S. 1970. Studies in the physiology of obligate parasitism VII. The effect of a curved thigmotropic stimulus. Phytopathologische Zeitschrift. **69**, 115-124.

Dixon R.A., Bolwell, G.P., Hamdan, M.A.M.S. and Robbins, M.P. 1987. In Genetics and Plant Pathogenesis. pp 245-259, eds P.R. Day and G.J. Jellis, Blackwells.

Ellingboe, A.H. 1976. Genetics of host-parasite interactions. Encyclopedia of Plant Physiology, N.S., **4,** 761-778.

Ellingboe, A.H., 1982. Genetical aspects of host defence. In: Active Defense Mechanisms in Plants. ed. R.K.S. Wood, pp 179-192. Plenum Press, New York.

Epstein, L., Laccetti, L., Staples, R.C., Hoch, H.C. and Hoose, W.A.,1985. Extracellular proteins associated with induction of differentiation in bean rust uredospore germlings. Phytopathology, **75**, 1073-1076.

Firmin, J.L., Wilson,K.E., Rossen, L., and Johnston, A.W.B., 1986. Flavonoid activation of nodulation genes in *Rhizobium* reversed by other compounds present in plants. Nature, **324**, 90-92.

Gabriel, D.W., Burges, A., & Lazo, G.R. 1986. Gene-for-gene interactions of five cloned avirulence genes from *Xanthomonas campestris* pv. *malvacearum* with specific resistance genes in cotton. Proceedings of the National Academy of Science, USA, **83**, 6415-6419.

Gollin, D.J., Darvill, A.G. & Albersheim, P. 1985. Plant cell walls inhibit flowering and promote vegetative growth in *Lemna gibba*. Biol. Cell, **51**,

Grove, S.N. and Bracker, C.E. 1978, Protoplasmic changes during zoospore encystment and cyst germination in *Pythium aphanidermatum*. Experimental Mycology, **2**, 51-98.

Hadwiger, L.A., Daniels, C., Fristensky B.W., Kenfra D.F. and Wagoner, W. 1986. Pea genes associated with the non-host resistance to *Fusarium solani* are also induced by chitosan and in race-specific resistance by *Pseudomonas syringae*. In Biology and Molecular Biology of Plant-Pathogen Interactions. pp263-269, ed. J.A. Bailey, Springer-Verlag.

Hahn M.G., Darvill, A.G. & Albersheim, P. 1981. Host-pathogen interactions XIX: The endogenous elicitor, a fragment of a plant cell wall polysaccharide that elicits phytoalexin accumulation in soybeans. Plant Physiology. **68**, 1161-1169.

Hardham, A.R. and Susaki, E. 1986. Encystment of zoospores of the fungus *Phytophthora cinnamomi* is induced by specific lectin and monoclonal antibody to the cell surface. Protoplasma. **133**. 165-173.

Hardham, A.R., Susaki, E. and Perkin, J.L. 1985. The detection of monoclonal antibodies specific for the surface components on zoospores and cysts of *Phytophthora cinnamomi*. Experimental Mycology, **9**, 264-268.

Hardham, A.R., Susaki, E. and Perkin, J.L. 1986. Monoclonal antibodies to isolate species- and genus-specific components on the surface of zoospores and cysts of the fungus, *Phytophthora cinnamomi*. Canadian Journal of Botany, **64**, 311-312.

Hemmes D.E. & Hohl, H.R. 1971. Ultrastructural aspects of encystation and cyst germination in *Phytophthora parasitica.* Journal of Cell Science, **9**, 175-191.

Hinch, J.M. & Clarke, A.E. 1980. Adhesion of fungal zoospores to root surfaces is mediated by carbohydrate determinants of the root slime. Physiological Plant Pathology, **16**, 303-307.

Irving. H.R. & Grant, B. 1984. The effects of pectin and plant root surface carbohydrates on encystment and development of *Phytophthora cinnamomi* zoospores. Journal of General Microbiology. **130**, 1015-1018.

Keen, N.T. 1982. Specific recognition in gene-for-gene host-parasite systems. Advances in Plant Pathology, **1**, 35-81.

Kondorosi, E., & Kondorosi, A. 1986. Nodule induction on plant roots by *Rhizobium.* Trends in Biochemical Science, **11**, 296-299.

Kota, D.A. & Stelzig, D.A.1977. Electrophysiology as a means of studying the role of elicitors in plant disease reactions. Proceedings of the American Phytopathological Society, **4**, 216-217.

Larkin P.J. 1977. Plant protoplast agglutination and membrane-bound ß-lectins. Journal of Cell Science **26**, 31-46.

Longman D. and Callow, J.A. 1987. Specific saccharide residues are involved in the recognition of plant root surfaces by zoospores of *Pythium aphanidermatum.* Physiological and Molecular Plant Pathology. **30**, 139-150.

Mendgen, K., Lange, M., & Bretschneider, K. 1985. Qualitative estimation of the surface carbohydrates on the infection structures of rust fungi with enzymes and lectins. Archives of Microbiology, **140**, 307-311.

Mitchell, J.E. 1976. The effects of roots on the activity of soil-borne plant pathogens. In: Encyclopedia of Plant Physiology, N.S., Physiological Plant Pathology, eds. R. Heitefuss and P.H. Williams, pp 94-128, Springer-Verlag, Berlin, Heidelberg, New York.

Mulligan J.T. and Long, S. 1985. Induction of *Rhizobium meliloti* nodC expression by plant exudate requires nodD. Proceedings of the National Academy of Sciences, U.S. A., **82,** 6609-6613.

Nothnagel, E.A., McNeil, M., Albersheim, P., and Dell, A. 1983. Host-pathogen interactions XXII. A galacturonic acid oligosaccharide from plant cell walls elicits phytoalexins. Plant Physiology, **71**, 916-926.

Ossowski P., Pilotti A., Garegg P.J. & Lindberg B. 1983. Synthesis of a branched hepta- and octasaccharide with phytoalexin-elicitor activity. Angewandte Chemie **22**, 793-795.

Peters B.M., Cribbs D.H. & Stelzig D.A. 1978. Agglutination of plant protoplasts by fungal cell wall glucans. Science **201**,364-365.

Pillai, C.G.P., & Weet,J.D. 1975. Sterol-binding polysaccharides of Rhizopus arrhizus, *Penicillium roquefortii,* and *Saccharomyces carlsbergensis.* Phytochemistry, **14**, 2347-2351.

Rossen, L., Johnston, A.W.B., Firmin, J.L., Shearman, C.A., Evans, I.J. and Downie, J.A. 1986. Structure, function and regulation of nodulation genes of *Rhizobium.* Oxford Surveys of Plant Molecular and Cell Biology. **3**, 441-447.

Ryder, T.B., Bell, J.N., Cramer, C.L., Dildine, S.L., Grand, C., Hedrick, S.A., Lawton, M.A., and Lamb, C.J. 1986. Organization, structure and activation of plant defence genes. In Biology and Molecular Biology of Plant-Pathogen Interactions. 207-219, ed. J.A. Bailey, Springer-Verlag.

Schell, J. 1988. The New Chimeras. In: New Perspectives in Plant Science, Annals of Botany Centenary Symposium (in press).

Sharp J.K., McNeil M. & Albersheim P. 1984a. The primary structures of one elicitor-active and seven elicitor-inactive hexa(ß-D-glucopyranosyl)-D-glucitols isolated from the mycelial walls of *Phytophthora megasperma* f.sp. *glycinea..* Journal of Biological Chemistry **259**,11321-11336.

Sharp J.K.,Valent B. & Albersheim P. 1984b. Purification and partial characterisation of a ß-glucan fragment that elicits phytoalexin accumulation in soyabean. Journal of Biological Chemistry **259**,11312-11320.

Sing, V.O. and Bartnicki-Garcia S. 1975. Adhesion of *Phytophthora palmivora* zoospores.

Electron microscopy of cell attachment and cyst wall fibril formation. Journal of Cell Science, **18**, 123-132.

Stachel, S.E., Messens, E., Van Montague, H., and Zambryski, P. 1985. Identification of the signal molecules produced by wounded plant cells that activate T-DNA transfer in *Agrobacterium tumefaciens*. Nature, **318**, 624-629.

Staples, R.C., Yoder O.C., Hoch, H.C., Epstein, L., and Bhairi, S., 1986.Gene expression during infection structure development by germlings of the rust fungi. In Biology and Molecular Biology of Plant-Pathogen Interactions, pp 331-341, ed. J.A. Bailey, Springer-Verlag.

Staskawicz, B.J., Dahlbeck, D. & Keen, N.T. 1984. Cloned avirulence gene *Pseudomonas syringae* pv. *glycinea* determines race-specific incompatibility on *Glycine max*. Proceedings of the National Academy of Science, USA, **81**, 6024-6028.

Tran Thanh Van, K., Toubart, P., Cousson, A., Darvill, A.G., Gollin, D.J., Chelf, P. & Albersheim, P. 1985. Manipulation of the morphogenetic pathways of tobacco explants by oligosaccharins. Nature, **314**, 615-617.

Ward, E.W.B. 1983, Effects of mixed or consecutive inoculations on the interaction of soyabeans with races of *Phytophthora megasperma* f.sp. *glycinea*. Physiological Plant Pathology, **23**,281-294.

Ward, E.W.B. 1986. Biochemical mechanisms involved in resistance of plants to fungi. In Biology and Molecular Biology of Plant-Pathogen Interactions, pp 107-131, ed. J.A. Bailey, Springer-Verlag.

West, C.A. 1981. Fungal elicitors of the phytoalexin response in higher plants. Naturwissenschaften, **68**, 447-457.

Wingate V.P.M., Norman P.M., Keen N.T., Staskawicz B.J. & Lamb C.J. 1984. Monoclonal antibodies to surface epitopes of *Pseudomonas syringae* pv. *glycinea* (PSG). Abstracts of the 6th John Innes Symposium, The Plant Cell Surface in Growth and Development, p 57.

Yamazaki, N., Fry, S.C., Darvill, A.G. & Albersheim, P. 1983. Host-pathogen interactions XXIV. Fragments isolated from suspension-cultured sycamore cell walls inhibit the ability of the cells to incorporate [^{14}C] leucine into proteins. Plant Physiology, **75**, 295-297.

York, W.S., Darvill, A.G., & Albersheim, P. 1984. Inhibition of 2,4-dichlorophenoxyacetic acid-stimulated elongation of pea stem segments by a xyloglucan oligosaccharide. Plant Physiology. **75**, 295-297.

Yoshikawa M., Keen N.T. & Wang M.C. 1983. A receptor on soybean membranes for a fungal elicitor of phytoalexin accumulation. Plant Physiology **73**, 497-506.

EARLY RECOGNITION SIGNALS IN THE *RHIZOBIUM TRIFOLII*-WHITE CLOVER SYMBIOSIS

Frank B. Dazzo, Rawle I. Hollingsworth, Saleela Philip-Hollingsworth, Kathryn B. Smith, Margaret A. Welsch, Michael Djordjevic[+], and Barry G. Rolfe[+]
Department of Microbiology
Michigan State University
East Lansing, Michigan 48824 U.S.A.
and
[+]Genetics Department
Australian National University
Canberra City, Australia

Rhizobium is a bacterial symbiont which infects root hairs of legumes, inducing the formation of symbiotic root nodules which fix atmospheric nitrogen into ammonia fertilizer for the host plant. A hallmark of the root nodule symbiosis is a high degree of host specificity which restricts the range of legume hosts infected by the bacterium. We are studying surface and extracellular molecules of *R. trifolii* which interact with white clover root hairs as a model of cell-cell communication in this plant-bacterial interaction.

We have identified several molecules which participate in cell recognition and/or communication between the bacterial and clover root symbionts during the early stages of infection. The molecules and their known activities are summarized as follows:

1. Trifoliin A, a 53,000 molecular weight glycoprotein lectin synthesized by white clover roots (Dazzo et al. 1978, Gerhold et al. 1985, Sherwood et al. 1984) which accumulates on root hair tips (Dazzo et al. 1978) where the receptor sites for the acidic capsular polysaccharide and lipopolysaccharide of the microsymbiont are located (Dazzo and Brill 1977, Dazzo et al. 1983). Much of the newly synthesized root lectin is excreted into the external environment where it can interact with *R. trifolii* (Dazzo et al. 1984, Sherwood et al. 1984).

2. Saccharides on the bacterial symbiont (capsular polysaccharide and oligosaccharide repeat unit fragments, lipopolysaccharide at a certain culture age) which specifically bind to trifoliin A and display an infection-related

NATO ASI Series, Vol. H17
Cell to Cell Signals in Plant, Animal and
Microbial Symbiosis. Edited by S. Scannerini et al.

biological activity resulting in stimulation of microsymbiont infection of clover root hairs (Abe et al. 1984, Dazzo and Brill 1979, Dazzo et al. 1983, Sherwood et al. 1984a, Sherwood et al. 1984b).

3. A class of low molecular weight aromatic signal molecules excreted by the bacteria which promote root hair differentiation and development in clover (Hollingsworth et al. 1986). One of these compounds, called Bacterial Factor-1 (BF-1 for short), may possibly be related to the bacterial factors which cause the thick, short-root response as described by others (Cantercremers et al. 1986, van Brussel et al. 1986). We believe that BF-1 affects biosynthesis of clover root hair walls and its action may be required for successful infection.

To determine whether production of the bacterial components involved in symbiont recognition requires functional nodulation genes, we have examined a collection of mutant and recombinant strains of wild type R. trifolii 843. These have alterations in genes encoding 5 contiguous "nodulation regions" on a 14 kb HindIII fragment of the symbiotic plasmid from the wild type strain. These strains include: a heat-cured pSym-minus strain, mutants with a single Tn5 insertion in nodA, nodB, nodC, nodD, nodE, nodF, nodI, nodJ, nodL, a deletion in nodulation region V (nodM + 2 kb), and recombinants of the pSym-minus strain containing all of the nodulation genes on the 14 kB HindIII fragment. Using a variety of microscopic and biochemical techniques, we have found that some of these nod genes on the symbiotic plasmid are required for wild type expression of certain surface components which interact with the white clover root hair.

The binding of trifoliin A to the bacteria has been examined by fluorescence microscopy after their incubation in situ in the clover root environment. Under these conditions, the pSym nod genes would be highly expressed in the wild type strain due to the presence of the clover flavone inducers (Redmond et al. 1986), thus amplifying the differences in phenotype between the wild type and the mutant strains. We have established that binding of trifoliin A to the bacteria in this host environment is significantly reduced or abolished by mutation of the common genes in the nodABCIJ operon, nodD, nodFE, and nodM and/or other genes in region V. Interestingly, mutation of nodL did not affect binding of trifoliin A. We have been able to define the chemical changes in capsular polysaccharide of the nodA, nodD, nodE, and nod region V (nodM + 2 kb) mutations and these are alterations in the non-carbohydrate substitutions (acetate, pyruvate, and 3-hydroxybutyrate). The nodE mutant is particularly interesting since it represents a "host specificity" mutant which lost the ability to infect white clover, retained the nodulation ability of subterranean clover, and gained pea nodulation capability (Djordjevic et al. 1985). Transfer of R. trifolii pSym nodFELM + 2 kb of region V to a wild type R. leguminosarum strain resulted in a hybrid which now can efficiently infect white clover and bind trifoliin A extensively in the clover root environment. We conclude that certain pSym nodulation and host specificity genes of R. trifolii are required for production of surface polysaccharide receptors which participate in molecular recognition by trifoliin A. These results strongly support a role for the interaction between acidic heteropolysaccharides of R. trifolii and trifoliin A in infection of white clover root hairs.

Trifoliin A participates in attachment of R. trifolii to clover root hairs if the bacterial inoculum can bind this clover lectin, and nonspecific mechanisms result in attachment of non-trifoliin A binding heterologous rhizobia. Thus, as anticipated, all mutants could attach to clover root hairs. We measured attachment of selected mutant strains in the presence of the hapten, 2-deoxy-D-glucose, which can block trifoliin A-mediated specific attachment (Dazzo and Brill 1979, Dazzo et al. 1976, Dazzo et al. 1984). A comparison of the lectin-binding wild type R. trifolii, the nodD::Tn5 mutant derivative which has a 95% reduction in trifoliin A binding ability, and the heat-cured pSym-minus mutant strain in which trifoliin A binding is totally abolished, showed that with the loss of lectin binding there was a corresponding decrease in the relative proportion of 2-deoxyglucose-inhibitable attachment to clover root hairs and an increase in the proportion of non-specific attachment. Detailed microscopic studies which distinguishes specific from nonspecific attachment during the reversible Phase 1 stage (Dazzo et al. 1984) showed that the pattern of attachment most closely correlating with symbiont specificity was displayed only with strains that could bind trifoliin A. Furthermore, mutants which bound trifoliin A poorly or not at all displayed a corresponding increase in the proportion of non-specific attachment during this phase. These results further reinforce a role of trifoliin A receptors in the specific attachment of the bacteria to the root hairs prior to infection. Scanning electron microscopic studies indicated that functions encoded by nodABCD genes are required for expression of the extracellular microfibrils associated with the attached bacteria during the firm Phase 2 adhesion stage (Dazzo et al. 1984).

Further work is in progress to identify the molecular signals permitting communication between R. trifolii and white clover root hairs during early stages of symbiotic infection.

Acknowledgments — Portions of the work described here were supported by NIH Grant GM 34331-03, USDA Grant 85-CRCR-1-167, and the Michigan Agricultural Experiment Station.

Literature Cited

Abe, M., Sherwood, J. E., Hollingsworth, R. I., Dazzo, F. B. 1984. Stimulation of clover root hair infection by lectin-binding oligosaccharides from the capsular and extracellular polysaccharides of Rhizobium trifolii. J Bacteriol 160: 517-520.

Cantercremers, H. C., van Brussell, A. N., Plazinski, J., Rolfe, B. G. 1986. Sym plasmid and chromosomal gene products of Rhizobium trifolii elicit developmental responses on various legume roots. J Plant Physiol 12: 25-40.

Dazzo, F. B., Brill, W. J. 1977. Receptor site on clover and alfalfa roots for Rhizobium. Appl Environ Microbiol 33: 132-136.

Dazzo, F. B., Brill, W. J. 1979. Bacterial polysaccharide which binds Rhizobium trifolii to clover root hairs. J Bacteriol 137: 1362-1373.

Dazzo, F. B., Napoli, C. A., Hubbell, D. H. 1976. Adsorption of bacteria to roots as related to host specificity in the Rhizobium-clover symbiosis. Appl Environ Microbiol 32: 166-172.

Dazzo, F. B., Truchet, G. L., Hrabak, E. M. 1983. Specific enhancement of clover root hair infections by trifoliin A-binding lipopolysaccharide from Rhizobium trifolii. 5th International Symposium on Nitrogen Fixation, Nordwegerhout, The Netherlands.

Dazzo, F. B., Truchet, G. L., Sherwood, J. E., Hrabak, E. M., Abe, M., Pankratz, H. S. 1984. Specific phases of root hair attachment in the Rhizobium trifolii-clover symbiosis. Appl Environ Microbiol 48: 1140-1150.

Dazzo, F. B., Yanke, W. E., Brill, W. J. 1978. Trifoliin: a Rhizobium recognition protein from white clover. Biochim Biophys Acta 539: 276-286.

Djordjevic, M. A., Shofield, P. R., Rolfe, B. G. 1985. Tn5 mutagenesis of Rhizobium trifolii host-specific nodulation genes results in mutants with altered host range ability. Molec Gen Genet 200: 463-471.

Gerhold, D. L., Dazzo, F. B., Gresshoff, P. M. 1985. Selective removal of seedling root hairs for studies of the Rhizobium-legume symbiosis. J Microbiol Methods 4: 137-142.

Hollingsworth, R. I., Philip, S., Smith, K., Welsch, M., Morris, P., Dazzo, F. B. 1986. Requirement of the nodulation region of pSym for production and excretion of a novel aromatic compound from Rhizobium trifolii which interacts with white clover root hairs. 3rd International Symp. on Molecular Genetics of the Plant-Microbe Interaction, Montreal, Canada.

Hrabak, E. M., Urbano, M. R., Dazzo, F. B. 1981. Growth-phase dependent immunodeterminants of Rhizobium trifolii lipopolysaccharide which bind trifoliin A, a white clover lectin. J Bacteriol 148: 697-711.

Redmond, J. W., Batley, M., Djordjevic, M. A., Innes, R. W., Kuempel, P. L., Rolfe, B. G. 1986. Flavones induce expression of nodulation genes in Rhizobium. Nature 323: 632-635.

Sherwood, J. E., Truchet, G. L., Dazzo, F. B. 1984a. Effect of nitrate supply on in vivo synthesis and distribution of trifoliin A, a Rhizobium trifolii-binding lectin in Trifolium repens seedlings. Planta 162: 540-547.

van Brussel, A. A., Zaat, S. A., Cantercremers, H. C., Wijffelman, C. A., Pees, E., Tak, T., Lugtenberg, B. J. 1986. Role of plant root exudate and sym plasmid-localized nodulation genes in the synthesis by Rhizobium leguminosarum of Tsr factor, which causes thick and short roots on common vetch. J Bacteriol 165: 517-522.

FLAVONOID COMPOUNDS AS MOLECULAR SIGNALS IN *RHIZOBIUM* - LEGUME SYMBIOSIS.

Sebastian A.J. Zaat, Herman P. Spaink, Carel A. Wijffelman, Anton A.N. van Brussel, Robert J.H. Okker and Ben J.J. Lugtenberg.
Leiden University, Department of Plant Molecular Biology, Nonnensteeg 3, 2311 VJ Leiden, The Netherlands

INTRODUCTION

The bacterium *Rhizobium* nodulates leguminous and some non-leguminous plants and establishes a symbiotic relationship with its host plant in which the bacterium fixes nitrogen, after differentiation to bacteroids. The symbiosis is host-specific, i.e. each species of *Rhizobium* nodulates only a limited set of host plants. *Rhizobium leguminosarum* nodulates plants like *Pisum* and *Vicia*, *R.trifolii* nodulates only *Trifolium*, and *R.meliloti* nodulates *Melilotus* and *Medicago*. Nodulation is a complex process in which probably many signals from the plant to the bacterium and vice versa are involved. The molecular mechanism of most of the steps in this process are still unknown, but recent results provide some insight into the first signals necessary for nodulation.

In *R.leguminosarum*, 10 genes or loci essential for nodulation have been localized in a 13 kb region on the Sym(biosis) plasmid pRL1JI (Fig. 1) (Evans and Downie 1986, Spaink et al. 1986, Spaink et al, in press). A nearly homologous organization of *nod* genes is found in other fast-growing *Rhizobium* species (Fig. 1) (Debellé et al 1986, Djordjevic et al 1986, Göttfert et al 1986, Kondorosi and Kondorosi 1986, Long et al 1986, Shearman et al 1986). Some of these *nod* genes are determinants of host-specificity, as pRL1JI confers the host specificity of *R.leguminosarum* when transferred to other species of *Rhizobium* cured for their own Sym plasmid (Johnston et al 1978). Most of the *nod* genes are not transcribed in a *Rhizobium* cultured in the usual laboratory media, but

NATO ASI Series, Vol. H17
Cell to Cell Signals in Plant, Animal and Microbial Symbiosis. Edited by S. Scannerini et al.

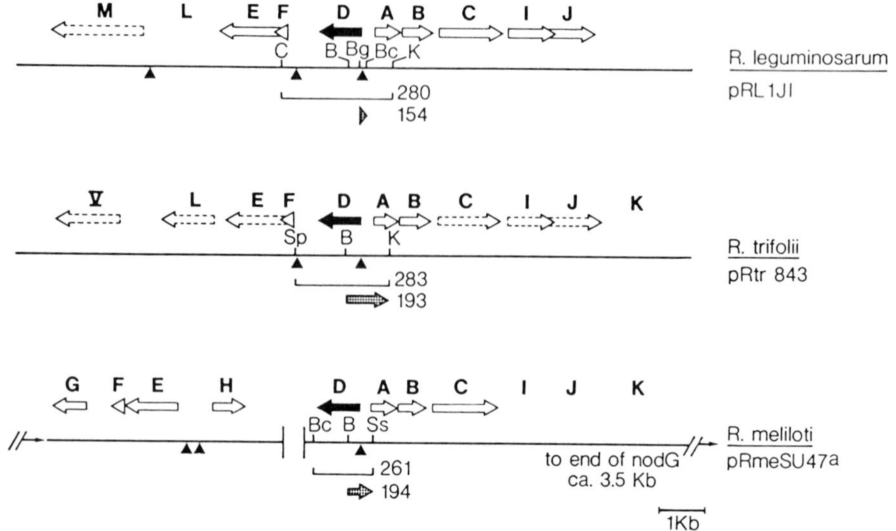

Figure 1. Genetic organization of nodulation regions of Rhizobium leguminosarum, trifolii, and meliloti. The nodulation genes localized on the Sym plasmids pRL1JI, pRtr843 and pRmeSU47a and the direction of their transcription are shown. Black arrows indicate constitutively transcribed genes, drawn arrows indicate inducible genes (see text) whose nucleotide sequence has been determined, dotted arrows indicate inducible genes whose nucleotide sequence has not been determined, and genes without arrows are only defined by mutation or complementation studies. The nod-boxes (see text) are indicated by triangles (R.leguminosarum: Evans and Downie 1986, Rossen et al 1984, Shearman et al 1986, Spaink et al 1986, Spaink et al in press; R.trifolii: Djordjevic et al 1986, Innes et al 1985, Schofield and Watson 1986; R.meliloti: Debellé et al 1986, Egelhoff et al 1985, Jacobs et al 1985, Rostas et al 1986, Török et al 1984). The fragments 280, 283, and 261, containing the nodD genes, were cloned in the IncP vector pMP92 (i.e. pMP220 without lacZ gene, see Fig. 2). The fragments 154, 193, and 194, containing the inducible promoters of the nodABCIJ operon, were cloned in the indicated direction towards lacZ in the IncQ expression vector pMP190 (see Fig. 2).

require the presence of host plant roots or root exudate for their activation (Innes et al 1985, Mulligan et al 1985, Rossen et al 1985, Shearman et al 1986, Spaink et al 1986, van Brussel et al 1986). This paper describes the plant signals that activate nod genes whose products have an essential, although as yet unknown function in the nodulation process.

TRANSCRIPTION OF NOD GENES

To study the expression of nod genes, expression probes have been widely used (for a discussion see Okker et al 1986). These probes are genetic constructs in which the promoters of nod genes are fused to the Escherichia coli lacZ gene in such a way that transcription of nod genes can be monitores as beta-galactosidase activity. We have developed a set of such expression probes (Fig. 2) which allow (i) easy

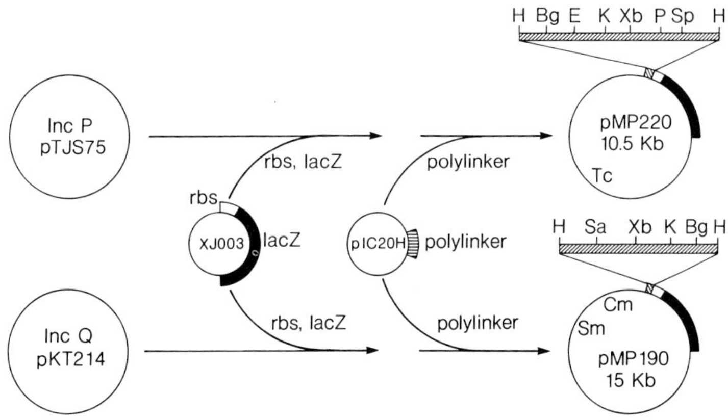

Figure 2. Construction of expression vectors. The plasmids pTJS75 (Schmidhauser and Helsinki 1985) and pKT214 (Bagdasarian et al 1979), of the incompatibility classes IncP and IncQ and with the selectable markers tetracycline resistance (Tc) and chloramphenicol plus streptomycin resistance (Cm,Sm), respectively, can replicate in a wide range of Gram-negative bacteria such as Escherichia coli and Rhizobium. The plasmids are suitable for cloning experiments in E.coli and subsequent mobilization to Rhizobium strains. The E.coli lacZ gene, coding for β-galactosidase, was cloned from plasmid XJ003 (Legocki et al 1984) into these plasmids together with a DNA fragment which contains a ribosome binding site (rbs), assuring an efficient start of translation. After removal of some undesired restriction sites (not shown) the polylinker of pIC20H (Marsh et al 1984) was cloned in front of the lacZ gene to create multiple unique cloning sites. The resulting plasmids pMP220 and pMP190 do not express the lacZ gene due to the absence of a promoter, so that any transcription-promoting activity of DNA fragments cloned into the polylinker can be detected as β-galactosidase activity. Restriction sites: H, HindIII; Bg, BglII; E, EcoRI; K, KpnI ; Xb, XbaI; P, PstI; Sp, SphI; Sa, SalI.

cloning of promoters, (ii) transfer to and maintenance in Rhizobium, (iii) recording of transcriptional activity of the cloned promoter as β-galactosidase activity, and (iv) a choice of incompatibility classes. The 13 kb nod region of pRL1JI was cloned as small fragments in vector pMP190 (Fig. 2). In the absence of plant products only one promoter, of nodD, was detected. In addition to this promoter, three inducible promoters were detected. The expression of these promoters requires the presence of host plant root exudate as well as a functional nodD gene. The promoters were designated as pr.(omoter) nodA, pr.nodF, and pr.nodM, according to their positions (Fig. 1). Together they activate all genes of the nod region except nodD. Two of these inducible nod promoters were also described by Shearman et al (1986) for R.leguminosarum, and analogous inducible promoters have been reported for R.trifolii (Schofield and Watson 1986) and R.meliloti (Rostas et al 1986). Nucleotide sequencing revealed highly conserved sequences of 49 bp long in these promoters (Shearman et al 1986, Spaink et al 1986, Spaink et al in press), which were designated as nod boxes (Rostas et al 1986). Deletion analysis showed that the inducible pr.nodA of R.leguminosarum consists of this nod box and the next 20 bp downstream of this sequence (Spaink et al in press). In conclusion, the nod region appears to be one regulon in which nodD acts as a positive regulator, and which is activated by one or more signal (s) from the plant.

RHIZOBIUM NOD GENE INDUCERS ARE FLAVONOIDS.

Recently the chemical nature of some of the inducers has been established. Inducers for R.meliloti, R.trifolii and R.leguminosarum were reported for the first time and simultaneously at a NATO workshop in 1986 (Lugtenberg 1986). Purifications started from either seed exudate, seed extracts or root exudates. Seed exudate of Medicago sativa (alfalfa) contains several inducers, the most active being the flavone (Fig. 3) luteolin (Table 1)(Peters et al 1986). Seed extract of Trifolium repens contains three inducers of which

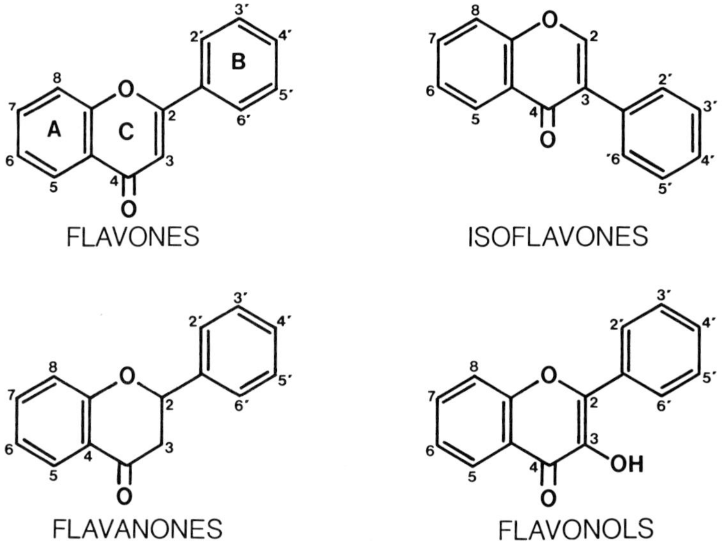

FLAVONES

ISOFLAVONES

FLAVANONES

FLAVONOLS

Figure 3.Skeletons and numbering schemes of flavonoids. See Table 1 for substitution patterns of compounds mentioned in the text.

7,4'-dihydroxyflavone (Table 1) was the most active (Redmond et al 1986a). The seed exudate of *Pisum sativum* contains four inducers, one of which was tentatively identified as apigenin-7-O-glucoside (Table 1) (Firmin et al 1986). The root exudate of *Vicia sativa*, another host plant of *R.leguminosarum*, contained a flavanone (Fig. 3) as the major inducer (Wijffelman et al 1986, Zaat et al 1987). This flavanone has not been characterized in detail until now, but a number of commercially available flavanones and flavones (Table 1) are able to activate the inducible *R.leguminosarum* *nod* promoters. Figure 4 shows a dose-response curve with the flavanone naringenin, and shows that an inducer concentration as low as 2.5 nM is sufficient for detectable induction. The amount of inducer secreted by host plants is also very low. In root exudates of *V.sativa* *V.hirsuta*, and *Pisum sativum* produced under standard conditions inducers are present in concentrations equivalent to 200, 250, and 55 nM naringenin, respectively (Fig. 4). The induction is a very rapid process: significant induction by naringenin as well as by *V.sativa* root exudate is already observed after 5 min (Zaat et al 1987). As the characteristics of induction by naringenin and by *V.sativa* root exudate are very similar

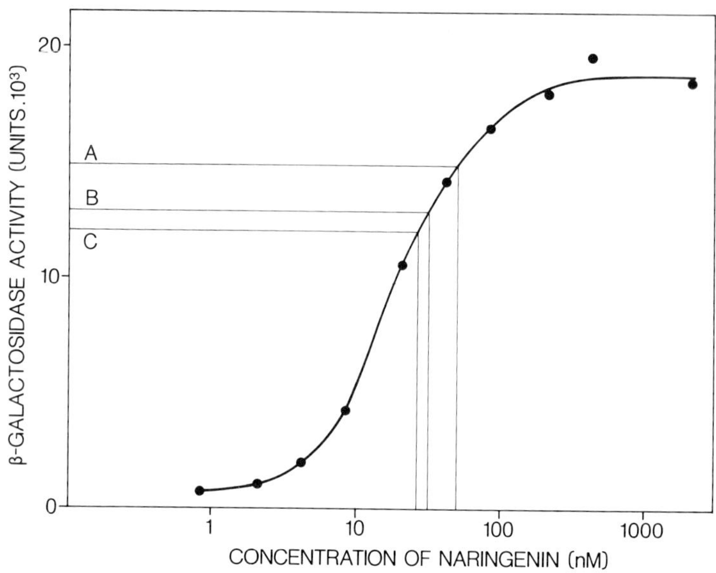

Figure 4. Induction of the R.leguminosarum nodA promoter in relation to the naringenin concentration. A, induction obtained with 4-fold diluted Vicia sativa root exudate; B, induction by 8-fold diluted Vicia hirsuta exudate; C, induction by 2-fold diluted Pisum sativum exudate. Induction was quantified by the β-galactosidase activity in the bacteria after 20 hours of incubation. Exudates were prepared as described in the legend of Table 2.

Zaat et al 1987), naringenin is used routinely in our laboratory to study nod gene related bacterial responses to the host plant (Zaat et al in press).

STRUCTURAL REQUIREMENTS FOR INDUCERS, A SURVEY

In addition to the research on the natural inducers, a large number of commercially available flavonoids has been tested for nod gene-inducting activity (see Table 1 for a summary). It appeared that induction is not accomplished by isoflavones, flavonols and flavone or flavanone glycosides, except apigenin-7-0-glucoside. The structural requirements for the induction of R.leguminosarum, R.trifolii and R.meliloti are not the same. The most active inducers for R.leguminosarum are tri- and tetrahydroxyflavones or flavanones with hydroxyl-groups at C5, C7, C3', or C4'. C4' may also be methoxylated (Table 1). R.trifolii is induced most effi-

Table 1. Structural requirements for *nod* gene inducers

Flavonoid	Substitution pattern						Induction reported for [a]			
	3	5	6	7	3'	4'	*R. leguminosarum* (1)	*R. leguminosarum* (2)	*R. trifolii* (3)	*R. meliloti* (4)
Flavones										
Primuletin		OH						0		
7-Hydroxyflavone				OH				52	100	
Chrysin		OH		OH				10	23	0
7,4'-Dihydroxyflavone				OH		OH			100	
7-Hydroxy-4'-methoxyflavone				OH		OCH_3	0			
4'-Hydroxy-7-methoxyflavone				OCH_3		OH			10	
Galangin	OH	OH		OH			0			
Apigenin		OH		OH		OH	56	98	100	20[b]
7,3',4'-Trihydroxyflavone				OH	OH	OH	74			
Apigenin-7-O-glucoside		OH		O-glu		OH	65			
Geraldone				OH	OCH_3	OH			83/77	
Kaempferol	OH	OH		OH		OH		3	19	
Fisetin	OH			OH	OH	OH		0	5	0
Scutellarein		OH	OH	OH		OH		0		
Luteolin		OH		OH	OH	OH	55	96	57	100
5,7-Dihydroxy-3',4'-dimethoxy-flavone		OH		OH	OCH_3	OCH_3	0			
Quercetin	OH	OH		OH	OH	OH		0	7	0
Flavanones										
5-Hydroxyflavanone		OH					0			
Pinocembrin		OH		OH				25		
Liquiritigenin				OH		OH			29	
Naringenin		OH		OH		OH	49	100	74	0
Eriodictyol		OH		OH	OH	OH	77	97		
Hesperetin		OH		OH	OH	OCH_3	100	98		

[a] Values given are percentages of the induction observed with the most active inducer in each paper, which induces 100%. References: 1, Firmin et al 1986; 2, Wijffelman et al 1986, Zaat et al 1987, Zaat unpublished results; 3, Redmond et al 1986a and 1986b; 4, Peters et al 1986.

[b] Apigenin had to be used at 100 times the concentration of luteolin required for maximal induction.

---

ciently by flavones hydroxylated at C7, C4' and C5, in order of decreasing importance. Methoxylation at C4' decreases inducing activity in this species considerably, and flavanones are less active than the corresponding flavones. Only a few flavonoids were tested on *R.meliloti*. Most strikingly, naringenin, which efficiently induces *R.leguminosarum* and *R.trifolii*, is not active on *R.meliloti* (Table). Whether the structural features of the inducers are important for uptake, for direct or indirect interaction with the bacterial *nodD* gene product, or for possible further processing by the bacteria remains to be determined.

THE *nodD* GENE DETERMINES SPECIES-SPECIFIC RESPONSES TO VARIOUS INDUCERS

Since *nodD* is the positive regulator of the *nod* regulon and because the structural requirements for the inducers of the *Rhizobium* species tested are different, we investigated the role of *nodD* in the response to inducers more closely. To eliminate differences between the *Rhizobium* species studied we cloned the *nodD* genes of *R.leguminosarum*, *R.trifolii*, and *R.meliloti* on IncP plasmids and transferred these constructs, pMP280, 283, and 261, respectively (Fig. 1), to the strain LPR5045, which lacks its own Sym plasmid. Subsequently, plasmid pMP154, on which the *R.leguminosarum* pr.*nodA* is cloned in front of *lacZ* (Fig. 1), was introduced and the resulting strains RBL5280, 5283, and 5261, respectively, were tested with several inducers for the activation of pr.*nodA*. Induction data for five inducers are presented in table 2. The main differences in induction behaviour of the three strains, only differing in their *nodD* gene, are: (i) RBL5280 (R.leguminosa-

TABLE 2. Induction of the R.leguminosarum nodA promoter in the presence of the R.leguminosarum, R.trifolii, or R.meliloti nodD gene.

Inducer[a]	Level of induction[b] with		
	RBL5280 (nodD R.leg.)	RBL5283 (nodD T.trif.)	RBL5261 (nodD R.mel.)
No inducer	0.26	0.84	0.36
Apigenin	24.0	29.0	2.6
Luteolin	19.0	22.0	6.0
Naringenin	21.0	28.0	0.54
Eriodictyol	21.0	4.6	0.90
7-Hydroxyflavone	7.6	29.0	0.55
Root exudate[c]			
Vicia sativa	22.7	18.3	2.0
Vicia hirsuta	18.8	18.7	6.0
Pisum sativum	18.3	22.5	9.3
Trifolium repens	14.0	30.0	7.5
Trifolium pratense	0.41	8.0	0.40
Medicago sativa	22.4	26.9	9.2
Melilotus alba	12.0	28.3	8.7

[a] Flavonoids were tested at concentration of 400 nM.

[b] Induction was recorded after 20 h and is expressed as units of β-galactosidase x 10^{-3}.

[c] Sterile root exudates of 20-30 plants grown in 100 ml medium for three days were used.

rum nodD) is highly induced by eriodictyol, and only moderately by 7-hydroxyflavone (ratio eriodictyol/7-hydroxyflavone = 2.8); (ii) RBL 5283 (R.trifolii nodD) is highly induced by 7-hydroxyflavone and not by eriodictyol (ratio erio./7-hf. = 0.16); and (iii), RBL5261 (R.meliloti nodD) is only efficiently induced by luteolin, and not by the other tested inducers. Similar response patterns were observed when pr.nodA of

R.leguminosarum had been replaced by pr.nodA of R.trifolii or R.meliloti, although the quantitative data of induction were different.

Various Sym plasmids of R.leguminosarum strains and two Sym plasmids of R.trifolii strains were introduced into strains RBL5045.pMP154. Testing of the resulting strains with the same set of inducers revealed that the induction followed the same pattern as was found with the nodD clones: all strains with R.leguminosarum Sym plasmids showed a ratio for induction with eriodictyol/7-hydroxyflavone of 2.2 - 2.8, and strains with the R.trifolii Sym plasmids showed a ratio of 0.16 - 0.3. Therefore it is tempting to speculate that the response to a given inducer is a species-specific trait of Rhizobium, determined by its nodD gene.

EXUDATES CONTAIN A MIXTURE OF INDUCING COMPOUNDS

The finding that response to flavonoid inducers is Rhizobium species-specific was not consistent with the general assumption that induction of nod genes by exudates is not specific, i.e. that exudates of host as well as non host leguminous plants are capable of induction (Peters et al 1986, Redmond et al 1986, Firmin et al 1986). We therefore investigated the induction of strain RBL5280, 5283, and 5261 by root exudates of host plants of R.leguminosarum, R.trifolii, and R.meliloti. The data (Table 2) revealed a few exceptions to the assumption described above. V.sativa exudate induced only marginally in the presence of the R.meliloti nodD gene (RBL5261), and T.pratense exudate induced only with the nodD gene of R.trifolii, its natural symbiont.

The low levels of induction observed could also have been caused by the presence of inhibitors of induction, as isoflavones and flavonols closely resembling the active inducers (Fig. 3) have been reported to inhibit induction of R.leguminosarum nod genes (Firmin et al 1986). However, no indications for the presence of inhibitors in the exudates were found in the cases described above.

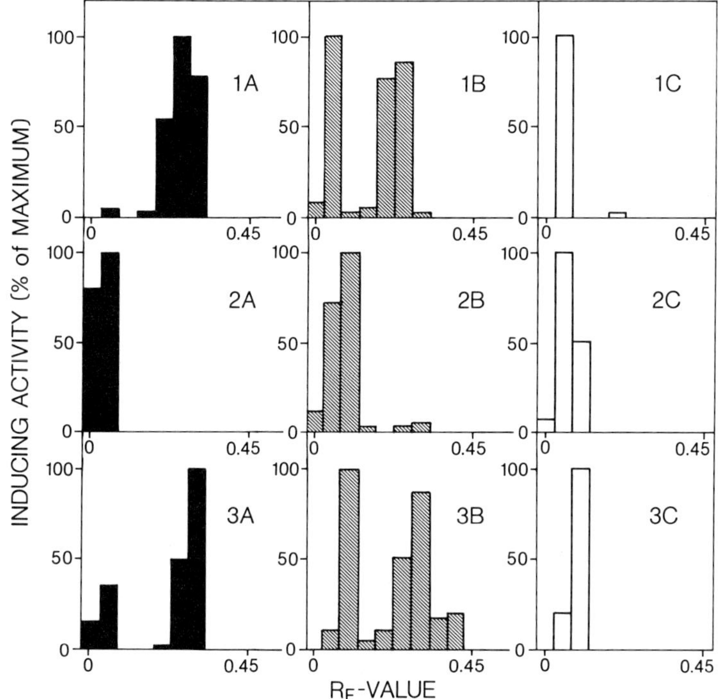

Figure 5. Induction of RBL5280 (nodD R.leguminosarum) (A), RBL5283 (nodD R.trifolii) (B), and RBL5261 (nodD R.meliloti) (C), by chromatogram fragments of root exudates of V.sativa (1), V.hirsuta (2), and P.sativum (3), after thin layer chromatography on cellulose with 15% acetic acid ($^{v}/v$) as the solvent. Levels of induction were measured as β-galactosidase activity after 20 h of incubation, and are expressed as percentages of the induction recorded for the most active fragment in each exudate/test strain combination. Note that the maximum induction in panel 1C (100%) represents only a moderate induction (see Table 2).

Thus, with only one exception, T.pratense, the induction by root exudates is not species-specific. However, separation of root exudate compounds and subsequent testing of the various fractions with RBL 5280, 5283, and 5261 revealed that the exudates contain a variety of inducers, most of which are able to induce in the presence of only one of the three nodD genes (Fig. 5). Furthermore, for a given test strain the compounds responsible for induction can differ between exudates. Thus for these exudates of the pea cross-inoculation group there is only a seeming contradiction between the nodD-specific response to commercially available inducers and the non-specific induction by root exudates: each exudate contains inducers for all three test strains, but most of these inducers only efficiently activate the nod genes in the presence of one particular nodD gene, in the concentrations present in the exudates.

nodD IS A DETERMINANT OF HOST-SPECIFICITY OF NODULATION

The nodD gene is generally considered as a common nod gene, i.e. nodD mutants can be complemented by the wild type nodD allele of another Rhizobium species. Yet, some exudates do not or only marginally activate the inducible nod genes in the presence of a particular nodD gene. To test whether this low level of induction also prevents nodulation, plasmids pMP280, 283, and 261 were transferred to R.leguminosarum strain RBL5561 (nodD::Tn5). The resulting strains were tested for nodulation on V.sativa. All three plasmids were able to restore nodulating ability of the nodD mutant strain, indicating that the low level of induction observed with V.sativa exudate and RBL5261 (with the R.meliloti nodD), is sufficient for normal nodulation. Subsequently, the three nodD containing plasmids were transferred to R.trifolii ANU851 (nodD::Tn5) and strains were tested for nodulation on T.repens, the exudate of which induces each nodD strain (Table 2), and on T.pratense, the exudate of which only induces the R.trifolii nodD strain. The results in Table 3 show that T.repens was nodulated by all three strains, but that T.pratense was only nodulated effectively by the strain with the R.trifolii nodD. Thus, in this bacterium-plant combination the nodD gene is a determinant of host-specificity, and acts as such at the level of induction by plant products. To test whether T.pratense is an exception in its specific requirements for a nodD of R.trifolii, forty other species of Trifolium were tested for nodulation with ANU851.pMP280 or 283 or 261. Half of the Trifolium species were nodulated by all three strains, whereas the other 50% reacted like T.pratense and were only nodulated by ANU851. pM283 (Spaink et al, unpublished). This means that in many Rhizobium-clover interactions the nodD gene is a determinant

TABLE 3. Nodulation of *T.pratense* and *T.repens* by *R.trifolii* derivatives[a] containing *nodD* genes of various *Rhizobium* species.

nodD source (plasmid)	% nodulated plants (average number of nodules)	
	T.pratense	*T.repens*
R.leguminosarum (pM280)	3 (1)	100 (4)
R.trifolii (pMP283)	90 (2)	100 (4)
R.meliloti (pMP261)	0	70 (2)
None	0	0

[a] The *nodD* containing plasmids were transferred to *R.trifolii* strain ANU851 (*nodD*::Tn5).

of host-specificity of nodulation.

DISCUSSION AND CONCLUSIONS

The symbiosis between *Rhizobium* and its host plant is a very complicated process. Recent developments have clarified the outlines of (one of) the first interactions of bacterium and plant. The plant exudes a mixture of flavonoid compounds, the composition of which is different for different plant species, even within one cross-inoculation group. The bacteria respond by recognizing one or more compounds as inducer and with the help of the *nodD* gene, express the other *nod* genes. In order to activate these *nod* genes, the flavonoid must fullfill a number of structural requirements, differing for *R.leguminosarum*, *R.trifolii*, and *R.meliloti*. This is also reflected in the *nodD*-specific response to inducers in plant root exudates, indicating a high degree of adaptation of the *nodD* gene to the host plant.

The mechanism of induction of the inducible *nod* genes is unknown but one may speculate that, by analogy to other bacterial regulation systems, the flavonoid inducer binds to the *nodD* protein and that this activated protein stimulates transcription, presumably by interaction with (parts of) the *nod*-box. This model is supported: (i) by the different responses to flavonoids by strains only differing in the nodD gene; and (ii) by the homology (27%) of the NH_2 terminal parts of the predicted nodD proteins, from start to amino acide 90, with the regulatory and probably DNA-binding lysR protein of E.coli. The predicted nodD proteins of the three Rhizobium species share at least 76% homology. The implicit assumption for this model is that the flavonoid inducers can enter the bacterial cell, which is still under investigation.

The role ofspecificity in nod gene induction in determining the host-specificity of nodulation is of varying importance. For a series of clover species, a clear correlation exists. The root exudates of these plants are only active with a R.trifolii nodD gene, and the plants are nodulated exclusively by R.trifolii. A number of other plants can activate nod genes in the presence of any of the three nodD genes, but are not normally nodulated by all three Rhizobium species. Therefore in these cases the host-specificity of nodulation must be determined in the stages following nod gene induction.

ACKNOWLEDGEMENTS

We thank Ine Mulders and Teun Tak for their skilfull technical assistence.

This study was partly supported by the Foundation for Fundamental Biological Research, which is subsidized by the Netherlands Organization for the Advancement of Pure Research.

REFERENCES

Bagdasarian, M., Bagdasarian, M. M., Coleman, S., Timmis, K. N. 1979. New vectors for gene cloning in Pseudomonas. In: Timmis, K. N., Pühler, A. (eds) Plasmids of medical,environmental, and commercial importance. Elsevier/North Holland, Amsterdam pp. 411-422.

Debellé, F., Sharma, S. B. 1986. Nucleotide sequence of Rhizobium meliloti RCR2011 genes involved in host specificity of nodulation. Nucl Acids Res 14: 7453-7472.

Djordjevic, M. A., Innes, R. W., Wijffelman, C. A., Schofield, P. R., Rolfe, B. G. 1986. Nodulation of specific legumes is controlled by several distinct loci in Rhizobium trifolii. Plant Mol Biol 6: 389-401.

Egelhoff, T. T., Fischer, R. F., Jacobs, T. W., Mulligan, J. T., Long, S. R. 1985. Nucleotide sequence of Rhizobium meliloti 1021 nodulation genes: nodD is read divergently from nodABC. DNA 4: 241-248.

Evans, I. J., Downie, J. A. 1986. The nodI gene product of Rhizobium leguminosarum is closely related to ATP-binding bacterial transport proteins; nucleotide sequence analysis of the nodI and nodJ genes. Gene 43: 95-101.

Firmin, J. L., Wilson, K. E., Rossen, L., Johnston, A. W. B. 1986. Flavonoid activation of nodulation genes in Rhizobium reversed by other compounds present in plants. Nature 324: 90-92.

Göttfert, M., Horvath, B., Kondorosi, E., Putnoky, P., Rodrigues-Quinones, F., Kondorosi, A. 1986. At least two nodD genes are necessary for efficient nodulation of alfalfa by Rhizobium meliloti. J Mol Biol 191: 411-420.

Innes, R. W., Kuempel, P. L., Plazinski, J., Canter-Cremers, H., Rolfe, B. G., Djordjevic, M. A. 1985. Plant factors induce expression of nodulation and host-range genes in Rhizobium trifolii. Molec Gen Genet 201: 426-432.

Jacobs, T. W., Egelhoff, T. T., Long, S. R. 1985. Physical and genetic map of a Rhizobium meliloti nodulation gene region and nucleotide sequence of nodC. J. Bacteriol 162: 469-476.

Johnston, A. W. B., Beynon, J. L., Buchanon-Wollaston, A. V., Setchell, S. M., Hirsch, P. R., Beringer, J. E. 1978. High frequency transfer of nodulating ability between strains and species of Rhizobium. Nature 276: 634-636.

Kondorosi, E., Kondorosi, A. 1986. Nodule induction in plant roots by Rhizobium. TIBS 11: 296-299.

Legocki, R. P., Yun, A. C., Szalay, A. A. 1984. Expression of β-galactosidase controlled by a nitrogenase promoter in stem nodules of Aeschynomene scabra. Proc Natl Acad Sci USA 81: 5806-5810.

Long, S. R., Peters, K., Mulligan, J. T., Dudley, M. E., Fischer, R. F. 1986. Genetic analysis of Rhizobium-plant interactions. In: Lugtenberg, B. (ed) Recognition in microbe-plant symbiotic and pathogenic interactions. Springer Verlag, Berlin, pp. 1-15.

Lugtenberg, B. (ed) 1986. Recognition in microbe-plant symbiotic and pathogenic interactions. Springer Verlag, Berlin.

Marsh, J. L., Erfle, M., Wykes, E. J. 1984. The pIC plasmid and phage vectors with versatile cloning sites for recombinant selection by insertional inactivation. Gene 32: 481-485.

Mulligan, J. T., Long, S. R. 1985. Induction of Rhizobium meliloti nodC expresion by plant exudate requires nodD. Proc Natl Acad Sci USA 82: 6609-6613.

Okker, R. J. H., Spaink, H., Wijffelman, C. A., van Brussel, A. A. N., Lugtenberg, B. 1986. Regulation of bacterial genes involved in plant-bacterium interactions by plant signal molecules. In: Bailey, J. (ed) Biology and molecular biology of plant-pathogen interactions. Springer Verlag, Berlin, pp. 307-314.

Peters, N. K., Frost, J. W., Long, S. R. 1986. A plant flavone, luteolin, induces expression of Rhizobium meliloti nodulation genes. Science 233: 977-980.

Redmond, J. W., Batley, M., Djordjevic, M. A., Innes, R. W., Kuempel, P. L., Rolfe, B. G. 1986a. Flavones induce expression of nodulation genes in Rhizobium. Nature 323: 632-635.

Redmond, J. W., Batley, M., Innes, R. W., Kuempel, P. L., Djordjevic, M. A., Rolfe, B. G. 1986b. Flavones induce expression of the nodulation genes in Rhizobium. In: Lugtenberg, B. (ed) Recognition in microbe-plant symbiotic and pathogenic interactions. Springer Verlag, Berlin, pp. 115-121.

Rossen, L., Johnston, A. W. B., Downie, J. A. 1984. DNA sequence of the Rhizobium leguminosarum nodulation genes nodAB and C required for root hair curling. Nucl Acids Res 12: 9497-9508.

Rossen, L., Shearman, C. A., Johnston, A. W. B., Downie, J. A. 1985. The nodD gene of Rhizobium leguminosarum is autoregulatory and in the presence of plant exudate induces the nodABC, genes. EMBO J 4: 3369-3373.

Rostas, K., Kondorosi, E., Horvath, B., Simoncsits, A., Kondorosi, A. 1986. Conservation of extended promoter regions of nodulation genes in Rhizobium. Proc Natl Acad Sci USA 83: 1757-1761.

Schmidhauser, T. J., Helinski, D. R. 1985. Regions of broad-host range plasmid RK2 involved in replication and stable maintenance in nine species of gram-negative bacteria. J. Bacteriol 164: 446-455.

Schofield, P. R., Watson, J. M. 1986. DNA sequence of Rhizobium trifolii nodulation genes reveals a reiterated and potentially regulatory sequence preceeding nodABC and nodFE. Nucl Acids Res 14: 2891-2903.

Shearman, C. A., Rossen,L., Johnston, A. W. B., Downie, J. A. 1986. The Rhizobium leguminosarum nodulation gene nodF encodes a polypeptide similar to acyl-carrier proteins and is regulated by nodD plus a factor in pea root exudate. EMBO J 5: 647-652.

Spaink, H. P., Okker, R. J. H., Wijffelman, C. A., Pees, E., Lugtenberg, B. 1986. Promoters and operon structure of the nodulation region of the Rhizobium leguminosarum Symbiosis plasmid pRL1JI. In: Lugtenberg, B. (ed) Recognition in microbe-plant symbiotic and pathogenic interactions. Springer Verlag, Berlin, pp. 55-68.

Spaink, H. P., Okker, R. J. H., Wijffelman, C. A., Pees, E., Lugtenberg, B. J. J. Promoter in the nodulation region of the Rhizobium leguminosarum Sym plasmid pRL1JI. Plant Mol Biol, in press.

Török, I., Kondorosi, E., Stepkowski, T., Pósfai, J., Kondorosi, A. 1984. Nucleotide sequence of Rhizobium meliloti nodulation genes. Nucl Acids Res 12: 9509-9523.

Van Brussel, A. A. N., Zaat, S. A. J., Canter-Cremers, H. C. J., Wijffelman, C.A., Pees, E., Tak, T., Lugtenberg, B. J. J. 1986. Role of plant root exudate and Sym plasmid localized nodulation genes in the synthesis by Rhizobium leguminosarum of Tsr factor, which causes thick and short roots on common vetch. J. Bacteriol 165: 517-522.

Wijffelman, C., Zaat, B., Spaink, H., Mulders, I., Van Brussel, T., Okker, R., Pees, E., De Maagd, R., Lugtenberg, B. 1986. Induction of Rhizobium nod genes by flavonoids: differential adaptation of promoter, nodD gene and inducers for various cross-inoculation groups. In: Lugtenberg, B. (ed) Recognition in microbe-plant symbiotic and pathogenic interactions. Springer Verlag, Berlin, pp. 123-135.

Zaat, S. A. J., Wijffelman, C. A., Spaink, H. P., Van Brussel, A. A. N., Okker, R. J. H., Lugtenberg, B. J. J. 1987. Induction of the nodA promoter of the Rhizobium leguminosarum Sym plasmid pRL1JI by plant flavanones and flavones. J. Bacteriol 169: 198-204.

Zaat, S. A. J., Van Brussel, A. A. N., Tak, T., Pees, E., Lugtenberg, B. J. J. Flavonoids induce Rhizobium leguminosarum to produce nodABC gene-related factors that cause thick, short roots and root hair responses on common vetch. J. Bacteriol, in press.

SOREDIA FORMATION OF COMPATIBLE AND INCOMPATIBLE LICHEN SYMBIONTS

M. Galun and J. Garty*
Department of Botany and *Institute for
Nature Conservation Research,
The George S. Wise Faculty of Life Sciences,
Tel Aviv University, Tel Aviv 69978, Israel.

Introduction

Since 1887, when De Bary claimed that lichen algae have become adapted to symbiosis with fungi and cannot exist in the free-living state, the lichenological literature has been full of controversy as to whether the lichen components encounter each other in nature, and the existence of these components, as free-living organisms in nature, was held questionable.

On the other hand, it was assumed that the very many lichen species which do not form vegetative diaspores (containing both components) and among them species with a very wide distribution, depend on resynthesis between

NATO ASI Series, Vol. H17
Cell to Cell Signals in Plant, Animal and
Microbial Symbiosis. Edited by S. Scannerini et al.

germinated spores and free-living photobionts (Bowler and Rundel, 1978; Tehler, 1982).

Although still infrequently observed, there is now unequivocal evidence for the presence of germinated mycobiont spores (Garty and Delarea, 1987) and the common photobionts *Trebouxia*, *Pseudotrebouxia* and *Myrmecia* in the free-living form (Slocum et al. 1980; Tschermak-Woess, 1978; Jahns et al. 1979).

Bubrick et al. 1984 documented the presence of both the mycobiont and photobiont of *Xanthoria parietina* on tree bark surfaces. The same substrates were examined for *in situ* recombination of the two symbionts, and here we demonstrate some of the further steps in the development of *X. parietina* thalli. The formation of soredial clusters was found as an intermediate process during thallus development.

Materials and Methods

The bark surfaces of oak, plum and citrus trees which were examined for free-living components of the lichen *X. parietina*, were also screened for the various stages of thallus development, by one of three methods :

1. Very thin, transparent surface layers were removed from the bark, as previously described (Bubrick et al. 1984), mounted on slides, stained with cotton blue and examined with a light microscope.

2. Scotch-tape was gently applied to the bark surface, lifted, mounted on SEM stubs - with the adhesive side upwards - coated with gold and observed with a Jeol-35 scanning electron microscope, operating at 25 KV.

3. Pieces of bark were examined directly with the SEM, after the coating procedure as above.

Observations were carried out on trees upon which only *X. parietina* and no other lichen species were growing, either on branches not colonized with *X. parietina* or between mature lichen colonies.

Specific areas, at the foothills of the Judean Mountains, inhabited by non-worediious, terricolous lichen species were searched for terricolous, soredial mats.

Results

In a previous study, Bubrick et al. (1984) have documented completely free- living components of the lichen X. parietina on tree bark surfaces. Further examination of these bark surfaces revealed several different stages of thallus development.

One of the early stages observed was algal cells tightly enveloped by fungal hyphae (Figs. 1 and 2) as well as dividing algal cells also tightly enveloped by the fungus (Fig. 3). The process of cell division and fungal encroachment apparently continues until soredial clusters are formed and both symbionts become encrusted by a mucilaginous matrix (Figs. 4 and 5). At this stage the colonies have a pale yellow-silver colour, are hydrophobic and react purple with KOH, indicating the content of anthraquinones, characteristic for X. parietina. From these soredial primordia thallus differentiation proceeds. Fig. 6 shows the beginning of a thallus lobe of X. parietina with an upper cortex formation.

Generally, during the screening, soredial clusters (or masses) of different shapes and sizes, were the phases most often encountered.

Soredial mats are often found in nature. Fig. 7 is an example of a soredial mat on soil in a habitat where no soredia-bearing species exist. The terricolous species in this habitat are Squamarina crassa, Lecidea decipiens, Fulgensia fulgens, Dermatocarpon hepaticum and Toninia coeruleonigricans.

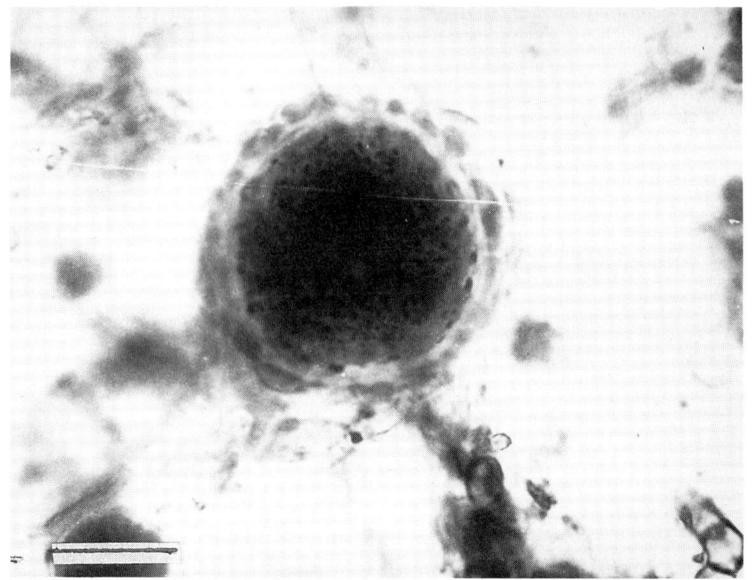

Fig. 1. Single algal cell enveloped by fungal hyphae. Bar = 5 um.

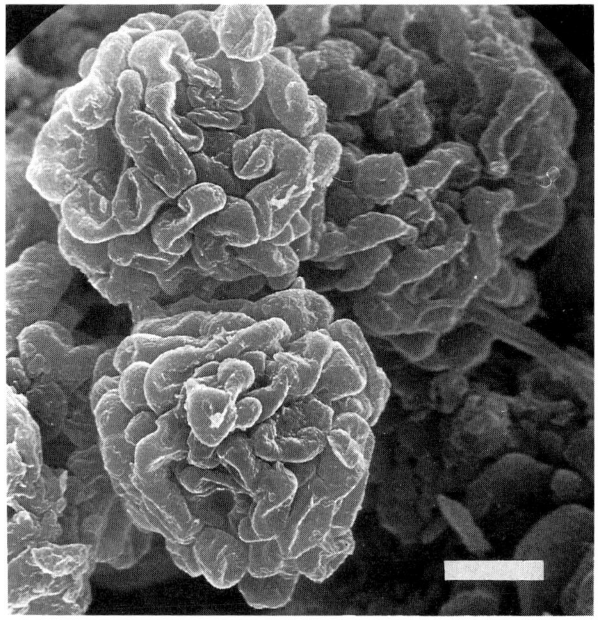

Fig. 2. An *in situ* aggregate of algal cells tightly enveloped by fungal hyphae. Bar = 5 um.

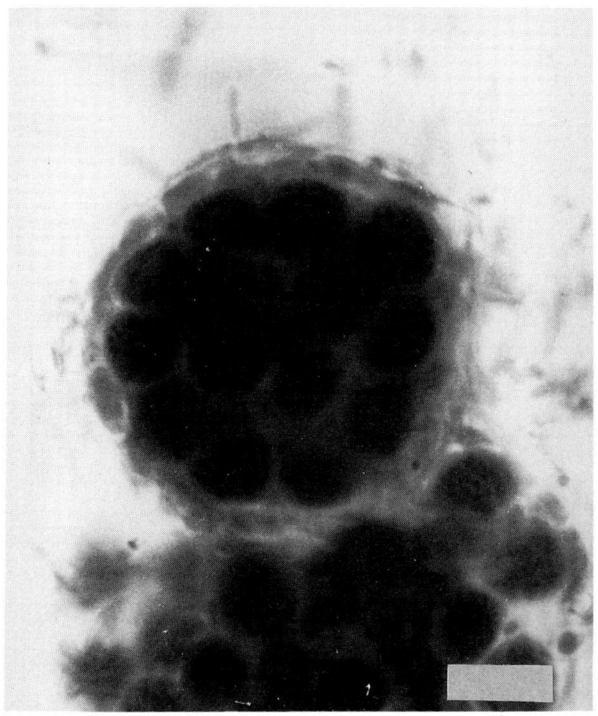

Fig. 3. Fungal hyphae around a dividing algal cell and between the daughter cells. Bar = 10 um.

Discussion

The existence of free-living mycobionts and photobionts *in situ* are the prerequisite for lichen re-establishment in nature. By morphological and highly sensitive immunological examinations, Bubrick et al. (1984) have provided evidence for the presence of potential candidates for the re-establishment of *X. parietina* on tree bark surfaces.

Consequently, it was natural to search these same substrate surfaces for algal-fungus resynthesis and stages of thallus development.

Fig. 4. Soredial mat. Bar = 100 um.

The absolute identity of the components at the various early stages of envelopment (Figs. 1-3) may be considered uncertain. It could, however, be demonstrated that the small soredial clusters formed by further development (Figs. 4 and 5) were those of _X. parietina_, by the purple positive reaction with KOH indicating the presence of lichen substances characteristic for this species. Thalline lobe formation from the soredial masses (Fig. 6) could clearly be defined as a stage in _X. parietina_ development.

In an _in vitro_ spore-to-spore resynthesis of _X. parietina_ (Bubrick and Galun, 1986) the centre of the colonies was still a soredial mass, while along the periphery lobes with mature apothecia, containing asci and spores had developed.

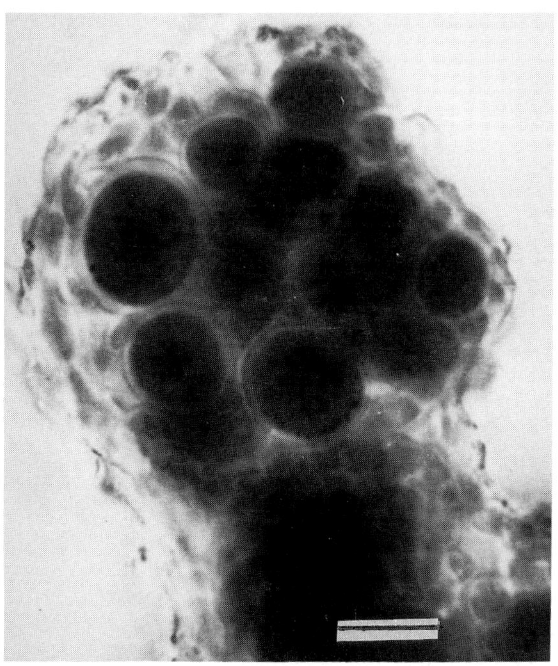

Fig. 5. Young primordium, surrounded by fungal hyphae and incorporated in a mucilinagous matrix. Bar = 5 um.

Many artificial combinations between the mycobiont of Cladonia cristatella with heterologous photobionts (5 species of Trebouxia, 12 of Pseudotrebouxia, 3 of other photobionts and a further 12 species of free-living, unicellular green algae) formed soredia and did not develop further (Ahmadjian, 1982; Ahmadjian et al. 1980; Culberson and Ahmadjian, 1980). Lallemant (1983) also observed soredial clusters during thallus formation of X. parietina and Parmelia furfuraceae.

During numerous field excursions, we often found unidentifiable soredial mats. Those appear in habitats inhabited by soredia-forming species, but are found also in habitats inhabited by non-soredia forming species, as in the case described here.

All these observations lead to the assumption that both compatible and incompatible combinations develop up to a soredial stage and from this stage onwards only preferred combinations continue to develop up to a mature and functional lichen, whereas non-preferred combinations do not undergo any further development.

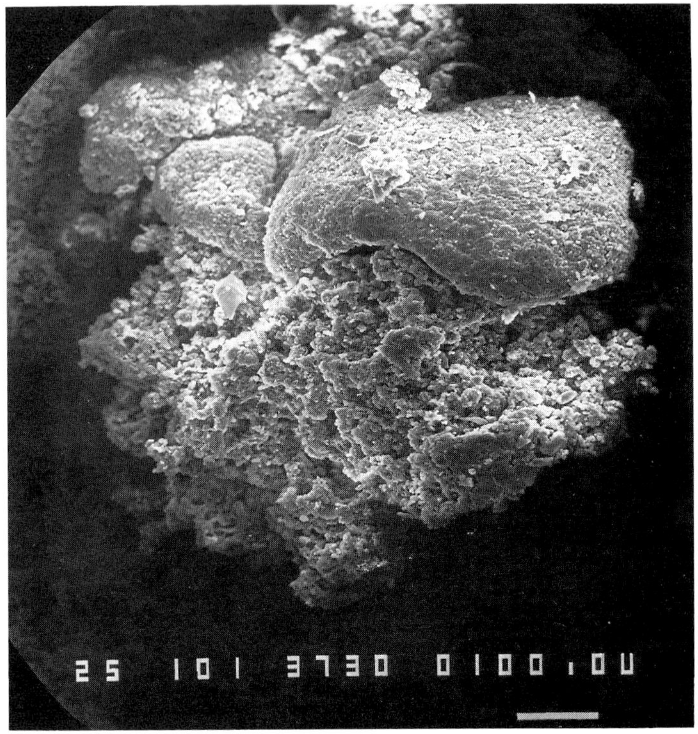

Fig. 6. Primordial differentiation into a lobe, upper cortex beginning to form. Bar = 100 um.

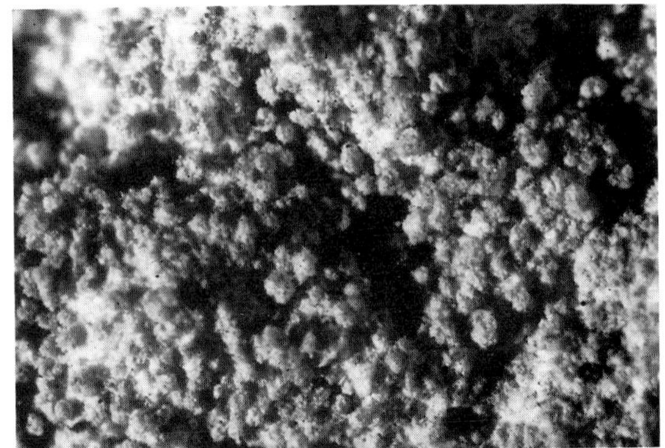

Fig. 7. Soredial mat on soil in nature, where no soredia-bearing terricolous species exist.

In conclusion, we suggest that :

1. In soredia-forming and non-soredia-forming species an intermediate stage of development consists of soredial clusters or soredial masses (*in situ* and *in vitro*).
2. Compatible and incompatible combinations form soredial masses (*in situ* and *in vitro*).
3. Soredial clusters are the final stage of development in incompatible associations.
4. Soredial clusters of compatible components further develop to form a mature and functional lichen.

There are certainly also prerequisites for soredia formation of ultimately incompatible partners. Besides physical conditions - such as size, shape, surface properties - the availability of the preferred partner, plays perhaps an important role.

This situation does not contradict the hypothesis that there are selective events at the initial contact phases and that "signals", as to whether lichenization should continue or not are elicited at earlier stages of interaction (Lockhart et al. 1978; Petit, 1982; Bubrick and Galun, 1980, Bubrick et al. 1981, 1985).

In other symbiotic associations, incompatible combinations may also develop up to a certain stage, often even up to a very advanced morphological and physiological stage, which nevertheless, do not proceed to the development of the final, functional symbiotic unit. For example in the case of the legume-*Rhizobium* symbiosis, there can be infection and nodule formation of non-functional nodules, where no N_2-fixation takes place (Leigh et al. 1985; Shantharam, 1986). Similarly, certain *Frankia* isolates form nodules which are totally inefficient in nitrogen fixation (Moiroud and Gianinazzi- Pearson, 1983). Rahat (pers. comm.) could infect aposymbiotic *Hydra viridis* with *Trebouxia* cells of *X. parietina*. Infection was complete after 24 h and lasted for a week during which the digestive cells hosted *Trebouxia*.

References

Ahmadjian, V. 1982. Algal/fungal symbioses. In : Round, F., Chapman, D. J. (eds.). Progress in Phycological Research, Vol. 1. Elsevier/North-Holland Biomedical Press, Amsterdam, New York, pp. 179-223.

Ahmadjian, V., Russel, L. A. and Hildreth, K. C. 1980. Artificial re-establishment of lichens. I. Morphological interactions between the phycobiont of different lichens and the mycobionts of Cladonia cristatella and Lecanora chrysoleuca. Mycologia 72 : 73-89.

Bowler, P. A. and Rundel, P. W. 1975. Reproductive strategies in lichens. Bot. J. Linn. Soc. 70 : 325-340.

Bubrick, P. and Galun, M. 1980. Proteins from the lichen Xanthoria parietina which bind to phycobiont cell walls. Correlation between binding patterns and cell-wall cytochemistry. Protoplasma 104 : 167-173.

Bubrick, P., Galun, M. and Frensdorff, A. 1981. Proteins from the lichen Xanthoria parietina which bind to phycobiont cell walls. Localization in the intact lichen and cultured mycobiont. Protoplasma 105 : 207-211.

Bubrick, P., Galun, M. and Frensdorff, A. 1984. Observations on free-living Trebouxia de Puymaly and Pseudotrebouxia Archibald, and evidence that both symbionts from Xanthoria parietina (L.) Th. Fr. can be found free-living in nature. New Phytol. 97 : 455-462.

Bubrick, P., Frensdorff, A. and Galun, M. 1985. Proteins from the lichen Xanthoria parietina (L.) Th. Fr. which bind to phycobiont cell walls. Symbiosis 1 : 85-95.

Bubrick, P. and Galun, M. 1986. Spore to Spore resynthesis of Xanthoria parietina. Lichenologist 18 : 47-49.

Culberson, C. F. and Ahmadjian, V. 1980. Artificial re-establishment of lichens II. Secondary products of resynthesized Cladonia cristatella and Lecanora chrysoleuca. Mycologia 72 : 90-109.

Garty, J. and Delarea, J. 1987. Some initial stages in the formation of epilithic crustose lichens in nature : a SEM study. Symbiosis 3 : 49-57.

Jahns, H. M., Mollenhauer, D., Jenninger, M. and Schonborn, D. 1979. Die Neubesiedlung von Baumrinde durch Flechten I. Nat. Mus. 109 : 40-51.

Lallemant, R. 1983. Quelques problemes de morphogenese dans la symbiose lichenique : etude descriptive et experimentale. These de doctorat d'etat. Universite Pierre et Marie Curie, Paris.

Leigh, J. A., Signer, E. R. and Walker, G. C. 1985. Exopolysaccharide- deficient mutants of Rhizobium meliloti form ineffective nodules. P.N.A.S. 82 : 6231-6235.

Lockhart, C. M., Rowell, P. and Stewart, W. D. P. 1978. Phytohaemagglutinins from the nitrogen-fixing lichens *Peltigera canina* and *P. polydactyla*. FEMS Microbial. Lett. 3 : 127-130.

Moiroud, A. and Gianinazzi-Pearson, V. 1984. Symbiotic relationships in Actinorrhizae. In : Genes Involved in Microbe-Plant Interactions. Verma, D. P. S. and Hohn, T. (Eds.). Plant Gene Research, Springer-Verlag, Vienna, N. Y.

Petit, P. 1982. Phytolectins from the nitrogen-fixing lichen *Peltigera horizontalis* : the binding pattern of the primary protein extract. New Phytol. 91 : 705-710.

Shantharam, S. 1986. Anomalous nodulation of soybean by a promiscuous strain of *Rhizobium phaseoli*. Cytobios 46 : 53-64.

Slocum, R. D., Ahmadjian, V., Hildreth, K. C. 1980. Zoosporogenesis in *Trebouxia gelatinosa* : ultrastructural potential for zoospore release and implications for the lichen association. Lichenologist 12 : 173-187.

Tehler, A. 1982. The species pair concept in lichenology. Taxon 31 : 708-714.

Tschermak-Woess, E. 1978. *Myrmecia reticulata* as a phycobiont and free-living *Trebouxia* - the problem of *Stenocybe septata*. Lichenologist 10 : 69-79.

THE ROLE OF THE CELL WALL AS A SIGNAL IN MYCORRHIZAL ASSOCIATIONS

Paola Bonfante-Fasolo

Dipartimento di Biologia Vegetale and Centro di Studio sulla Micologia del Terreno del CNR - Viale Mattioli 25, 10125 Torino, Italy

INTRODUCTION

The dynamic succession of a variety of cell interactions between two highly integrated eucaryotes - i.e. soil fungi and plant root cells - makes mycorrhizal symbioses extremely complex. Transmission electron microscopy (TEM) offers some clues to the understanding of the numerous cell interactions leading to the mutualistic symbiotic state (Scannerini and Bonfante-Fasolo, 1983).

In mycorrhizae, as in most symbiotic associations, major events such as adhesion, penetration and cellular compatibility involve the surfaces of interacting cells. In plant and fungal cells, the surface is a dynamic complex, consisting of cell walls, plasma membrane and cortical cytoskeleton (Roberts et al., 1985).

Recent researches have shown that the cell wall is not an inert and passive envelope for cells, but, on the contrary, it represents an active, adaptable device, sensitive to the environment. When analyzed at the ultrastructural level, cell walls show a three-dimensional structure that is the expression of morphogenetic and physiological events regulating growth and differentiation (Vian and Roland, 1987). At the molecular level, cell walls possess complex carbohydrates and oligosaccharide fragments displaying biological regulatory activities as active components in plant defense responses against microbial invasion (Albersheim et al., 1986).

NATO ASI Series, Vol. H17
Cell to Cell Signals in Plant, Animal and Microbial Symbiosis. Edited by S. Scannerini et al.

These recent views on cell walls stress the importance of studying symbiont walls in mycorrhizal systems in more detail, both at the cellular and molecular level in order to understand how the symbiotic state is reached. Moreover, unlike associations where the symbiont is a unicellular organism, endomycorrhizal fungal symbionts are far more complex. The extraradical and intraradical mycelia are connected, so that the cell walls become the surfaces interacting between partner cells as well as acting as a link between the ectosymbiontic and endosymbiontic phases.

This report will focus on two current projects on the role of cell walls in two different mycorrhizal systems (vescicular-arbuscular and ericoid mycorrhizae) followed by a brief discussion of a working hypothesis for further research.

VESICULAR-ARBUSCULAR MYCORRHIZAE

Fungal wall. Vesicular-arbuscular mycorrhizal (VAM) endophytes occur in about 80% of plants and generally lack host specificity (Powell and Bagjaray, 1984; Bonfante-Fasolo, 1984). Their development pattern is very constant inside the roots of the different hosts, as they form: (i) an extraradical phase represented by chlamydospores and extramatrical mycelia; and (ii) an intraradical phase consisting of hyphal coils, intercellular hyphae, vesicles and arbuscules.

During the complex morphogenesis, a detailed description of which is beyond the aim of this report, the fungus shows a striking modification of its cell wall both as regards morphology and composition (Fig. 1).

Chlamydospores of many VAM species possess a composite and thick wall. Under electron microscopy, two layers, each with a distinct architecture can often be seen. One of the most significant observations is that some VAM spores have a typical helicoidal pattern. Helicoidal walls occur both in the primary walls of elongating cells and in the secondary walls of fully differentiated cells in higher plants as well as in some algae. This organization is widespread in the animal world among crustacea and insects. It results from an oscillatory system and a rhythmical activity and can be interpreted as a time register of cell metabolism (Vian and Roland, 1987). Interestingly this type of organization has never been reported in fungi, except for VAM fungi (Bonfante-Fasolo and Vian, 1984; Bonfante-Fasolo and Schubert, 1987; Mosse, 1986). Layer deposition is very regular in Glomus versiforme and G. macrocarpum, where up to 12-20 layers can be laid down during spore maturation. In the former an electron dense line represents a natural starting point of the twisted layers, while this line is missing in the latter. G. caledonium is another example of twisted organization, because the twisted pattern is limited to the outer part of its wall.

The cell wall of the G. versiforme spore is rich in chitin representing 27% of the cell wall fraction (Bonfante-Fasolo and Grippiolo, 1984). We localized the polysaccharide both in primary and secondary wall by using a new ultrastructural technique based on the molecular affinity both between gold and chitinase, an endoenzyme able to link to the glycosidic linkages of chitin, and between colloidal gold and wheat germ agglutinin (WGA), the lectin specific for N-acetylglucosamine (GlcNAc) residues (Bonfante-Fasolo et al., 1986a). The distribution on the thin sections was not the same for the two layers and was weaker on the helicoidal wall after chitinase gold. This is due to differences in the structural organization of chitin in the two walls and to a difference in binding of the two probes. Chitin is accessible to both probes when it is organized in fibrillar parallel chains, but when it has a helicoidal pattern, its accessibility to chitinase decreases, since only oligomers belonging to different fibrils occur on the sectioning plane. Thus combined morphological and cytochemical techniques can unravel the different three dimensional organizations of the chitin molecule in a way in which mere biochemical analysis cannot (Fig. 1).

The complex helicoidal organization disappears when the spore germinates, leading to an active mycelium that represents the extraradical phase, together

Fig. 1 - Schematic drawing showing i) the various VAM fungal structures, ii) the wall ultrastructure, iii) the macromolecular chitin organization, that can be fibrillar (helicoidal or parallel) or amorphous. The last one might correspond to a different number of GlcNAc residues in the chain.

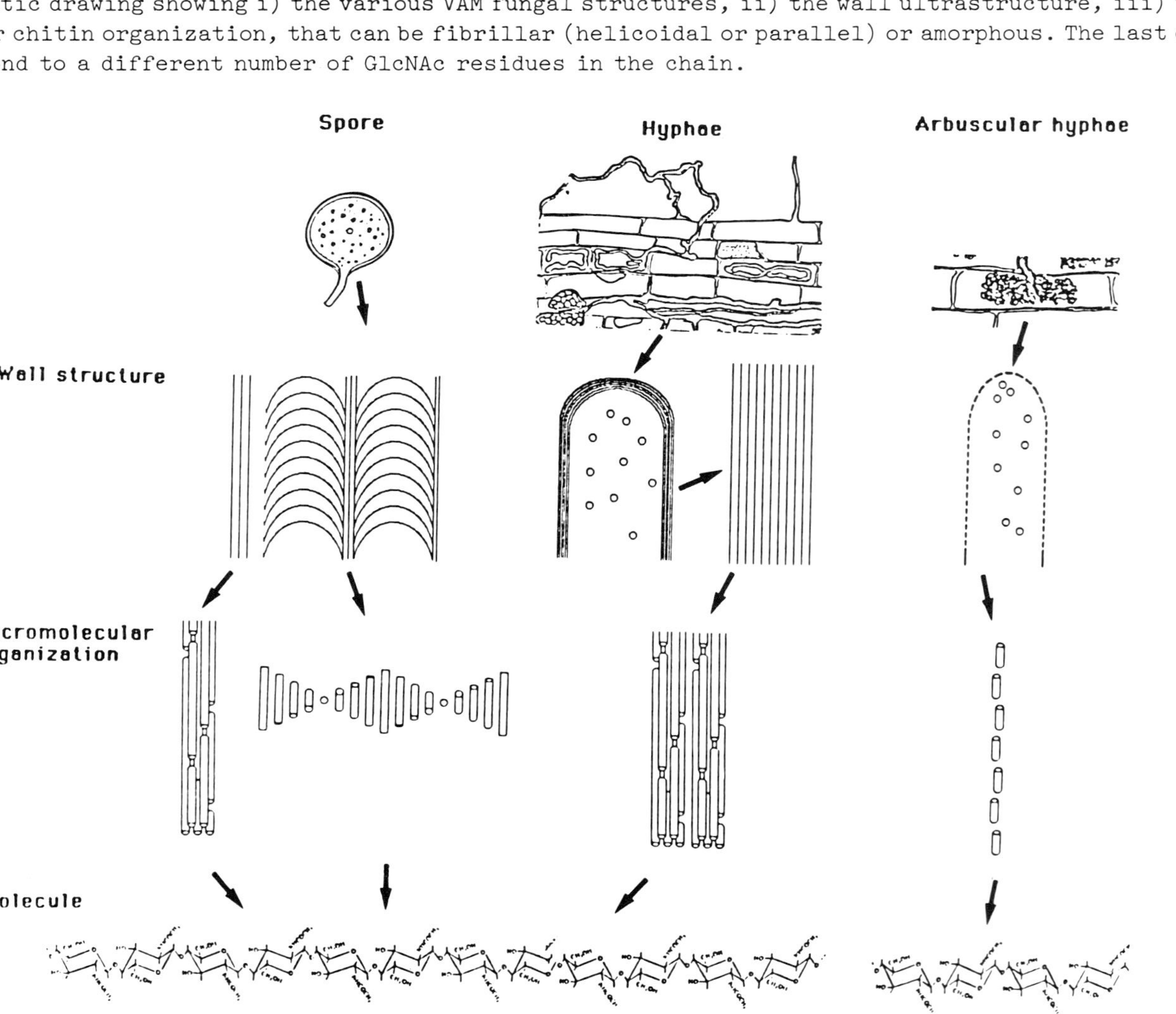

with the mycelium which surrounds the root when the symbiosis is fully established. In the soil, the fungal wall is thick, characterized by a fibrillar texture, and biochemical analyses show the presence of chitin. The polysaccharide can be localized both by fluorescence microscopy using WGA linked to FITC or by electron microscopy using gold linked to WGA or chitinase. Both probes give a positive reaction, suggesting that chitin has a fibrillar and polymerized organization.

During the intraradical phase (see review by Bonfante-Fasolo, 1984), the fungal wall shows striking changes correlated with the different types of structure. When the fungus forms coils, the wall is thick, with a fibrillar texture, rich in alkali soluble polysaccharides. Chitin appears as polymerised fibrils, as suggested by cytochemical observations. The wall thins when intercellular hyphae run through the roots, mostly during their active growth and near the apical zone. However, when the intercellular hyphae are old, their wall shows the same thick and fibrillar texture.

When VAM endophytes penetrate cortical host cells, they stretch the host wall and cause the host plasmalemma to invaginate, forming a regularly dichotomously branched structure called the «arbuscule». The position of arbuscules as well of that of coils is still being debated: these fungal structures are usually regarded as being «intracellular», while *Rhizobium* bacteroids surrounded by the peribacteroid membrane are regarded as being extracellular (Robertson et al., 1985), since they are enclosed in an envelope derived from the outer membrane of the host cell.

Arbuscules are considered the physiological key to plant-fungus interactions, representing the site where nutritional exchanges between the partners occur.

Arbuscular branches can be described as the apical structures of the infection unit, being living and active, with a regular nuclear distribution (Bonfante-Fasolo et al., 1987).

When observed by TEM, the arbuscular wall appears very distinctive: it is much thinner, about 50 nm (up to ten times thinner than the extraradical hyphal wall); it loses its fibrillar texture and becomes amorphous in structure (Bonfante-

Fasolo, 1982). Chitin chains can no longer be detected by using the chitinase-gold technique. On the contrary, oligomers of N-acetylglucosamine can be detected after the use of WGA linked to FITC, ferritin or gold (Bonfante-Fasolo, 1982; Bonfante-Fasolo et al., 1986). One can therefore assume in the apical part of the infection unit (the arbuscule) chitin polymerization is either not accomplished or chitin does not yet occur in a crystalline form, according to recent views on the plasticity of the fungal wall at the growing apex (Wessels, 1986).

The results emerging from the combination of different techniques suggest that fungal morphogenesis can be traced step by step both in the diversified wall organization at an ultrastructural level and in the different number of GlcNAc residues in the chitin chain at the molecular level (Fig. 1). Wall structures become simplified, as a VAM fungus passes from the extraradical to the intradical phase.

There may be more than one cause behind this. According to Margulis (1981), in animal and plant symbioses of long standing, redundancy seems to have been selected against. A complex cell wall may be redundant in a cellular habitat: many algal symbionts are accompanied by a reduction of their cell wall (Trench, 1979); and rhizobia also lose components of their cell walls when they become bacteroids. This general phenomenon fits in well with VAM fungi, but we also have to remember that VAM fungi show an apical growth. New wall material that remains thin and plastic may be actively deposited at the apex, as in ascomycetes and basidiomycetes in pure culture (Wessels, 1986). However, there might be a more complex host-governed regulation mechanism.

As a preliminary hypothesis, one could suggest a host-controlled fungal development by modulation - among other mechanisms - of the production of hydrolytic enzymes. Chitinase might be important because of its role in various processes. In fact, fungal chitinase is known to play a role in fungal growth, since it is associated with chitin-synthase (Gooday et al., 1986). Moreover the same enzyme seems to control branch initiation and degradation of newly synthesized chitin at apices, where it appears nonfibrillar (Wessels, 1986).

Higher plants also often produce chitinase and this capability can be interpreted as a defense against pathogens (Boller, 1986). These observations suggest that fungal morphogenesis may be affected by chitinase activity, from the fungus and from the host plant. Some observations show an increase in chitinase activity in mycorrhizal plants (Dehne and Schönbeck, 1978) or a modulation of this activity (Boller and Spanu, personal communication). The effect of a hypothetical mechanism involving chitinase could be recognized in the morphological responses at different levels during arbuscule formation: non polymerized GlcNAc residues inside the wall, amorphous wall texture and repeated apical branchings.

Host wall structure in mycorrhizal and non mycorrhizal plants. Observations involving the host wall structure and the localization of its components have so far been limited. Two cellular types are usually affected by the fungal penetration in VAM: epidermal or exodermic cells, and cortical cells. Observations on *Ginkgo biloba* (Bonfante-Fasolo and Fontana, 1985) and on *Allium porrum* (Vian, unpublished results) show that the basic structural architecture does not change. The primary wall is thin, with a fibrillar well recognizable texture, where cellulose as well as pectin components can be localized. The most important modification concerns a deposition of new wall material, laid down by the host between the host plasmalemma and the fungal branches both during arbuscule and coil formation. This wall material is usually called interfacial material: it remains thick and compact around the coil, its texture being modified during arbuscule development. Cytochemical results suggest the presence of Thiery- reactive polysaccharides as well as proteinaceous material: however, very little is still known about the structural organization of this material and how it differs from the native cell wall.

In plant-pathogen interactions, pathogens penetrate the walls of their hosts by secreting a mixture of cell-wall degrading enzymes, while plant cells can activate defense responses leading to the synthesis of cell wall components such as cutin, suberin, phenols and lignin (Halbrock et al., 1986). When the possibilities of the VAM fungi evoking defense responses at the host cell wall level were studied, the results were negative: callose does not occur in interfacial material; suberin occurs in *Allium porrum* epidermal cell walls, whether or not fungi are present, phenols change neither in quantity or in quality: syringic, vanillic, ferulic and caffeic acid are found in *Allium porrum* and *Ginkgo biloba* in comparable quantity in mycorrhizal and non-mycorrhizal plants (Maffei et al., 1986). Peroxidase is another important biochemical marker of plant pathogen interactions: in addition to other capabilities, this enzyme is able to stiffen the wall by forming bridges between peptydiltyrosin (Fry, 1986). The result is a rigid wall, much more resistant to pathogen penetration. Our experiments performed during mycorrhizal establishment show that activity peaks differ in mycorrhizal and non mycorrhizal plants. In mycorrhizal plants, the peak occurred early in the infection, when the fungus is mostly in the intercellular phase. When the fungus penetrates inside cells to form arbuscules, the activity decreases and to reach same value as for non-mycorrhizal plants. Peroxidase activity has not been detected in the interfacial material by cytochemical techniques, but only on the residues of the middle lamella (Spanu and Bonfante, in preparation).

These biochemical findings added to the previous morphological observations that development of VAM fungi changes according to the host cell wall composition, confirm our previous hypothesis for the regulatory role of the host cell wall (Bonfante-Fasolo and Fontana, 1985). Cell wall components can regulate structural features through their different distribution in the root tissues. This distribution is unconnected with symbiotic status, but reflects the strong microheterogenity of the cell wall components (McNeil et al., 1984). The cell wall of VAM plants does not stiffen, it allows fungal development, and does not evoke defense responses at least at wall level.

Speculations. The picture of the fungal and host walls suggested by cytochemical and biochemical researches creates a scenario of cellular compatibility between the partners, where the fungus maintains a thin and plastic wall when it

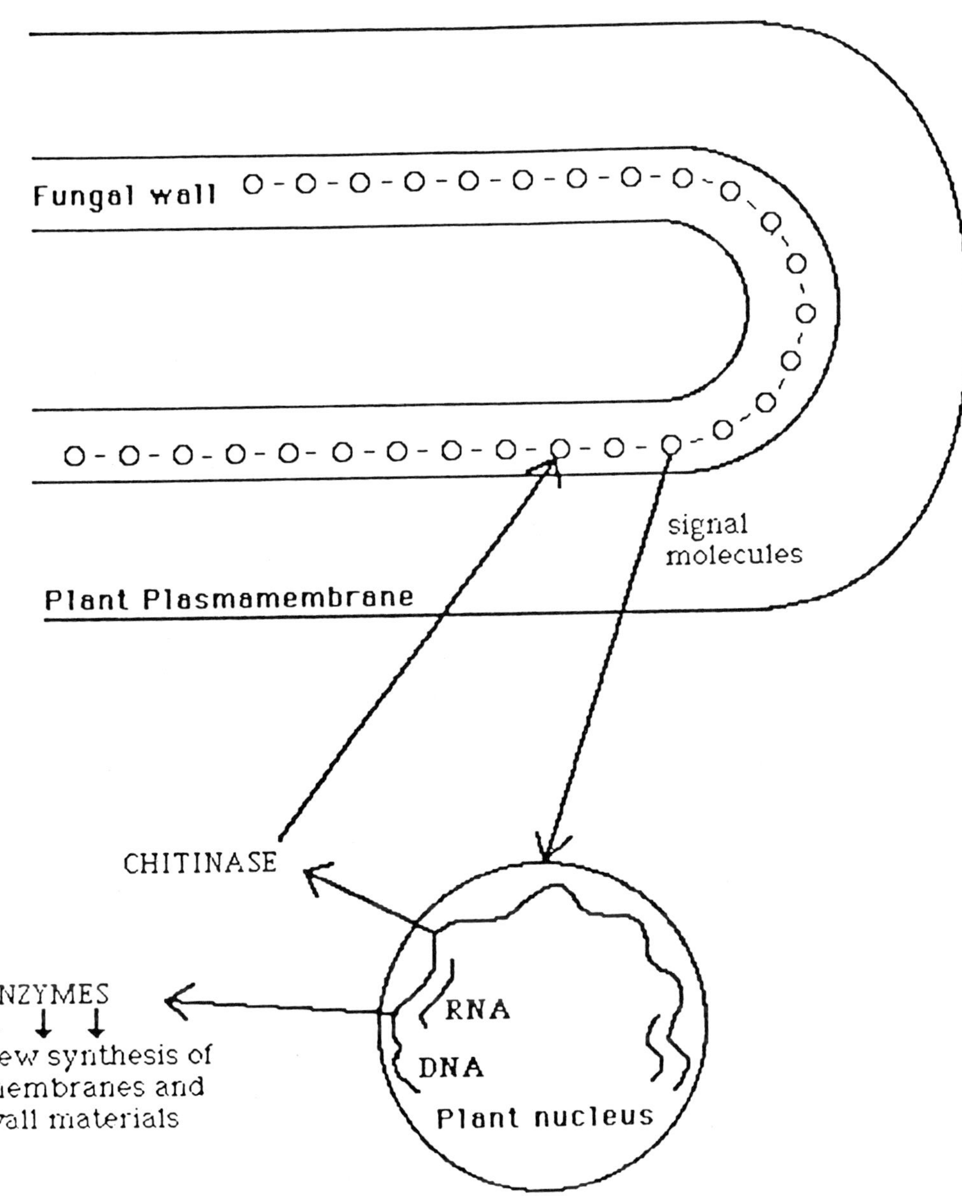

Fig. 2 - A proposed mode of action of signal molecules during the VAM fungal-plant interaction (modified from Hadwiger et al., 1986). Signal molecules (among which GlcNAc oligomers may occur) may influence chromatin structure of the host, and activate some genes, products of which lead to new cell components and new enzymatic activities such as - for example - hydrolytic enzymes. They might be able to act on the fungal wall.

penetrates the host cells, and then grows under the control of the host plasmamembrane.

As in other plant-microorganism interactions, a regulation mechanism between host and fungus can be suggested (Fig. 2). The fungal symbiont can produce signal molecules, able to reach the host nucleus and stimulate some inducible genes, the expression of which leads to enzymes provoking a new synthesis of membranes and cell wall materials that can be observed at ultrastructural level. We can speculate that chitin may act as one of these signal molecules as has already been suggested both for chitin and chitosan in some pathogen/plant interactions (Pearse and Ride, 1982; Hadwiger et al., 1986). Chitin or N-acetylglucosamine oligomers occurring on the fungus surface may act as enzyme-stimulating elicitors. The morphological observations by Berta et al. (1986) suggest the presence of a host nucleus in arbuscule-containing cells with an intense transcriptional activity. The result of this induced activity may be typical inducible enzymes, which is what hydrolitic enzymes are.

ERICOID MYCORRHIZAE

Ericoid mycorrhizae are characterized by a strong host specificity, since some fungi related to *Hymenoscyphus ericae* species or *Clavaria* infect a limited number of plant species within the *Ericales*. Cellular interactions between the partners develop inside thin hair roots, giving rise to heterologous relationships, well described in different associations (Bonfante-Fasolo and Gianinazzi-Pearson, 1982). The molecular basis of this specificity is so far unknown, even if a role of surface structures has been recently suggested (Gianinazzi-Pearson et al., 1986).

Fungal walls. Some fungal symbionts possess an extracellular weft of material which is particularly abundant in presence of the host: we suggested that it may anchor the fungus to the host during initial contact. By using different fungal strains, striking correlations between capacity for extracellular material production and infection capability were observed (Gianinazzi-Pearson et al., 1986). Moreover, the amount of fibrillar material can be quantified by using computer-aided image analysis: results show a strong correlation between the ability of the fungus to infect a plant and production of extracellular material. The amount of fibrillar material also significantly increases in the presence of the plant.

Correlations between the different surface organizations of the fungal strains and their infectivity were also sought (Fig. 3). The biochemical composition of the extracellular material is so far not known: however, cytochemical observations suggest the presence of Thiery- reactive polysaccharides as well as soluble sugars, and which can only be kept in situ by low temperature techniques such as cryoultramicrotomy or Lowicril embedding (Bonfante-Fasolo et al., 1987a). The fungal strains with different infective capabilities showed a dissimilar distribution of their surface sugar residues both in pure culture and in presence of the plant host, while extensive cell surface modifications did not occur during the passage from the saprophytic to the symbiotic phase (Bonfante-Fasolo and Perotto, 1986; Bonfante-Fasolo et al., 1987a). GlcNAc residues belonging to chitin chains were present in septa and longitudinal walls of all the strains, both in pure culture, in the extraradical and in the intracellular phase. However, in the infective strains, where extracellular material occurred, the residues were only localized at the TEM level, since they were not accessible to the probe WGA/FITC which was blocked by the extracellular material. Moreover, if the chitinase gold probe was used, the labeling was almost the same both in pure culture and in the presence of the plant, suggesting that there were no modifications in the molecular organization of chitin in ericoid fungi. Con A specific for mannose and glucose residues exhibited great differences in binding according the strains and throughout the fungal life cycle. Binding was scanty on the non-infective

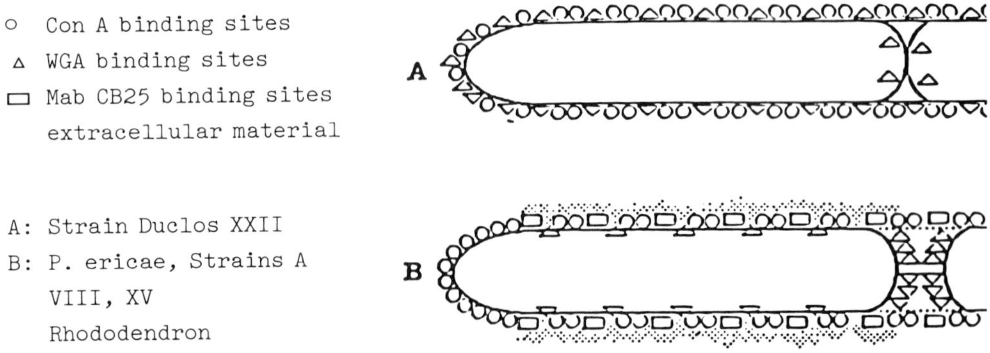

○ Con A binding sites
△ WGA binding sites
▭ Mab CB25 binding sites
extracellular material

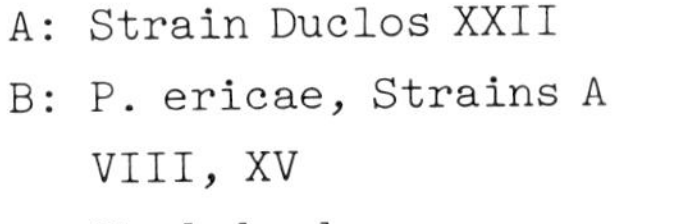
A: Strain Duclos XXII
B: P. ericae, Strains A
VIII, XV
Rhododendron

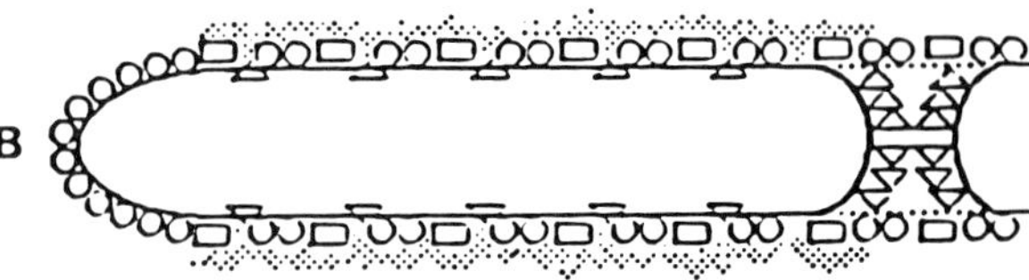

Fig. 3 - Schematic drawing plotting the different distribution of some molecules identified by immuno- and cytochemical methods on the surface of mycorrhizal ericoid isolates (modified from Bonfante-Fasolo and Perotto, 1986).

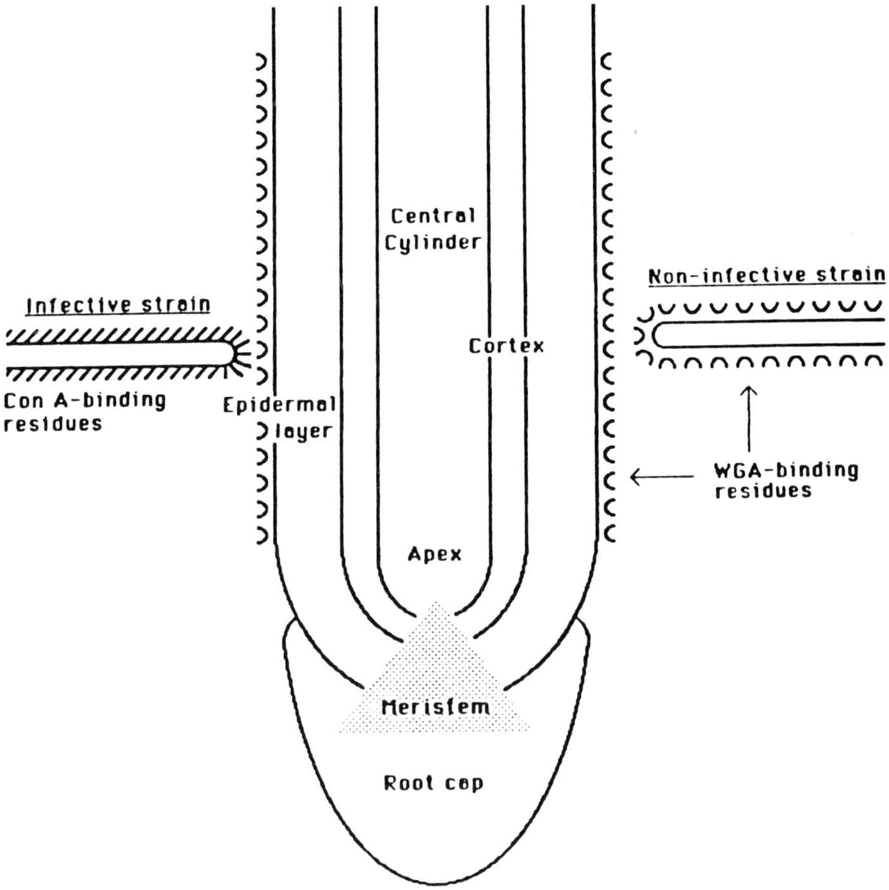

Fig. 4 - Schematic drawing of the partner surfaces in an ericoid mycorrhiza. Infective and non-infective strains as well as the thin hair roots of Calluna are drawn. WGA binding residues only occur on the elongation zone of the root.

strains, abundant on the infective strains and precisely localized on the extracellular material which was sensitive to mannosidase digestion. In presence of the plant binding of Con A seemed to improve further.

These results clearly suggest that the extracellular material rich in mannose residues is actually important since mycorrhizal establishment only occurs in its presence. Moreover, when the fungus is inside the cell, its deposition seems to decrease.

In order to give a deeper insight into the role of surface molecules, monoclonal antibodies were obtained. Hyphae of H. ericea were used as the first immuniser agent.

An immunological approach had already been used by Mueller et al. (1986), who raised a polyclonal antibody in order to clarify the taxonomic position of ericoid fungi. By contrast, monoclonal antibodies (Mab) offer the possibility of analysing a mixture of molecules, among which there may be some with antigenic properties, and to study their expression during cell differentiation. The different antibodies secreted by hybridomas also include one surface antibody, studied in detail (Perotto in preparation). The Mab not only recognizes the first immuniser, but also other strongly infective strains, while it does not link to the surface of non-infective ones. Preliminary results of pre-embedding experiments demonstrate the presence of such antigens on the extracellular material.

Host wall. The epidermal cells of Calluna vulgaris and Vaccinium myrtillus show a complex wall with the typical helicoidal organization, rarely observed in root cells. The complex plywood architecture does not represent a barrier against fungal penetration. The root is surrounded by a well developed mucilage layer. Labelled lectins as probes for sugar residue localization demonstrated that Con A bound to the apical zone, and RCA to the differentiated zone, demonstrating the presence of galactose residues (Perotto, unpublished results). WGA binding was also specifically localized on the hair roots, although not at the apex, exactly in the zone where fungal attachment does not occur. Experiments showed that C. vulgaris hair roots could take up labelled glucosamine, and they possessed the metabolic pathways leading to the formation of N-acetylglucosamine and to its incorporation into a polymeric material (Dalessandro et al., in preparation).

Speculations. Current researches on specificity between our strains of ericoid fungi and Calluna hosts suggest there may be a specific molecular basis for recognition. GlcNAc residues occurring on specialized sites of the hair roots may link to the fungal extracellular material. It is rich in mannose residues, and covers the cell wall, where GlcANc residues are organized as polymerized chains of chitin (Fig. 4). As such, carbohydrates are involved, as in many other recognition processes such as Rhizobia-Legumes, sperm binding to eggs, phagocytosis processes, Dictyostelium development (Monsigny, 1984).

CONCLUSION

Notwithstanding the great morphological and physiological differences, cell walls actually play an important role during the contact between partners in endomycorrhizae. They represent the cell surface of the interacting partners, where molecular signals and their receptors are located.

In ericoid mycorrhizae there seems to be a selective mechanism acting on the surface and preceding mycorrhizal establishment. In this system, chitin has an exclusively skeletal function, while GlcNAc residues in polymeric material on the part of the host, and mannose residues on the part of the fungus, may play a regulatory role.

By contrast, since we were not able to highlight selection filters acting on the surface of VAM, at least in our laboratory system, various intraradical mechanisms work, whereby the fungus is accepted inside the host cell as «not different from self». Chitin and its oligomers may act as signals playing a regulatory as well as a structural role.

ACKNOWLEDGEMENT

Research work was supported by CNR, Italy.

REFERENCES

Albersheim, P., Darvill, A.G., Sharp, J.K., Davis, K.R., Doares, S.H. 1986. Studies on the role of carbohydrates in host-microbe interactions. In: Recognition in Microbe-Plant Symbiotic and Pathogenic Interactions. B. Lugtenberg ed. Springer-Verlag, Berlin, pp. 297-309.

Berta, G., Fusconi, A., Sgorbati, S., Trotta, A. and Scannerini, S. 1986. Preliminary results on the ploidy and fine structure of the nuclei of the host cells in a VA mycorrhiza. Giorn. Bot. Ital. 120: 84-86.

Boller, T. 1986. Chitinase, a defense of higher plant against pathogens: In: Chitin in Nature and Technology. R. Muzzarelli, C. Jeuniaux and G.W. Gooday, eds. Plenum Press, New York and Londong, pp. 223-230.

Bonfante-Fasolo, P. 1982. Cell wall architectures in a mycorrhizal association as revealed by cryoultramicrotomy. Protoplasma 11: 113-120.

Bonfante-Fasolo, P. 1984. Anatomy and morphology In: VA mycorrhizas. CL. Powell and D.J. Bagyaraj eds. CRC Press, Boca Raton, Florida, pp. 5-33.

Bonfante-Fasolo, P. and Fontana, A. 1985. VAM Fungi in Ginkgo biloba roots: Their interaction at cellular level. Symbiosis 1: 53-67.

Bonfante-Fasolo, P. and Gianinazzi-Pearson, V. 1982. Ultrastructural aspects of endomycorrhiza in the Ericaceae. III. Morphology of the dissociated symbionts and modifications occurring during their reassociation in axenic culture. New Phytol. 91: 691-704.

Bonfante-Fasolo, P. and Grippiolo, R. 1984. Cytochemical and biochemical observation on the cell wall of the spore of Glomus epigaeum. Protoplasma 123: 140-151.

Bonfante-Fasolo, P. and Perotto, S. 1986. Visualization of surface sugar residues in mycorrhizal ericoid fungi by fluorescein conjugated lectins. Symbiosis 1: 269-288.

Bonfante-Fasolo, P. and Schubert, A. 1987. Spore wall architecture of Glomus spp. Can. J. Bot. 65: 539-546.

Bonfante-Fasolo, P. and Vian, B. 1984. Wall texture in the spore of a vesicular-arbuscular mycorrhizal fungus. Protoplasma 120: 51-60.

Bonfante-Fasolo, P., Berta, G. and Fusconi, A. 1987. Distribution of nuclei in a VAM fungus during its symbiotic phase. Trans. Br. Mycol. Soc. 88: 263-266.

Bonfante-Fasolo, P., Marzachì, C. and Testa, B. 1986 - Structural modifications of the fungal wall before and during VAM symbiosis. In: Recognition in Microbe-Plant Symbiotic and Pathogenic Interactions. B. Lugtenberg ed. Springer-Verlag, New York, pp. 283-286.

Bonfante-Fasolo, P., Vian, B. and Testa, B. 1986a. Ultrastructural localization of chitin in the cell wall of a fungal spore. Biol. Cell 57: 265-270.

Bonfante-Fasolo, P., Perotto, S., Testa, B. and Faccio, A. 1987a. Ultra-structural localization of cell surface sugar residues in ericoid mycorrhizal fungi by gold labeled lectins. Protoplasma 139:25-35.

Dehne, H.W., Schönbeck, F. 1978. Untersuchungen zum Einfluss der endotrophen Mykorrhiza auf Pflanzenkrankheiten. 3. Chitinase-Aktivitat und Ornithin-zyklus. Zeitschrift für Pflanzenkrankheiten und Pflanzenschutz. 85: 666-678.

Fry, S.C. 1986. Cross-linking of matrix polymers in the growing cell walls of angiosperms. Ann. Rev. Plant Physiol. 37: 165-186.

Gianinazzi Pearson, V., Bonfante-Fasolo, P. and Dexheimer, J. 1986. Ultrastructural studies of surface interactions during adhesion and infection by ericoid endomycorrhizal fungi. In: Recognition in Microbe-Plant Symbiotic and Pathogenic Interactions. B. Lugtenberg ed. Springer-Verlag, New York, pp. 273-281.

Gooday, G.W., Humphreys, A.M. and McIntoch, W.H. 1986. Roles of chitinases in fungal growth. In: Chitin in Nature and Technology. R. Muzzarelli, C. Jeuniaux and G.W. Gooday eds. Plenum Press, London, pp. 83-91.

Hadwigwer, L.A., Kendra, D.F., Fristensky, B.W. and Wagoner, W. 1986. Chitosan both activates genes in plants and inhibits RNA synthesis in fungi. In: Chitin in nature and Technology. R. Muzzarelli, G. Jeuniaux and G.W. Gooday eds. Plenum Press, New York, pp. 209-214.

Halbrock, K., Cuypers, B., Douglas, C., Fritzemeier, K.H., Hoffman, H., Rohwer, F., Scheel, D. and Schulz, W. 1986. Biochemical interactions of plants with potentially pathogenic fungi. In: Recognition in Microbe-Plant Symbiotic and Pathogenic Interactions. B. Lugtenberg ed. Springer-Verlag, New York, pp. 311-323.

Maffei, M., Codignola, A., Spanu, P., Scannerini, S. e Bonfante-Fasolo, P. 1986. Costituenti fenolici in pareti cellulari di radici in piante axeniche e micorrizate. Giorn. Bot. It. 120: (suppl. 2), 22.

Margulis, L. 1981. Symbiosis in Cell Evolution W.H. Freeman and Company, San Francisco.

McNeil, M., Darvill, G.A., Fry, C.S. and Albersheim, P. 1984. Structure and

function of the primary cell wall of plants. Ann. Rev. Biochem. 53: 625-663.
Monsigny, M. 1984. The role of carbohydrates in cell recognition. Endogenous lectins. Biol. Cell. 51, 2.

Mosse, B. 1986. Ultrastructure of the spore wall of some VA Mycorrhizal Fungi. In: Physiological and Genetical Aspects of Mycorrhizae. V. Gianinazzi-Pearson and S. Gianinazzi eds. INRA Press Paris, pp. 615-620.

Mueller, W.C., Tessier, B.J. and Englander, L. 1986. Immunocytochemical detection of fungi in the roots of Rhododendron. Can. J. Bot. 64: 718-723.

Pearce, R.B. and Ride, J.P. 1982. Chitin and relate compounds as elicitors of the lignification response in wounded wheat leaves. Physiol. Plant Pathol. 20: 119-123.

Powell, C.L., Bagyaraj, D.J. 1984. VA Mycorrhiza. CRC Press, Boca Raton, Florida.

Roberts, K., Johnston, A.W.B., Lloyd, C.W., Shaw, P. and Woolhouse, H.W. 1985. The cell surface in plant growth and development. J. Cell. Sci. Suppl. 2.

Robertson, J.G., Wells, B., Brewin, N.J., Wood, E., Knight, C.D. and Downie, J.A. 1985. The legume-Rhizobium symbiosys: a cell surface interaction. J. Cell Sci., Suppl. 2: 317-331.

Scannerini, S. and Bonfante-Fasolo, P. 1983. Comparative ultrastructural analysis of mycorrhizal associations. Can. J. Bot. 61: 917-943.

Trench, R.K. 1979. The cell biology of plant-animal symbiosys. Ann. Rev. Plant Physiol. 30: 485-532.

Vian, B. and Roland, J.C. 1987. The helicoidal cell wall as a time register. New Phytol. 105: 345-357.

Wessels, J.G.H. 1986. Cell wall synthesis in apical hyphal growth. Intern. Rev. Cytol. 104: 37-79.

PEPTIDE AND CARBOHYDRATE MOIETIES AS MOLECULAR SIGNALS IN ANIMAL CELL RECOGNITION

M. MONSIGNY, A.C. ROCHE, C. KIEDA, R. MAYER and P. MIDOUX
Laboratoire de Biochimie Cellulaire et Moléculaire des Glycoconjugués
Centre de Biophysique Moléculaire, C.N.R.S.
et Université d'Orléans
1, Rue Haute - 45071 - ORLEANS Cedex 2, France

Cell to cell and to substratum interactions are usually mediated by specific ligands and receptors. During the last decade, various cell surface receptors were evidenced, they include carbohydrate binding and peptide binding proteins. Cell surface carbohydrate specific receptors may be lectins (Goldstein et al, 1980), ectoglycosyltransferases or glycosidases. Glycolipids, glycoproteins and proteoglycans have been shown to interact with lectins on the surface of a large number of animal cells as well as parasites, bacteria and viruses (for reviews, see Monsigny et al, 1979, 1983, 1984 a, Monsigny 1984, Ashwell and Harford 1982, Liener et al, 1986, Lis and Sharon 1986, Mirelman, 1986, Sharon, 1984). Conversely, a given cell expresses on its surface a set of specific glycoconjugates which may be involved in recognition mechanisms. These glycoconjugates include glycolipids, glycoproteins and glycosaminoglycans (see Schrével et al., 1981 for a review on the cytochemistry of cell surface glycoconjugates). Their biosynthesis involves a number of different glycosyltransferases, glycosidases and/or epimerases and sulfotransferases. The nature and the number of the glycoconjugates are modulated depending on the type, on the maturation or on the transformation state of cells. During the last 15 years, the structure of the sugar moiety of many glycoconjugates have been elucidated (for reviews, see Montreuil (1982, 1984) for glycoproteins, Hakomori (1981) for glycolipids and Poole (1986), Roden and Horowitz (1982) for glycosaminoglycans). The main biosynthetic pathways have also been elucidated (see Schachter, 1984 ; Kornfeld and Kornfeld, 1985).

More recently, specific peptide receptors have been shown to mediate the binding of the protein moiety of proteins or glycoproteins such as fibronectin, fibrinogen, vitronectin, von Willebrand factor... (for reviews see Ruoslahti and Pierschbacher, 1986, Hynes, 1987). In some cases, a dual mechanism involving both peptide and carbohydrate receptors may be involved.

NATO ASI Series, Vol. H17
Cell to Cell Signals in Plant, Animal and Microbial Symbiosis. Edited by S. Scannerini et al.

In this paper, the properties of cell adhesion molecules of embryonic cells (CAM), of neural cells (N-CAM) or of liver cells (L-CAM) which were recently summarized by Cunningham (1986) and by Edelman (1985), will not be presented. In this report, we intend to present recent results of our current work on recognition mechanisms together with some recent studies from others.

Peptide receptors in animal cell recognition

Peptides as such are known to be signal molecules acting as biological modifiers : adrenocorticotropin, angiotensin, bradykinin, opioïd peptides : (dynorphin, endorphin, enkephalin), growth hormone releasing factors, luteinizing hormone releasing factors, melanocyte stimulating hormone, neurotensin, oxytocin, vasopressin, parathyroïd hormone, somatostatin, substance P. Other peptides induce chemotaxis, macrophage activation, immunosuppression, etc. Recently, it was shown (see Ruoslahti and Pierschbacher (1986) and references therein) that short peptide fragments of large molecules such as fibronectin are involved in the binding to specific cell surface receptors (Table 1).

a - Arg Gly Asp as cell recognition signal

(See Dufour *et al*, 1986, Hynes 1987, Ruoslahti and Pierschbacher, 1986).

A short peptide in the cell-binding domain of fibronectin has been shown to inhibit the binding and the spreading of cells incubated on a fibronectin coated surface. The smallest peptide showing such activity is the tripeptide Arg-Gly-Asp.

Subsequently, this tripeptide and various tetrapeptides were found to be present in several other proteins interacting with cell surface components and to be able to inhibit cell recognition phenomena. Interestingly, Arg-Gly-Asp-Ser inhibits the binding of viruses such as Sindbis virus or foot and mouth disease virus to target cells. Furthermore, small peptides containing this sequence inhibit lung colonization by tumor cells in mice (Humphries, *et al*., 1986).

b - Other peptides binding cell surface receptors

Pert *et al*. (1980) found that an octapeptide Ala-Ser-Thr-Thr-Thr-Asn-Tyr-Thr present in the glycoprotein gp 120 of HIV virus was a very efficient inhibitor of the binding of HIV to CD4 of T lymphocytes. This peptide is now being tested as an anti AIDS drug, Kolatag (1987). A pentapeptide corresponding to the C-terminal of the above peptide Thr-Thr-Asn-Tyr-Thr and related pentapeptides found in other HTLV or Epstein Barr virus bind a specific receptor on the surface of monocytes and induce chemotaxis (Ruff *et al*. (1987).

TABLE I

Peptide specific receptors involved in cell recognition

Peptide	Protein
RGD	Von Willebrand factor
RGDS	Fibronectin
	Sindbis virus
	Foot and mouth virus
	Fibrinogen alpha chain
RGDA	Collagen α-1
	Thrombin
RGDT	Collagen α-2
	Influenza A virus
RGDY	Vitronectin
SDGR	MHC-I α-2 domain
ASTTTNYT	HIV - gp 120
TTSYT	LAV
STNYT	HTLV1
TENYT	EBV

Characterization of cell surface lectins

The first membrane lectin of animal cells has been characterized about 20 years ago by Ashwell et al. Working on the metabolism of ceruleoplasmin, these authors treated this glycoprotein with neuramidase labeled the galactose terminal and found that this asialo-glycoprotein had a clearance drastically shorter than the native glycoprotein. subsequently, they showed that asialo-glycoproteins were very efficiently taken up by hepatocytes. Since, membrane lectins have been evidenced in many other animal cells (Table II).

Membrane lectins may be directly evidenced (see Table III) by measuring the binding of labeled glycoconjugates to the cell surface. The glycoconjugates used are either native glycoproteins or glycosidase-treated glycoproteins or synthetic glycoproteins called neoglycoproteins (Lee and Lee, 1982). Neoglycoproteins are obtained by adding an activated sugar or oligosaccharide to a protein such as serum albumin. The activated sugar is either a glycosyl alkylimidate, a glycosylphenyl diazonium or a glycosylphenyl isothiocyanate. Neoglycoproteins usually contain 25 sugar residues. When the number of linked sugar residues is too small the binding efficacy is poor. When the number of sugar residues is much higher than 25, the specificity is poor. On a sugar molar basis, the apparent affinity of a neoglycoprotein to a membrane lectin is 100 to 1000 times higher than the corresponding free glycoside. Neoglycoproteins can easily be labeled by substitution with radioactive iodide or with fluorescein-isothiocyanate. A series of fluorescent neoglycoproteins have been prepared (Monsigny et al., 1984 c) and have been used to visualize (by fluorescence microscopy) and to quantitate (by flow cytometry) cell surface lectins of a variety of animal cells, (Midoux et al., 1986, 1987).

TABLE II

Membrane lectins of animal cells

Cell	Sugar specificity
Hepatocytes	β- Gal, β- GalNAc (mammals)
	β- GlcNAc (avians)
Monocytes	6-P-Man
Macrophages	Man/GlcNAc
	6-P-Man
	β- Gal
Fibroblasts	6-P-Man
	β- Gal
Lymphocytes	
T Helper	β- Gal
T Suppressor	α- Rha
B	α- Man
Tumor cells	
Teratocarcinoma	β- Gal
L1210	β- Gal/α Fuc
3LL	α- Glc
B 16	β- Gal

TABLE III

Characterization of membrane lectins (Tools)

Ligands : Glycoproteins and hydrolase-treated derivatives
Neoglycoproteins
Glycolipids embedded in liposomes
Polysaccharides
Red blood cells and coated red blood cells
Bacteria, yeasts, etc
Antilectin antibodies

Labeling : Radioactive elements
^{125}I or ^{131}I : using chloramine T, iodogen, lactoperoxidase
^{3}H, ^{35}S, ^{14}C using metabolic precursors
Glycosylated enzymes
Glycosylated ferritin
Gold granules coated with glycoproteins or neoglycoproteins

Use of neoglycoproteins to detect lectins and to study their function.
Binding measured at 4°C
Endocytosis measured at 37°C
Inhibition of cell adhesion, *in vitro*
Inhibition of organ colonisation by tumor cells
Inhibition of lymphocyte homing
Isolation of cell subpopulations
Purification of membrane lectins by affinity chromatography
Toxic drug targeting
Immunomodulator targeting

Characterization of membrane lectins (Approaches)

Molecular biology and biochemical approach

- Number of binding sites and their affinity
- Sugar specificity
- Cell agglutination between homologous cells
- Rosettes by incubating native or coated red blood cells and target cells
- Cell adhesion to plated cells or glycoconjugate coated cells

- Isolation :
  - solubilization in the presence of surfactants
  - purification by conventional methods including affinity chromatography
- Molecular characteristics :
  - molecular mass, pI, composition
  - primary and 3 D structure
- Insertion in liposomes and/or cell membrane
- Biosynthesis and catabolism :
  - post-translational modifications
  - turnover
- Preparation of cDNA and production in host cells

Cell biology approach

Lectin mediated endocytosis

Intracellular traffic

Internalization of toxic drugs or of biological response modifiers

Occurrence related to cell cycle

Modulation of occurrence related to cell differentiation

Cytological and histological approach

Fluorescence microscopy

Flow cytofluorometry

Modulation related with metabolic inhibitors, with cytoskeleton specific drugs

In vitro adhesion on coated surfaces
on plated cells
on sections

Homing and colonization *in vivo* and their inhibition by glycoproteins or neoglycoproteins.

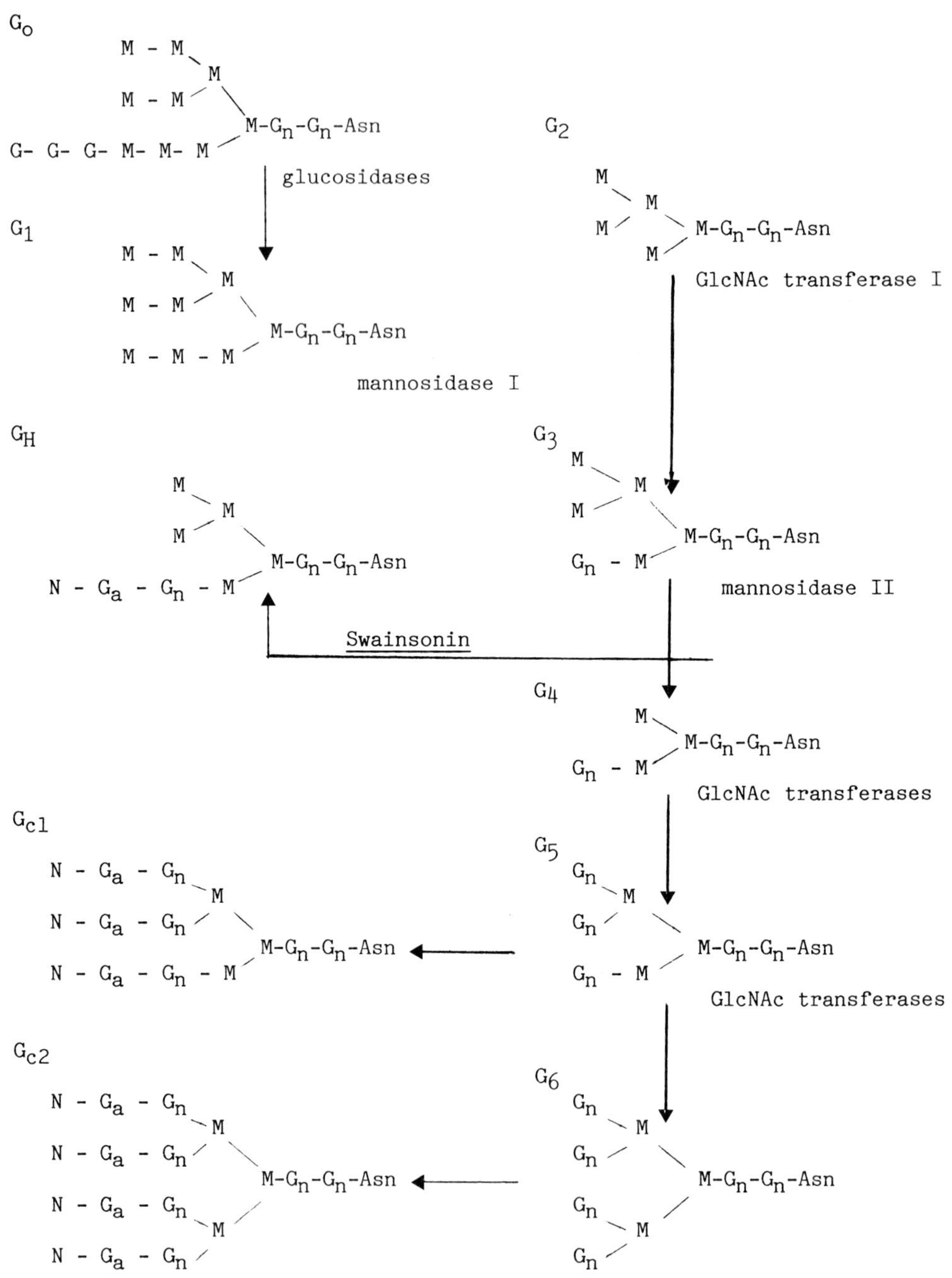

Figure 1 : Schematic pathway of Asn-linked oligosaccharides. G : glucose ; M : mannose ; G_n : N-acetyl glucosamine ; G_a : galactose ; N : N-acetyl or N-glycoloyl neuraminic acid ; G_o : common tetradecasaccharide ; G_1 and G_2 : oligomannoside - type oligosaccharides ; G_5 to G_6 : intermediate complex-type oligosaccharides ; G_H : Hybrid type oligosaccharide ; G_{c1} and G_{c2} : complex-type or lactosamine-type oligosaccharides (see Kornfeld and Kornfeld, 1985 ; Schachter, 1984).

Tumor cells colonization and metastasis

The synthesis and expression of cell surface carbohydrates are developmentally regulated and subpopulations of tumor cells may have unusual carbohydrate structures which may control cell-cell interactions. Various clones have been isolated and several variants have been induced. Particularly, several lectin resistant tumor cells have been obtained and were shown to have defects in the biosynthesis of oligosaccharides. When oligosaccharides were deficient in both sialic acid and galactose (G5 in Fig. 1), the cells showed a greatly attenuated metastatic phenotype compared to the parent cells (Dennis and Laferte, 1987). In a similar way, when cells are treated with Swainsonin, an inhibitor of golgi α -mannosidase II, which blocks the synthesis of complex - type Asn-linked oligosaccharides : Gc1 or Gc2 in figure 1 and leads to hybrid type oligosaccharide G_H in figure 1, the incidence of lung colonies after i.v. inoculation was dramatically reduced.

In addition to the specific cell surface oligosaccharides of tumor cells which may interact with endogenous lectins in various target organs, tumor cells contain also membrane lectins (see Monsigny et al., 1984 a). The specificity and the expression of tumor cell lectins depend on the cell type and/or the metastatic capacity of tumor cells. Raz and Lotan (1987) have identified a β-galactose specific lectin on B16 melanoma cells and other tumor cells, and have shown that the expression of this cell surface lectin is associated with transformation and metastasis (Meromsky et al., 1986 ; Raz et al., 1987). Gabius et al (1987 a) have identified β-galactose specific lectins and lectins with other specificities on a large number of tumor cells of various origins. Paietta et al. (1986) evidenced a lectin on Hodgkin disease cells.

In our laboratory, we have identified an α-glucose specific lectin on mouse 3LL (Lewis lung carcinoma) cells (Roche et al, 1983) by a spectrofluorimetric method using fluoresceinylated neoglycoproteins and a α-fucose-specific lectin on mouse L1210 leukemic cells (Monsigny et al., 1984 c) by a quantitative flow cytometry method. In both cases, these lectins induce the internalization of their ligand ; when specific neoglycoproteins were substituted with an anticancer drug such as methotrexate or with an inhibitor of the protein biosynthesis, it was shown that such carried drugs are delivered inside the cells and are able to block their growth. Recently, Gabius et al. (1987 b) showed that human colon carcinoma cells have a membrane lectin and that FUdR bound to a relevant neoglycoprotein was more efficient than FUdR bound to an irrelevant neoglycoprotein or sugar-free bovine serum albumin.

The cell surface membrane lectins of 3LL cells was shown to be directly involved in the binding to α-glucosylated serum albumin bound to hydrophilic trisacryl beads and to lung cells either in culture or in sections. The binding of 3LL cells to plated lung cells was poorly inhibited by α-glucosylated serum albumin, but was strongly inhibited by a mixture of α-glucosylated serum albumin and of α-fucosylated serum albumin. It was shown that plated lung cells have a fucose-specific binding lectin. So, in this case, the binding involves the lectins and the glycoconjugates of both cells in a dual heterotypic way (Monsigny et al., 1984 a, Kieda and Monsigny, 1986).

Lymphocyte adhesion via membrane lectins

Lymphocytes possess membrane lectins (Kieda et al., 1978, 1979 ; Monsigny et al., 1979, 1983). Recently, it was found that subpopulations of lymphocytes identified by using specific monoclonal antibodies have defined membrane lectins. Human B cells bind mannose terminated glycoconjugates (Barzilay et al., 1982). Human T suppressor cells bind specifically α-rhamnosylated serum albumin (Kieda et al., 1982). T suppressor cells were isolated by plating, on anti-serum albumin antibody coated plates, peripheral blood T cells which have been preincubated in the presence of α-rhamnosylated serum albumin and washed. The isolated T suppressor cells were released by addition of serum albumin and were found to contain less than 15 p 100 of cells not labeled by the Leu2a specific monoclonal antibody. Furthermore, supernatants of cultured T suppressor cells contain a protein which specifically binds to immobilized α-rhamnosylated serum albumin and induces suppression on B cells (Kieda et al., unpublished data).

The cell surface lectin of lymphocytes are directly involved in the binding of lymphocytes on lymph node veinules as shown by inhibition experiments using lymph node sections (Kieda and Monsigny, 1983). Fluorescein-labeled splenocytes were found to bind to mesenteric lymph node veinules and the binding is specifically inhibited by β-galactosylated serum albumin but not by unrelated neoglycoproteins. Recently, various subpopulations of lymphocytes from spleen, cervical, peripheral and mesenteric lymph nodes were isolated by using the method described above to isolate T suppressor cells. Cells containing either a β-galactose specific lectin or an α-fucose specific lectin were fluoresceinylated and injected intravenously to mice in the absence or in the presence of the relevant neoglycoprotein. The co-injection of neoglycoproteins and specific cells strongly inhibit the homing of lymphocytes in specific organs : splenocytes bearing a β-galactose specific lectin home to Peyer's patches or cervical cells bearing a α-fucose specific lectin home to cervical lymph nodes (Kieda et al.).

Given to the great diversity of the lymphocyte subpopulations and the continuous differentiation of lymphocytes (Butcher, 1986) on the one hand, and the number of different lymphoid organs on the other hand, it is not surprising that several types of cell-cell recognition mechanisms may be involved (Jalkanen et al., 1986). For instance, Stoolman et al. (1984)

showed that phosphomannosyl receptors may participate in the adhesive interaction between lymphocytes and high endothelial veinules, and Yednock et al. (1987) showed that the phosphomannose receptor is on the surface of lymphocytes. Rosen et al (1985) showed that sialic acids on endothelial cells are involved in organ specific lymphocytes recirculation. Others workers have also implied lymphocyte membrane lectins as potentiel receptors in cell recognition. Parish et al., (1984) demonstrated the presence of sulphated polysaccharide specific lectins on lymphocyte surface, Popisil et al. (1986) showed that lactosamine type asialo-oligosaccharides are involved in the recognition step of NK cell cytotoxicity, Powell et al. (1985) showed that asparagin-linked oligosaccharides on tumor cells are involved in the recognition step in the mixed lymphocytes reaction and Pimlott and Miller (1986) using tunicamycin, an inhibitor of the glycosylation of glycoprotein, showed that cytotoxic T cells require an interaction with the target cell that involves N-linked glycans on the target cell surface.

Lectins of monocytes and macrophages

The presence of membrane lectins on macrophages is known for ten years (see Stahl et al., 1984, Wileman et al., 1985). These membrane lectins bind primarily mannose and fucose terminated glycoconjugates, and secondarily N-acetylglucosamine terminated glycoconjugates. They are not present on freshly isolated monocytes or on macrophage-like cell lines. Fresly isolated human monocytes have a membrane lectin that binds and internalizes mannose-6 phosphate terminated glycoconjugate (Roche et al., 1985 b) as shown by a quantitative flow cytofluorometry method allowing to study the binding, the internalization and the degradation of endocytosed ligands (Midoux et al., 1986, 1987).

Upon culture, monocytes acquire the properties of macrophages and have both a mannose and a mannose-6-phosphate specific lectin. These membrane lectins are involved in the internalization of Leishmania donovani parasites (Wilson and Pearson, 1986).

Another lectin which is specific for β-galactose terminated receptors seems to be present on liver macrophages (Roos et al., 1985).

The endocytosis of mannose or mannose-6-phosphate specific terminated glycoconjugates is quite efficient and allows to efficiently target macrophage activators both in vitro and in vivo (Monsigny et al., 1984 b, Roche et al., 1985 a). However, the proteolytic activity of monocytes is low and acido-labile linkage between the carrier and the activator has to be preferentially used instead of peptide hydrolysable linkage, since the internalized material is contained in acidic endosomes (Roche et al., 1985 b). By using N-acetylmuramyldipeptide (MDP), it was found that this activator, bound to a mannose-6-phosphate substituted serum albumin or polypeptide, is 100 to 1000 times more active than free MDP in rendering macrophages tumoricidal. While free MDP is inactive in vivo, the above mentioned activator-carrier conjugates are very efficient as antitumor agents and allow more than 70 % of mice developing lung metastases from a 3LL primary tumor to be cured.

Invasion of red blood cells by Plasmodium merozoites

The invasion of erythrocytes by Plasmodium is mediated by a recognition mechanism involving a parasite receptor on one hand and erythrocyte membrane components on the other hand. Using neoglycoproteins, it was shown that fluoresceinylated neoglycoproteins bearing N-acetylglucosamine specifically bind on the surface of endoerythrocytic Plasmodium stages and merozoites (Schrével et al., 1986). This lectin supposedly interacts with transmembrane glycoproteins : glycophorin and/or Band 3. The binding is the first step of the reinvasion. A second step involves the activity of a specific parasite protease which cleaves synthetic peptides such as Val-Leu-GLy-Arg-X or Val-Leu-Gly-Lys-X between the basic amino acid and the substituent X. Peptide derivatives containing these sequences were shown to strongly inhibit the reinvasion of red blood cells (Schrével et al., French Patent 87-06785 and Schrével et al., unpublished results). It is interesting to note that the glycoprotein Band 3 contains an analogous sequence exposed on the surface of red blood cells.

Concluding remarks

The molecular signals involved in cell recognition which have been described above emphasize the role of peptide receptors and of sugar binding proteins. In all cases, the recognition between a ligand (or substrate) and a receptor (or enzyme) occurs at an optimal pH close to 7.4. The receptors could be enzymes that are devoid of enzymatic activity because of the high pH of the medium. Indeed, Rauvala and Hakomori (1981) showed that an α-manosidase at the fibroblast cell surface which is almost inactive at pH 7.4 but is active at pH values lower than 6, could be involved in cell recognition. These receptors could also be glycosyltranferases working as binders but unable to transfer any sugar on their ligand because of the absence of nucleotide sugar in the surrounding medium. Along this line, Shur (1982) showed that migrating embryonic cells have high levels of cell surface galactosyltransferases which bind to terminal N-acetylglucosamine within the extracelular matrix. However, in many cases, the membrane lectins bind their ligands at neutral pH (see Ashwell and Harford, 1982) but loose their affinity upon acidification. This mechanism is the basis of rapid turnover and recycling of membrane lectins. Conversely, in other cases, the reverse may also be found ; for instance, thyrocyte lectins bind N-acetylglucosamine terminated glycoconjugate at acidic pH but do not at neutral pH (Miquelis et al., and Alquier et al., 1987).

References

Alquier, C., Miquelis, R. and Monsigny, M. (1987), Direct fluorescence localization of an endogenous N-acetylglucosamine specific lectin in the Thyroid gland. Histochem. (submitted).

Ashwell, G. and Harford, J. (1982), Carbohydrate specific receptors of the liver. Ann. Rev. Biochem., 51: 531-534.

Barzilay, M., Monsigny, M. and Sharon, N. (1982), Interaction of soybean agglutinin with human peripheral blood lymphocyte subpopulations : evidence for the existence of a lectin-like substance on the lymphocyte surface. In Lectins Biology, Biochemistry, Chimical Biochimistry, Walter de Gruyter and Co, Berlin, Vol. 2, 67-81.

Butcher, E.C. (1986), The regulation of lymphocyte traffic Current Topics in Microbiol Immunol, 28: 85-122.

Cunningham, B.A. (1986) Cell adhesion molecules : a new perspective on molecular embyology. Trends in Biochemical Science, 11: 423-426.

Dennis, J.W. and Laferte (1987), Tumor cell surface carbohydrate and the metastatic phenotype. Cancer Metast. Rev., 5: 185-204.

Dufour, S., Duband, J.L. and Thiery, J.P. (1987), Role of a major cell substratum adhesion system in cell behavior and morphogenesis. Biol. Cell, 58: 1-14.

Edelman G.M. (1985), Cell adhesion and the molecular processes of morphogenesis. Ann. Rev. Biochem., 54: 135-169.

Gabius, H.J., Engelhardt, R. and Cramer, F (1987 a), Endogenous tumor lectins : overview and perspectives. Anticancer Research, in press.

Gabius, H.J., Engelhardt, R., Hellmann, T., Midoux, P., Monsigny, M., Nagel, G.A. and Vehmeyer K. (1987 b), Characterization of membrane lectins in human colon carcinima cells by flow cytofluorometry, drug targeting and affinity chromatography. Anticancer Res., 7: 109-112.

Goldstein, I.J., Hughes, R.C., Monsigny, M., Osawa, T. and Sharon, N. (1980), What should be called a lectin ? Nature, 285: 66.

Hakomori, S.I. (1981), Glycophinsgolipids in cellular interaction, differentiation and oncogenes. Ann. Rev. Biochem., 50: 733-764.

Humphries, M.J., Olden, K. and Yamada, K. (1986), A synthetic peptide from fibronectin inhibits experimental metastasis of murine melanoma cells. Science, 233: 467-470.

Hynes, R.O. (1987), Integrins : a family of cell surface receptors. Cell, 48: 549-554.

Jakalnen, S., Reichert, R.A., Gallatin, W.M., Weissman, I.L. and Butcher, E.C. (1986), Homing receptors and the control of lymphocyte migration. Immunological Reviews, 91: 39-60.

Kieda, C., Bowles, D.J., Ravid, A. and Sharon, N. (1978), Lectins in lymphocyte membranes. FEBS Lett., 94: 391-396.

Kieda, C. and Monsigny, M. (1983), The adhesion of mouse spleen cells to lymph node veinules involves endogenous membrane lectins (sugar receptors) of the lymphocytes. In Intercellular communication in Leucocyte J.W Parker and R.L. O'Brien ed., John Wiley and sons, 649-652.

Kieda, C. and Monsigny, M. (1986), Involvement of membrane sugar receptors and membrane glycoconjugates in the adhesion of 3LL subpopulations to cultured pulmonary cells. Invas. Metast., 6: 347-366.

Kieda, C., Monsigny, M. and Waxdal, M.J. (1982), Endogenous lectins on human peripheral mononuclear leukocytes. In Lectin Biology, Biochemistry and Chimical Biochemistry. Walter de Gruyter and Co, Berlin, Vol. 3, 427-433.

Kieda, C., Roche, A.C., Delmotte, F. and Monsigny, M. (1979), Lymphocyte membrane lectins. Direct visualization by the use of fluoresceinyl-glycosylated cytochemical markers. FEBS Lett., 99: 329-332.

Kolatag (1987), Chimical trials planned for new AIDS drug. Science, 235: 1138-1139.

Kornfeld, R. and Kornfeld S. (1985), Assembly of asparagine-linked oligosaccharides. Ann. Rev. Biochem., 54: 631-664.

Lee, Y.C. and Lee R.T. (1982), Neoglycoproteins as probes for binding and cellular uptake of glycoconjugates in the glycoconjugates. Horowitz, M.I. and Pigman, W., eds. vol 4, Academic Press, New York, 57-83.

Liener, I.E., Sharon, N. and Goldstein, I. J. eds., (1986) The lectins Properties, fonctions and applications in biology and medicine. Academic Press, New York, pp 600.

Lis, H. and Sharon, N. (1986), Lectins as molecules and as tools. Ann. Rev. Biochem., 55: 35-67.

Meromsky, L., Lotan, R. and Raz, A. (1986) Implications of endogenous tumor cell surface lectins as mediators of cellular interaction and lung colonization. Cancer Res., 46: 5270-5275.

Midoux, P., Roche, A.C. and Monsigny M. (1986), Estimation of the degradation of endocytosed material by flow cytofluorometry using two neoglycoproteins containing different number of fluorescein molecules. Biol. Cell, 58: 221-226.

Midoux, P., Roche, A.C. and Monsigny, M. (1987), Quantitation of the binding, uptake and degradation of fluoresceinylated neoglycoproteins by flow cytofluorometry. Cytometry, 8: 327-334.

Miquelis, R., Alquier, C. and Monsigny, M. (1987), GlcNAc specific receptor of the thyroid. Binding characteristics, identification and potential role. J. Biol. Chem., in press.

Mirelman, D., ed. (1986), Microbial lectins and agglutinins : Properties and biological activity. Wiley Intercience Pub., J. Wiley, New York, pp 443.

Monsigny, M., ed. (1984) The role of carbohydrates in cell recognition : endogenous lectins. Biol. Cell., 51: 113-294 (special issue).

Monsigny, M., Kiéda, C. and Roche, A.C. (1979), Membrane lectins, Biol. Cell, 86: 289-300.

Monsigny, M., Kieda, C. and Roche, A.C. (1983), Membrane glycoproteins, glycolipids and membrane lectins as recognition signals in normal and malignant cells. Biol. Cell, 47: 95-110.

Monsigny, M., Kiéda, C. and Roche, A.C. (1984 a) Des lectines endogènes dans la reconnaissance cellulaire. in Cellular and pathological aspects of glycoconjugate metabolism. Colloque INSERM-CNRS, INSERM, 126: 357-372.

Monsigny, M., Roche, A.C. and P. Bailly (1984 b). Tumoricidal activation of murine alveolar macrophages by muramyl dipeptide substituted mannosylated serum albumin. Biochem. Biophys. Res. Commun., 121: 579-584.

Monsigny, M., Roche, A.C. and Midoux, P. (1984 c) Uptake of neoglycoproteins, via membrane lectin(s) of L1210 cells evidenced by quantitative flow cytofluorometry and drug targeting. Biol. Cell., 51: 187-196.

Montreuil, J. (1982), Glycoproteins. In Comprehensive Biochemistry, A. Neuberger and L.L. M. Van Deenen, eds, Elsevier, Amsterdam 19B Part II, 1-188.

Montreuil, J. (1984) Spatial conformation of glycans and glycoproteins. Biol. Cell, 51: 115-132.

Paietta, E., Stockert, R.J., Morell, A.G., Diehl, V. and Wiernik, P.H. (1986), Lectin activity as a marker for Hodgkin disease cells. Proc. Natl. Acad. Sci., USA, 83: 3451-3456.

Parish, C.R., Rylatt, D.B. and Snowden, J.M. (1984), Demonstration of lymphocyte surface lectins that recognize sulphated polysaccharides, J. Cell, Sci., 65: 145-148.

Pert, C.D., Hill, J.M., Ruff, M.R., Berman, R.M., Robey, W.G., Arthur, L.O., Ruscetti, F.W. and Farrar, W.L. (1986) Octapeptides deduced from the neuropeptides deduced from the neuropeptide receptor-like pattern of antigen T4 in brain potentially inhibit human immunodeficency virus receptor binding and T cell invectivity Proc. Natl. Acad. Sci. USA, 83: 9254-9258.

Pimlott, N. J.G. and Miller, R.G. (1986) The use of tunicamycin to study the role of cell surface oligosaccharides in lymphocyte recognition, J. Immunol. (1986), 137: 2455-2459.

Poole, A.R. (1986), Proteoglycans in health and disease : structures and functions. Biochem. J., 236: 1-14.

Popisil, M., Kubrycht, J., Bezouska, K., Taborsky, O., Novak, M. and Kocourek, J. (1986) Lactosamine type of asialooligosaccharide recognition in NK cytotoxicity. Immunol. Lett, 12: 83-90.

Powell, L.D., Bause, E., Legler, G., Molyneux, R. and Hart, G.W. (1985) Influence of asparagine linked oligosaccharides on tumor cell recognition in the mixed lymphocyte reaction. J. Immunol., 135: 714-724.

Rauvala, H. and Hakomori, S.I. (1981) Studies on cell adhesion. III The occurence of α-mannosidase at the fibroblast cell surface and its possible role in cell recognition. J. Cell Biol., 88: 149-159.

Raz, A., Meromsky, L., Zvibel, I; and Lotan, R. (1987), Tranformation related changes in the expression of endogenous cell lectins. Int. J. Cancer, 39: 353-360.

Raz, A. and Lotan, R. (1987), Endogenous galactose-binding lectins : a new class of functional tumor cell surface molecules related to metastasis. Cancer Metast. Rev., 5: in press.

Roche, A.C., Bailly, P. and Monsigny, M. (1985 a), Macrophage activation by MDP bound to neoglycoproteins : Metastasis eradication in mice. Invas. Metast., 5: 218-222.

Roche, A.C., Barzilay, M., Midoux, P., Junqua, S., Sharon, N. and Monsigny, M. (1983), Sugar specific endocytosis of glycoproteins by Lewis lung carcinoma cells. J. Cell. Biochem., 22: 131-140.

Roche, A.C., Midoux, P., Bouchard, P. and Monsigny, M. (1985 b) Membrane lectins on human monocytes. Maturation dependent modulation of 6-phosphomannose and mannose receptors. FEBS Lett., 193: 63-68.

Roden, L. and Horowitz, M.I. (1982), Structure and biosynthesis of connective tissue proteoglycans. in The glycoconjugates Horowitz M.I. and Pigman W., eds. vol 2, Academic Press, New York, 3: 85.

Roos, P.H., Hartman, H.J., Schlepper-Schäfer J., Kolb, H. and Kolb-Bachofen, V. (1985), Galactose specific receptors on liver cells : characterization of the purified receptor for macrophages reveals no structural relationship to the hepatocyte receptor. Biochim. Biophys. Acta, 847: 115-121.

Rosen, S.D., Singer, M.S., Yednock, T.A. and Stoolman, L.R. (1985) Involvement of sialic acid on endothelial cells in organ-specific lymphocyte recirculation. Science, 228: 1005-1007.

Ruff, M.R., Martin, B.M., Ginns, E.I., Farrar, W.L. and Pert, C.B. (1987), CD4 receptor binding peptides that block HIV infectivity cause human monocyte chemotaxis. Relationship to vasoactive intertinal peptide. FEBS Lett., 271: 17-22.

Ruoslahti, E. and Pierschbacher, M.D. (1986) Arg-Gly-Asp : a versatile cell recognition signal. Cell, 44: 517-518.

Schachter, H. (1984) Coordination between enzyme specificity and intracellular compartimentation with control of protein bound oligosaccharide biosynthesis. Biol. Cell., 51: 133-146.

Schrével, J., Gros, D. and Monsigny, M. (1981), Cytochemistry of cell glycoconjugates, Progr. Histochem. Cytochem. Gustav Fischer Verlog, Stuttgart, New York. Vol. 13, n° 2, 1-269.

Schrével, J., Monsigny, M., Mayer, R., Bernard, F., Picard, I., Lawton P. et Grellier, P. (1987), Nouvelle protéase de plasmodium falciparum anticorps dirigés contre cette protéase, substrats peptidiques spécifiques de la dite protéase et leur utilisation comme médicaments contre le paludisme. French Patent : 87 06785.

Schrével, J., Philippe, M., Bernard, F. and Monsigny, M. (1986) Surface Plasmodium sugar binding components evidenced by fluorescent neoglycoproteins, Biol. Cell, 56: 49-55.

Sharon, N. (1984) Carbohydrates as recognition determinants in phagocytosis and in lectin mediated killing of target cells. Biol. Cell, 51: 239-245.

Shur, B.D. (1982) Cell surface glycosyltransferase activities during normal and mutant (T/T) mesenchyme migration, Dev. Biol., 91: 149-162.

Stahl, P.D., Wileman, T.E., Diment, S. and Shepherd, V.L. (1984) Mannose specific oligosaccharide recognition by mononuclear phagocytes Biol. Cell, 51: 215-218.

Stoolman, L.M., Tenforde, T.S. and Rosen, S.D. (1984) Phosphomannosyl receptors may participate in the adhesive interaction between lymphocytes and high endothelial vesicles, J. Cell Biol., 99: 1535-1540.

Wileman, T.E., Charding, C. and Stahl, P. (1985) Receptor mediated endocytosis, Biochem. J., 232: 1-14.

Wilson, M.E. and Pearson, R.D. (1986) Evidence that Leishinania donovani utilizes a mannose receptor on human mononuclear phagocytes to establish intracellular parasitism. J. Immunol., 136, 4681-4688.

Yednock, T.A., Stoolman, L.M. and Rosen, S.D. (1987) Phosphomannosyl derivatized beads detect a receptor involved in lymphocyte homing. J. Cell Biol., 104: 713-723.

GENETICAL AND BIOCHEMICAL INTERACTIONS BETWEEN THE HOST AND ITS ENDOCYTOBIOTES IN THE WEEVILS *Sitophilus* (COLEOPTERE, CURCULIONIDAE) AND OTHER RELATED SPECIES

P. NARDON and A.M. GRENIER
Institut National des Sciences Appliquées. Biologie 406 -
69621 Villeurbanne Cedex - France

Among invertebrates, insects offer good models of permanent heterospecific associations (=symbiosis). Three main types can be defined : ectosymbiosis (insects which cultivate fungi), endosymbiosis (gut symbiosis in Termites and some Scarabeid or Cerambycid beetles) and endocytobiosis, i.e. intracellular symbiosis. This last type has the highest degree of integration with the host, and is the only one which will be considered in this article.

Endocytobiosis is a fascinating problem, and an unsolved enigma. Its study is relevant to both ecology (interaction between genetically distinct organisms), cellular biology (symbiote integration with host cell metabolism) and evolutionary biology (the problem of the emergence and maintenance of a heterospecific association and its role in evolution.).

Among the best known symbiotic insects are the snout beetles *Sitophilus* : *S. oryzae*, *S. zea-mais* (two sibling species) and *S. granarius*. They are the most dangerous pests to stored cereals in the world. The female deposits the eggs inside the kernels after boring a hole with her rostris, and the larvae develop inside the grain from which the young adults emerge. All other cited Rhynchophorinae species live in roots or stems of monocotyledonous plants : *Metamasius hemipterus*, the sugar cane weevil ; *Cosmopolites sordidus*, the banana weevil ; and *Rhynchophorus palmarum* the palm weevil (Nardon et al, 1985).

The principal features concerning symbiosis in these species will be reviewed, with particular reference to *Sitophilus oryzae*, in order to emphasize the relationships between the two partners and to try to explain their mutual control.

I - The host-endocytobiote cycle

A) The endocytobiotes

Symbiosis in *Sitophilus* was first described by Pierantoni (1927), then by Mansour (1930, 1934, 1935) and Tarsia in Curia (1933). In the African *Rhynchophorus ferrugineus* it has been studied by Buchner (1965). It has also been discovered in *Cosmopolites sordidus* and *Metamasius hemipterus* (Nardon et al, 1985). Despite some contradictory reports, in all these species and in some other Rhynchophorinae (*Sphenophorus striatum* in particular : unpublished data) the endocytobiotes are Gram-negative non-sporulating and non-ciliated bacteria (Grinyer and Musgrave, 1966 ; Musgrave and Grinyer, 1968 ; Nardon, 1971). A complete description can be found in Nardon and Wicker (1981) and Dasch et al (1984). In *S. oryzae* the symbiote size varies

NATO ASI Series, Vol. H17
Cell to Cell Signals in Plant, Animal and
Microbial Symbiosis. Edited by S. Scannerini et al.

from 1 µm to 30 µm. They are bacilliform as in *S. granarius* where two kinds of bacteria have been reported (Singh and Musgrave, 1966), the secondary type lacking an outer membrane and being able to infect unusual tissues such as Malpighian tubules. Nevertheless, since Dasch (1975) reports only one sort of DNA in this species (50 % G + C content), secondary symbiotes may represent degenerating bacteria, as in the cockroach *Periplaneta americana* (Philippe, 1986). In *M. hemipterus* the bacteria are also bacilliform, with a length varying from 2 µm to 200 µm, but, in *Cosmopolites*, they are so curved that they become hieroglyph-like (Nardon et al, 1985).

In *S. zea-mais* the symbiotes are very pleomorphic, with all intermediate forms between C-shaped bacteria and spiral bacteria (Morris, 1979 ; Nardon and Wicker, 1981). Since Dasch (1975) found two DNA bands (50 % and 54.5 % G + C content) the possibility exists of two different species. But how can we explain (unpublished data) that each strain of *S. zea-mais* is characterized by a definite proportion of these two kinds of bacteria ? Furthermore, it is to be noted that the most useful characteristics to distinguish the two sibling species *S. zea-mais* and *S. oryzae* is the morphology of their symbiotes, since they are bacilliform in *S. oryzae* or only flexuous (Musgrave and Homan, 1962 ; Nardon and Wicker, 1981). This observation suggests a coevolution between insect and their endocytobiotes and the following question arises : *did the symbiote play a role in the evolution of the weevils and particularly in the isolation of the maize and rice weevils ?*

The endocytobiotes can be maintained alive *in vitro*, but have not been successfully cultured so far (Musgrave and Mc Dermott, 1961 ; Nardon, 1978a), so that they cannot be identified (Dasch et al, 1984).

B) Transmission and fate of symbiotes during host development

Some differences exist between the Rhynchophorinae (=Calandrini) so far studied , but only the common features will be reported here (for more details see Nardon and Wicker, 1981 ; Nardon and Grenier, 1987). *S. oryzae* is the best known species. In mature oocytes and oviposited eggs, bacteria are scattered among yolk spheres, but a great number of them are located at the posterior pole, mixed with ribonucleoproteic granules probably representing the oosome (Nardon, 1971 ; Nardon et al, 1985). Therefore, the first differentiated primordial germ cells, at the very beginning of embryogenesis, include both the oosome and the symbiotes (Scheinert, 1933) which then appear as *permanent organelles of female germ line*. For reasons which have not yet been explained, they are rapidly eliminated from the male germ line and the rudiments of testes never contain bacteria. The cytological argument is corroborated by genetical experiments showing an exclusive maternal heredity for symbiotes (Musgrave and Miller, 1956 ; Nardon and Wicker, 1981).

The fate of symbiotes during embryogenesis has been studied by several authors (Mansour, 1930, 1935b ; Tiegs and Murray, 1938), but the phenomenon is not yet completely understood in detail. The bacteriome is differentiated from gut cells, and especially from stomodeum, according to Buchner (1965). The larval bacteriome, in both sexes, is a unique ectodermal organ located at the beginning of the midgut, but unconnected to the lumen. The symbiotes have no digestive role.

During metamorphosis, the bacteriocytes, probably under the influence of the new hormonal balance, slip back along the intestine and are incorporated into the anterior midgut caeca (Tarsia in Curia, 1933 ; Murray and Tiegs, 1935 ; Schneider, 1956), but they do not persist in imaginal life and are eliminated after three weeks (Mansour, 1930 ; Nardon and Wicker, 1981) in *Sitophilus* species. Therefore, in the adult weevils, which live several months, the endocytobiotes are restricted to the ovaries.

In ovaries, each having only two ovarioles, they are only found in the germ cells, oocytes and trophocytes, and in the bacteriocytes located in four apical bacteriomes.

What is remarkable in *S. oryzae* and in the other related beetles is the fact that *the endocytobiotes are always intracellular*, no transmission forms being necessary since they behave as permanent organelles of female germ cells, just like mitochondria. Therefore, in *Sitophilus*, the endocytobiotes inhabit only two cell types : female germ cells and bacteriocytes of stomodeal origin. The reasons for such a specificity are unknown.

In these cells, *the endocytobiotes are not surrounded by a host membrane* and they lie free in the cytoplasm (Grinyer and Musgrave, 1966 ; Musgrave and Grinyer, 1968 ; Nardon, 1971 ; Nardon et al, 1985). This feature shows that the bacterium is perfectly tolerated by the insect and is recognized as a "self" organelle. It follows that exchanges between the two partners are direct.

C) The microecosystem *Sitophilus*

In *S. oryzae*, the endocytobiosis is a very simple one : only one symbiote species and no migratory forms. Moreover, despite the fact that the bacterium-like symbiote is perfectly integrated with the host, *aposymbiotic insects (= deprived of symbiotes) can survive.* Mansour (1935) was the first author to describe an asymbiotic strain of *S. granarius* . Aposymbiotic weevils have been obtained experimentally following X or Gamma irradiation (Nardon, 1973), or methyl bromide treatment (Musgrave et al, 1965). The best method to eliminate the symbiotes is heat-treatment. Schneider (1956) obtained aposymbiotic weevils (*S. granarius* and *S. oryzae*) following an exposure of adults to 35° C for 33 days. A F2 generation was only possible on sorghum. In *S. zea-mais* a temperature of 33° C (during 4 to 12 weeks) is lethal to C-shaped symbiotes but not lethal to all spiral symbiotes (Morris, 1979), which confirms the presence of two symbiotes in this species. In our laboratory we succeeded in obtaining an aposymbiotic strain of *S. oryzae* on wheat (Nardon, 1973), following a treatment for 28 days at 35° C and 80-90 % R.H. It was reared for 95 generations. Obtaining stable aposymbiotic strains of weevils is facilitated by feeding them with sorghum. After two generations, they can be generally put back on wheat. We possess five aposymbiotic strains of *S. oryzae* on wheat and two strains on sorghum. Following a treatment at

37° C (90 % RH) for one month, we obtained one aposymbiotic strain of *S. granarius* and two strains of *S. zea-mais*.

Since endocytobiotes are unculturable bacteria, the only way to understand their role in insects is to compare the performances of symbiotic and aposymbiotic strains. Nevertheless, in making such a comparison, we have to remember that the treatment used to obtain aposymbiotic insects may have affected the host itself, so the symbiotic insects are not always a real control. Fortunately, we possess one such true control strain ("R35") since it survived heat treatment but retained its symbiotes. An other difficulty is that any difference observed between symbiotic and aposymbiotic weevils can be due either to the absence of symbiotes and/or to a genetic drift of the strain following the symbiote elimination. *To take into account a possible genetic drift, we have compared reciprocal hybrids between symbiotic and aposymbiotic strains*, whose genotypes are the same and which only differ by the presence (if the mother was symbiotic) or the absence of symbiotes. Any influence of symbiotes would only be maternally inherited.

A final difficulty in comparing symbiotic and aposymbiotic insects is to be sure that no other microorganisms are present. In the case of *Sitophilus*, Gasnier-Fauchet and Nardon (1985) showed that this insect, in laboratory conditions, is free of intestinal flora, so that the only active bacteria are the symbiotes. It is quite different in *Cosmopolites* or *Metamasius* where an important microflora inhabits the gut.

II - Biological and biochemical interactions

A) Importance of symbiotes for the host

As we can rear aposymbiotic rice weevils (since 1973 for our oldest strain), it is evident that symbiosis is not absolutely necessary in this species, at least at the present time. Modern cereals have certainly a higher nutritional value than the wild strains of several million years ago, when the symbiosis was established. Although not necessary, the symbiotes are nevertheless very useful to the insect, despite the opinion of Mansour (1934).

1. Morphogenetical influence

The aposymbiotic adult weevils are softer and paler, as has been reported for *S. granarius* (Musgrave and Miller, 1956) and *S. oryzae* (Nardon, 1973). The same phenomenon has been described in aposymbiotic cockroaches and was attributed to a lack of tyrosine (Henry and Cook, 1964). Nevertheless this amino acid does not seem to be involved in the colour of *S. oryzae* (Wicker, 1984). The behaviour of aposymbiotic weevils is also modified since, in *S. oryzae* and *S. zea-mais,* they are no longer able to fly or only fly with difficulty (unpublished data).

In aposymbiotic rice weevils both the ovarial and the larval bacteriomes have disappeared (Nardon, 1973). Their differentiation seems therefore dependent on some inductive factor produced by the symbiotes during embryogenesis (see previous review in this book). This phenomenon is not general since in the asymbiotic Egyptian strain of Mansour (1935 b), rudiments of a bacteriome are present. The differentiation of these organs is probably under a double control, but in any case the symbiotes play a role of varying importance according to the host species.

2. Influence on growth and life cycle (Table 1)

For aposymbiotic insects reared on wheat, the mean weight only declines for the first six generations (Nardon, 1973) : 7.4 % in females and 10.3 % in males. In succeeding generations the difference is restricted to males (about - 4 %).

Strains	**R1**	**R2**	**S**	**L**	**A**
Symbiote number - per bacteriome		2 800 000 (± 125 000)	1 400 000 (± 100 000)	950 000 (± 125 000)	0
- per ovary	111 000 (± 4000)	88 800 (± 4 900)	56 500 (± 2 850)	41 900 (± 3 000)	0
- per egg		23 000 (± 2 700)	18 600 (± 3 000)	14 900 (± 2 250)	0
Development time (day)	27.303 (± 0.153)	29.037 (± 0.027)	32.260 (± 0.034)	35.291 (±0.071)	44.353 (± 0.124)
Fertility/female	79.47 (± 9.01)	61.23 (± 1.41)	42.32 (± 2.24)	60.87 (± 5.56)	31.87 (± 2.49)
Weight : - female (mg)	1.887 (± 0.071)	1.866 (± 0.020)	1.828 (± 0.024)	1.731 (± 0.024)	1.693 (± 0.031)
- male	1.764 (± 0.062)	1.792 (± 0.023)	1.737 (± 0.031)	1.647 (± 0.024)	1.558 (± 0.029)
Imaginal life (day)		98.90 (± 5.19)	119.24 (± 7.55)	81.65 (± 3.27)	132.85 (± 8.98)

Table 1 : Comparison of some biological characteristics according to the number of symbiotes in *S. oryzae*. Counts in ovaries concern only previtellogenic stages. Fertility is measured during 2 weeks (2 to 4 weeks old weevils). The weight of aposymbiotic (A) weevils is determined for the first six generations (Nardon, 1973 and 1978 a). All means are given with $\pm t\,\sigma / \sqrt{n}$ for $p = 0.05$.

On the contrary, *the developmental time* (from deposition of the egg to emergence of adults out of grains) *is permanently lengthened by the loss of symbiotes* : + 55 % during the first five generations, and about + 30 % afterwards. In the control strain "R35", development is the same as in non-treated symbiotic strains. The decrease of growth is therefore the direct consequence of symbiote elimination (Nardon, 1978b ; Nardon and Wicker, 1981, 1983). We have been able to demonstrate that the *developmental time is directly related to the number of symbiotes* (Table1). The more numerous are the bacteria, the faster is development. That the symbiotes have an influence on the rate of development was previously reported in other species and also in *S. oryzae* (Schneider, 1956 ; Baker and Lum, 1973), but no precise study has been made so far.

Aposymbiotic weevils have a longer life (133 days) than symbiotic ones (119 days), and they are more resistant to radiation (Nardon and Wicker, 1981).

3. Influence on fertility

The elimination of symbiotes generally decreases fertility in numerous species such as cockroaches (Brooks and Richards, 1955) or *Euscelis plebejus* (Schwemmler et al, 1973). In *S. oryzae*, the first three aposymbiotic generations had a very low fertility (F) : only 13 fertile eggs per female per two weeks (n = 577) compared with F = 58 (n = 70) in the control symbiotic strain. Nevertheless, from the 8th generation onwards fertility has increased and is stabilized at a value 20 % to 30 % below that of symbiotic weevils. This diminished fertility is the consequence of a higher egg mortality rather than of a lower fecundity (Nardon, 1973, 1978a). Fertility is not strictly correlated with the symbiote density (Table 1).

4. Role of symbiosis in population dynamics

The growth of two weevil populations (*S. oryzae*), one symbiotic (S), and the other aposymbiotic (A), has been compared (Grenier et al, 1986). From the life tables and the weekly fertility, a malthusian growth model was established. The intrinsic rate of increase is much more important in S insects : 0.61 against 0.46 (27.5° C). We have experimentally verified the validity of the model. Thus *endocytobiosis greatly enhances the growth of the population*, and is a very important ecological factor.

B) Trophic interactions

Since the symbiotic bacteria are intracellularly located, they receive their own nutrients from the host, and are probably heterotrophic for some of them.

The numerous data on symbiosis (Buchner, 1965) have shown that endocytobiotes are autotrophic for some nutrients necessary for host development and reproduction and for which the host itself is heterotrophic. In *S. oryzae* we have shown (Grenier et al, 1986) that aposymbiotic and symbiotic larvae eat the same quantity of wheat but symbiotic ones are able to convert it better, since ECD (efficiency conversion digested matter factor) is higher in S (9.2 for dry weight at 27.5° C) than in A weevils (8.8) (see also Baker, 1974 a). What are the factors involved ? Two suggestions have been made : the host is provided with growth factors, and the bacteria interact with some metabolic pathways.

In *Sitophilus oryzae*, strict requirement for choline cannot be demonstrated (Baker, 1975 ; Wicker, 1983). Nevertheless, the presence of choline increases the fertility of symbiotic weevils, and is needed for aposymbiotic insects, but in a very low concentration (Wicker, 1984a). Concerning vitamins, *the symbiotes provide their host with pantothenic acid, biotin and riboflavin* in sufficient quantities to promote development and reproduction (Wicker and Nardon, 1983). Pyridoxin and folic acid are also possibly supplied, but in too low a concentration to maintain development for more than one generation. The vitamins of wheat, except riboflavin, are in a sufficient concentration for *S. oryzae* (Wicker and Nardon, 1983). The concentration of riboflavin is superior in sorghum (1.7 more) and this would partly explain the best development of A weevils on this cereal. *S. oryzae* offers a good example of a trophic interaction : in a preliminary experiment, Wicker (1984a) showed that in the absence of riboflavin from the symbiotic weevil diet, the number of bacteria per ovary significantly increased (38 %).

In *S. oryzae* phenylalanine is not strictly required in the diet of symbiotic insects (Baker, 1979), while aposymbiotic weevils need more aromatic amino acids than symbiotic ones. The slower growth rate of aposymbiotic weevils could be due, in part, to a less efficient utilization of exogenous tyrosine (rather than a lack of this amino acid in the food), and to a lack of endogenous tyrosine (or phenylalanine) normally supplied by bacteria (Wicker and Nardon, 1982).

Moreover, the symbiotes of *S. oryzae* could supply a unidentified liposoluble substance (Wicker, 1984a) and a study suggests that *S. granarius* bacteria could desaturate fatty acids (Yadava and Musgrave, 1972).The role of bacteria in sterol utilisation is also probable, but not yet clear (Baker, 1974b ; Wicker, 1984a).

C) Metabolic interactions

Symbiotes not only supply the host with some necessary nutrients, but they interfere with its metabolism.

1. Amino acid metabolism

One of the consequences of the elimination of symbiotes in *S. oryzae* is a change in the free, peptide-bound and protein amino acid concentration (Wicker and Nardon, 1982 ; Gasnier et al, 1984 ; Wicker et al, 1985). All the observed differences do not result from symbiote activity. Thus, the free valine level (+ 35 % in old symbiotic larvae) is not affected by symbiotes, while the situation is not clear concerning free (glutamic acid + glutamine) level (+ 46 % in symbiotic larvae) and free proline level (+ 46 % in aposymbiotic larvae), since both a maternal and a chromosomic effect probably occur (Gasnier et al, 1984). *The most striking differences between symbiotic and aposymbiotic rice weevil larvae concern methionine sulfoxide* (Gasnier-Fauchet and Nardon, 1987) always at a higher level in S weevils, *and sarcosine* (Gasnier-Fauchet et al, 1986) whose concentration increases regularly during last instar in aposymbiotic larvae, reaching its highest value at the end of this instar (43.5 nmoles/insect), while it never exceeds 3.8 nmoles/insect in symbiotic larvae. The same situation prevails in *Sitophilus zea-mais* (Gasnier-Fauchet and Nardon, 1987) and, in the two species results directly from the absence of symbiotes as is shown in reciprocal crosses of A and S insects. From the experiments carried out, several conclusions emerge (Figure 1).

- Methionine is in excess in wheat for *Sitophilus*

- In symbiotic weevils methionine is tranformed into methionine sulfoxide by a non-energy consuming and reversible reaction. *In vitro* experiments show that the larval bacteriome is directly involved, the insect itself having a poor oxidation capacity. This reaction could be an adaptative response to diets high in methionine (Gasnier-Fauchet and Nardon, 1986).

- In aposymbiotic weevils, methionine is preferably demethylated via a glycine N-methyltransferase-like activity, leading to the accumulation of sarcosine, since this amino-acid (=methyl glycine) is not incorporated into proteins, and not excreted in the feces (Gasnier-Fauchet et al, 1986).

- In contrast to Mammals, the weevil mitochondria are not able to convert sarcosine into glycine. That is the reason why sarcosine accumulates and appears as a metabolic cul-de-sac. It is remarkable that this undesirable production is prevented by symbiotes (which is a role devolved to mitochondria in vertebrates). Furthermore, the oxidation of methionine spares ATP.

- Another method of methionine elimination is transformation into lanthionine, excreted in the feces. From preliminary experiments, symbiotes would favor this pathway (Gasnier-Fauchet, 1985) (Figure 1).

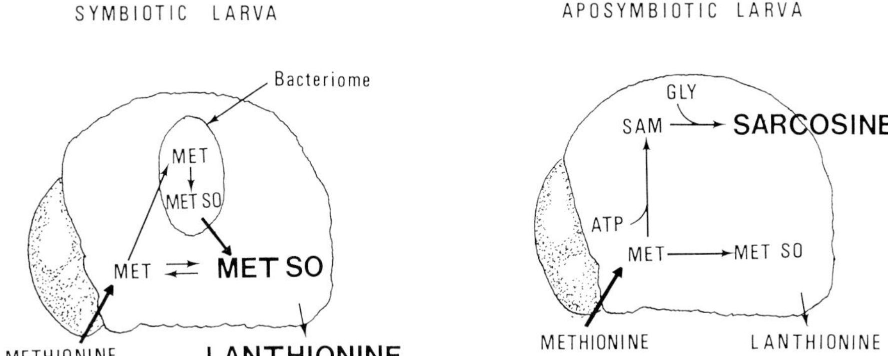

Figure 1 - Schematic pathways of methionine and sarcosine metabolisms in *S. oryzae* and *S. zea-mais*. From courtesy of F. Gasnier-Fauchet.

2. Methylation index

As a consequence of the influence of endocytobiotes upon methionine metabolism, in *S. oryzae* the *index of methylation* (SAM/SAH ratio) *is three times higher in the aposymbiotic larvae than in the symbiotic ones* (Gasnier-Fauchet et al, 1986). It is important to keep in mind that methylation probably plays a role in gene inactivation (Sneider, 1986).

3. Enzymatic interactions

In an unknown way the endocytobiotes influence some enzymatic activities in *S. oryzae*. Thus, phosphatase activity is diminished in aposymbiotic adults (Wicker, 1984b). The increasing rate of S-adenosyl methionine level without a combined effect on S-adenosyl homocysteine in aposymbiotic larvae suggests an inhibiting effect of bacteria upon ATP-L-methionine adenosyl transferase and on SAH-hydrolase. Concerning β-N-acetylglucosaminidase activity we have demonstrated that it is simultaneously controlled by the host genotype, epigenetic factors, and the inhibitory action of the symbiotes (Nardon et al, 1978 ; Wicker, 1980 ; Wicker and Nardon, 1980).

All these observations and experiments here reported lead us to conclude that the endocytobiotes of *S. oryzae* are perfectly integrated with the host metabolism, and that, in some respects, they fulfil the role of mitochondria. In cockroaches the endocytobiotes are also able to perform some fundamental mitochondrial functions (Laudani et al, 1973), and to interact with excretion metabolism (Valovage and Brooks, 1979).

III - Genetical interactions

A) Nucleic acid metabolism

In all Rhynchophorinae so far studied, as in most insects (Buchner, 1965), the *bacteriocytes are giant polyploid cells* (Nardon, 1971 ; Mansour, 1934 ; Nardon et al, 1985). In *S. oryzae*, thymidine ^3H is incorporated in the ovarial bacteriocyte nuclei and no mitotic figures are seen (Nardon, 1978a). In the aposymbiotic grain weevils studied by Mansour (1934), where rudiments of the gut bacteriome persist, the nuclei are smaller than in the corresponding infected cells of symbiotic insects. So it is tempting to consider that the polyploidy of host cells results from the presence of bacteria, which would then interfere with the nucleic metabolism of their host.

Quite interesting is the fact that a reciprocal effect could be possible : the gigantism of symbiotes has been emphasized in some species (e.g., the giant symbiote from the x-organ of the Fulgoroid *Myndus musivus* (Buchner, 1965)). In *Metamasius hemipterus* (Nardon et al, 1985), in the old larva, the bacteria lengthen but no longer divide, so they become extremely long, more than 200 mμ. In *E. coli* we know that cell division is controlled by specific genes (Donachie et al, 1984). We may suppose that in lengthening symbiotes such genes have been switched on by specific signals of the host.

B) Nucleo-cytoplasmic incompatibility

Reciprocal backcrosses between symbiotic (S) and aposymbiotic (A) strains have been made (Nardon, 1978a). After eight generations, the A genotype has been transfered into the S cytoplasm (type A_S) and vice versa (type S_A). These crosses have shown the genetic drift of the aposymbiotic strain (without their symbiotes the weevils are submitted to a different selection pressure). The most interesting phenomenon is the diminution of fertility in the successive back crosses, particularly in the aposymbiotic line S_A which became completely sterile after the 8th transfer. Such a nucleocytoplasmic incompatibility could be explained if we hypothesize that in the symbiotic weevil a harmony exists between nuclear (A1) symbiotic (S) and mitochondrial (M) genotypes (A_1,S,M). The balanced regulation system is modified in the aposymbiotic insect : (A_2,O,M) (A2 expresses the genetic drift of the strain). After eight successive back-crosses the initial systems are "recombined" since we then have symbiotic weevils (A_2,S,M), and aposymbiotic weevils (A_1,O,M). The latter leads to sterility. We note an intriguing resemblance with male cytoplasmic sterility in maize. We may imagine that the symbiote interferes also with mitochondrial genome.

C) Control of symbiotes by the host

The host controls not only the location of symbiotes but their density, constant for a given strain, but varying between geographical strains. We have demonstrated that such variations are due to both epigenetic factors and genetic control.

1. Epigenetic factors

As in every ecosystem, the symbiote is under the varied influences of its biotope i.e. the host. We have shown that the number of symbiotes varies according to the food (Grenier et al, 1983 ; Wicker, 1984). In cockroaches a similar phenomenon has been described (Brooks, 1963). Here, the balanced ratios of minerals (manganese, zinc and calcium) greatly influence the ovarial transmission of bacteroids. A very curious observation has been made in bacteriocytes cultivated *in vitro* (Philippe, 1986). While the symbiotes are rapidly degenerating in the absence of juvenile hormone, the depletion of cysteine, methionine and vitamin B12 in the culture medium, retards the lysis of bacteria. The author interprets these results by supposing that the symbiotes are regulated by the host cell according to its metabolic needs. Wicker has observed a similar phenomenon with *S. oryzae* symbiotes in the absence of riboflavine in the diet (see above).

2. Chromosomal factors

In *S. oryzae* the occurence of a genetic control is emphasized by three kinds of experiments. Firstly, when the oocytes are destroyed by X-irradiation (Nardon, 1978b), the apical bacteriocytes invade all the ovaries. This observation suggests that a local control is excerted by germ cells against bacteriocytes. Furthermore, in the invading bacteriocytes, there is no lytic area, contrary to that which is seen in non-irradiated weevils (Nardon, 1971).

Secondly, when irradiated males (5000 to 7000 Rads) are mated with control females some female offspring have no apical bacteriome and no endocytobiotes, an event never observed previously in control populations. Only the irradiation of the father's chromosomes can be responsible for the disappearance of bacteria and apical bacteriomes of ovaries. We have also irradiated virgin weevils of both sexes (4500 Rads) and mated them with control insects. A viable female progeny is obtained, presenting a reduced number of ovarial symbiotes (Nardon, 1978b ; Nardon and Wicker, 1983) :

♀ control	x	♂ control :	110 000 ± 6 500 (n = 96)
♀ irrad.	x	♂ control :	32 500 ± 3 600 (n = 128)
♀ control	x	♂ irrad. :	32 400 ± 4 000 (n = 96)

The observed effect is not the result of the treatment on the bacteria themselves (since they are not transmitted by the males), but is the consequence of the alteration of some chromosomal factor(s) controlling the symbiote density. This effect persists in the following generations but is progressively diluted.

In a third type of experiment we succeeded in obtaining selected strains with different symbiote densities, and we observed a strict relation between the number of symbiotes and the development time (Nardon, 1978a) (Table 1). The symbiote density is correlated in this way in ovaries, eggs and larval bacteriome. Furthermore, reciprocal crosses have been made between rapid (RR) and slow (LL) strains and the number of ovarial symbiotes counted :

Parents	RR : 111 430 ± 3 900		LL : 43 945 ± 2 400
F1	♀ RR x ♂ LL (RL)	and	♀ LL x ♂ RR (LR)
	RL : 76 500 ± 3 450		LR : 74 200 ± 3 700

Symbiotes are maternally inherited but their number is controlled by chromosomal factors and both sexes have an influence. We may assume the additive action of polygenes. In the successive backcrosses this conclusion is still verified : the introduction of LL genome in RR cytoplasm decreases the symbiote number towards that of LL strain (57 300 ± 4 500 at BC 5) and, vice versa, the introduction of RR genome in LL cytoplasm increases the symbiote density which reaches 134 600 ± 8 000 at BC 5. This experiment proves that the chromosomal structure of the weevil is the only active factor controlling the symbiote number. From selection experiments we could imagine that two types of bacteria were present in the unselected strain, each being revealed by selection. But the backcrosses show that the host genotype directly influences endocytobiote multiplication.

3. Mechanism of control

The question is : what physiological process(es) is(are) involved in the control of symbiote density ? We have no decisive answer. We can imagine at least three different mechanisms which are not mutually exclusive : cytolysis of symbiotes under the influence of either lytic enzymes or bacteriophages, or control of symbiote division by a specific signal. The elimination of the bacteriome occurring at metamorphosis could be the consequence of the new hormonal status, as has been described in cockroaches (Philippe, 1986). Lytic activity has been reported in *Sitophilus* (Musgrave and Grinyer, 1968 ; Nardon, 1971) and in other insects (Louis, 1980). We have hypothesized the role of the β-N-acetylglucosaminidase (Nardon et al, 1978) but further experiments did not clearly confirm this idea. Nevertheless, the symbiote-host cell interplay probably depends on resistance or inhibition mechanisms similar to those observed in pathogen-phagocyte interaction. In Mycobacteria, at least four known substances are produced to inhibit phagosome-lysosome fusion (Ryter and Chastellier, 1983). Other substances confer upon the bacteria a resistance to lysosomal enzymes, like mycoside which surrounds the bacteria to form a halo, as is shown in some endocytobiotes. Such protection is not always effective. In symbiosis they are perhaps best controlled by the host.

CONCLUSION

The *Sitophilus oryzae* model of endocytobiosis in insects is particularly interesting in several ways. Firstly, it is a very simple model since only one symbiote is involved, with no migratory forms, the bacteria being permanently located inside the host cells. Secondly, endocytobiosis can be disrupted and aposymbiotic strains obtained in some defined conditions, so that the role of symbiotes can be analysed precisely in comparing the two strains and their reciprocal crosses. Thirdly, despite the fact that endocytobiosis is not absolutely necessary, the symbiote is perfectly integrated into the host cell and organism. This is emphasized by the following three features : the bacteria lie free within the cytoplasm, they participate in insect nutrition, anabolism and catabolism, and they are perfectly controlled (in their location and density) by the host, in particular by chromosomal factors.

Concerning the biochemical interactions, one of the most interesting results is their role in methionine metabolism since it is demonstrated that, in some way, they replace deficient mitochondria (if we compare them with those of Mammals) to avoid sarcosine accumulation. Furthermore the presence of symbiotes leads to a three-fold decrease of the methylation index. Such an observation suggests that methylation could be one method used by symbiotes to interact with the host genome expression (the alternative way being a possible plasmid transfer, but we have no results in this field).

In conclusion, our study on weevil endocytobiosis confirms our interest in favor of the endosymbiotic theory with regard to the origin of mitochondria (Margulis, 1981). The study of endocytobiosis shows us that not only war and struggle for life are natural laws, but also harmony and cooperation.

Acknowledgements : we wish to thank Dr. D.C. Smith for correcting the English and C. Nardon for illustrations.

REFERENCES

Baker J.E. - 1974a - Differential net food utilization by larvae of Sitophilus oryzae and Sitophilus granarius. - *J. Insect Physiol.*, **20**: 1937-1942.

Baker J.E. - 1974b - Differential sterol utilization by larvae of Sitophilus oryzae and Sitophilus granarius. - *Ann. Entomol. Soc. Am.*, **67**: 591-594.

Baker J.E. - 1975 - Vitamin requirements of larvae of Sitophilus oryzae. - *J. Insect Physiol.*, **21**: 1337-1342.

Baker J.E. - 1979 - Requirements for the essential dietary amino acids of larvae of the rice weevil. - *Environ. Entomol.*, **8**: 451-453.

Baker J.E. & Lum P.T.M. - 1973 - Development of aposymbiosis in larvae of Sitophilus oryzae by dietary treatment with antibiotics. - *J. Stored Prod. Res.*, **9**: 241-245.

Bhatnagar R.D.S. & Musgrave A.J. - 1970 - Cytochemistry, morphogenesis and tentative identification of mycetomal microorganisms of Sitophilus granarius L. (Col.). - *Canad. J. Microb.* , **16**, 12 : 1357-1362.

Brooks M.A. - 1963 - Symbiosis and aposymbiosis in Arthropods - *in "Symbiotic associations". University Press Cambridge* , : 200-230.

Brooks, M. A. & Richards, A. G. - 1955 - Intracellular symbiosis in cockroaches. II - Mitotic division of mycetocytes - *Science* , **122**: 242-.

Buchner P. - 1965 - - *"Endosymbiosis of animals with plant microorganisms." Wiley J. and Sons (Ed.) Intersciences Publishers N.Y.* , : 1-909.

Dasch G.A. - 1975 - Morphological and molecular studies on intracellular bacterial symbiotes of insects. - *Thesis Ph D, University of Yale* , : 1-330.

Dasch G.A., Weiss E. & Chang K.P. - 1984 - *In"Bergey's manual of systematic Bacteriology", Krieg N.R. and Holt J.G. (Eds.), Williams and Wilkins, Baltimore / London* , : 811-836.

Donachie W.D., Begg K.J. & Sullivan N.F. - 1984 - Morphogenes of Escherichia coli - *In "Microbial Development" Losick R. and Shapiro L. (Eds.) Cold Spring Harbor Laboratory, New-York* , : 27-62.

Gasnier F., Nardon P. & Guillaud J. - 1984 - Influence of bacterial symbiotes on the amino acid composition of Sitophilus oryzae larvae (Coleoptera, Curculionidae) - *Endocyt. C. Res.* , **1**: 69-79.

Gasnier-Fauchet F. - 1985 - Etude du rôle des bactéries symbiotiques dans le métabolisme protéique de leur hôte Sitophilus oryzae (Coléoptère, Curculionide). - *Thèse de 3ème cycle, Microbiologie - Université Lyon I- ENV Lyon.* , : 1-130.

Gasnier-Fauchet F. & Nardon P. - 1985 - Les interactions biochimiques dans la symbiose endocellulaire chez le charançon Sitophilus oryzae : étude de la sarcosine - *Bull. Scient. Techn. INRA* , **19**: 23-30.

Gasnier-Fauchet F. & Nardon P. - 1986 - Comparison of methionine metabolism in symbiotic and aposymbiotic larvae of Sitophilus oryzae L.(Coleoptera : Curculionidae) II- Involvement of the symbiotic bacteria in the oxidation of methionine. - *Comp. Biochem. Physiol.* , **85B**, 1 : 251-254.

Gasnier-Fauchet F. & Nardon P. - 1987 - Comparison of sarcosine and methionine sulfoxide levels in symbiotic and aposymbiotic larvae of two sibling species, Sitophilus oryzae L. and S. zeamaïs Mots.(Coleoptera : Curculionidae) - *Insect Biochem.* , **17**, 1 : 17-20.

Gasnier-Fauchet F., Gharib A. & Nardon P. - 1986 - Comparison of methionine metabolism in symbiotic and aposymbiotic larvae of Sitophilus oryzae L. (Coleoptera : Curculionidae) I- Evidence for a glycine N-methyltransferase-like activity in the aposymbiotic larvae. - *Comp. Biochem. Physiol.* , **85B**, 1 : 245-250.

Grenier A.M., Nardon P. & Bonnot G. - 1986 - Importance de la symbiose dans la croissance des populations de Sitophilus oryzae L. (Coléoptère Curculionidae). Etude théorique et expérimentale. - *Acta Oecologica, Oecol. Applic.* , **7**, 1 : 93-110.

Grenier A.M., Nardon P. & Wicker C. - 1983 - Influence du changement de nourriture sur le développement, l'activité de la N-acétylglucosaminidase et les symbiotes de Sitophilus oryzae L. - *Bull. Soc. ent. Fr.* , **88**: 323-332.

Grinyer I. & Musgrave A.J. - 1966 - Ultrastructure and peripheral membranes of the mycetomal microorganisms of Sitophilus granarius - *J. Cell Sci.* , **1**: 181-186.

Henry S.M. & Cook T.W. - 1964 - Amino acid supplementation by symbiotic bacteria in the cockroach - *Contrib. Boyce Thompson Inst.* , **22**: 507-.

Laudani U., Frizzi G.P., Roggi C. & Montani A. - 1974 - The function of the endosymbiotic bacteria of Blattoidea - *Experientia* , **30**, 8 : 882-883.

Louis C. - 1980 - Recherches ultrastructurales sur les cycles de procaryotes intracytoplasmiques d'Arthropodes. - *Thèse de Doctorat ès Sciences, Montpellier* , : 1-191.

Mansour K. - 1930 - Preliminary studies on the bacterial cell-mass (accessory cell-mass) of Calandra oryzae (Linn.) : the rice weevil. - *Q. J. Microsc. Sc.* , **73**: 421-436.

Mansour K. - 1934 - On the so-called symbiotic relationship between coleopterous insects and intracellular micro-organisms. - *Q. J. Microsc. Sc.* , **77**: 255-272.

Mansour K. - 1935a - On the intracellular microorganisms of some Bostrychid beetles. - *Q. J. Microsc. Sci.* , **77**: 243-254.

Mansour K. - 1935b - On the micro-organism-free and the infected Calandra granaria (Linn.) - *Bull. Soc. Roy. Entomol. Egypte* , : 290-306.

Margulis L. - 1981 - - *"Symbiosis in cell evolution- Life and its environment on the early earth", Margulis L. (Ed.) - Freeman W.H. & Co.-San Francisco* , : 1-419.

Morris G.W. - 1979 - An attempt to produce aposymbiotic Sitophilus zeamais (Coleoptera : Curculionidae) by rearing at 33°C. - *Proc. Ent. Soc. Ontario* , **110**: 107-108.

Murray F.V. & Tiegs O.W. - 1935 - The metamorphosis of Calandra oryzae. - *Q. J. Microsc. Sci.* , **77**: 404-495.

Musgrave A.J. & Grinyer I. - 1968 - Membranes associated with the disintegration of mycetomal micro-organisms in Sitophilus zea-mais (Mots.). - *J. Cell. Sci.* , **3**: 65-70.

Musgrave A.J. & Homan R. - 1962 - Sitophilus sasakii in Canada : Anatomy and mycetomal symbiotes as valid taxonomic characters. - *Can. Ent.* , **94**: 1196-1197.

Musgrave A.J. & Mac Dermott L.A. - 1961 - Some media used in an attempt to isolate and culture the mycetomal microorganisms of Sitophilus weevils. - *Can. J. Microbiol.* , **7**: 842-843.

Musgrave A.J. & Miller J.J. - 1956 - Some microorganisms associated with the weevils Sitophilus granarius (L.) and Sitophilus oryzae (L.) (Coleoptera). II - Population differences of mycetomal microorganisms in different strains of S. granarius. - *Can. Ent.* , **88**: 97-100.

Musgrave A.J., Monro H.A.V. & Upitis E. - 1965 - Apparent elimination of symbiotes in successive generations of Sitophilus fumigated with methyl bromide. - *J. Invertebr. Pathol.* , **7**: 506-511.

Nardon P. - 1971 - Contribution à l'étude des symbiotes ovariens de Sitophilus sasakii : localisation, histochimie et ultrastructure chez la femelle adulte. - *C. R. Acad. Sci.* , **272D**: 2975-2978.

Nardon P. - 1973 - Obtention d'une souche aposymbiotique chez le charançon Sitophilus sasakii Tak. : différentes méthodes et comparaison avec la souche symbiotique d'origine. - *C. R. Acad. Sci.* , **277D**: 981-984.

Nardon P. - 1978a - Etude des interactions physiologiques et génétiques entre l'hôte et les symbiotes chez le Coléoptère Curculionide Sitophilus sasakii (= S. oryzae). - *Thèse de doctorat INSA-Université Lyon I* , : 1-285.

Nardon P. - 1978b - Etude de l'action des rayons X sur les symbiotes ovariens et l'ovogénèse chez Sitophilus oryzae L. (Col. Curculionide). - *Bull. Soc. Zool. Fr.* , **103**: 295-300.

Nardon P. & Grenier A.M. - 1987 - Endocytobiosis in Coleoptera : Biological, biochemical and genetical aspects - *In "Handbook of Insect Endocytobiosis : Morphology, Physiology, Genetics, Evolution" Schwemmler W. (Ed.), CRC Press Inc. (Boca Raton, Florida, USA)* , **(in press)**: -.

Nardon P. & Wicker C. - 1981 - La symbiose chez le genre Sitophilus (Coléoptère Curculionide). Principaux aspects morphologiques, physiologiques et génétiques. - *Ann. Biol.* , **20**: 327-373.

Nardon P. & Wicker C. - 1983 - Genetic control of symbiotes by the host in the insect Sitophilus oryzae L. (Coleoptera Curculionidae). - *In "Endocytobiology" De Gruyter & Co (Ed.), Berlin-New York* , **2**: 727-731.

Nardon P., Louis C., Nicolas G. & Kermarrec A. - 1985 - Mise en évidence et étude des bactéries symbiotiques chez deux charançons parasites du bananier : Cosmopolites sordidus (Germar) et Metamasius hemipterus (L.) (Col. Curculionidae). - *Ann. Soc. ent. Fr.* , **21**, 3 : 245-258.

Nardon P., Wicker C., Grenier A.M. & Laviolette P. - 1978 - Contrôle de la prolifération des symbiotes par l'hôte : étude préliminaire de l'hexo- - N - acétylglucosaminidase chez le Curculionide Sitophilus oryzae L. - *C. R. Acad. Sci.* , **287D**: 1157-1160.

Philippe C. - 1986 - Developpement d'un nouveau modèle in vitro pour la culture à long terme de tissus de Periplaneta americana : applications à l'étude de la symbiose intracellulaire et de la physiologie du corps gras et de l'ovaire - *Thèse de Doctorat d'etat, Université Pierre et Marie Curie, Paris VI* , : 1-247.

Pierantoni U. - 1927 - L'organo simbiotico nello sviluppo di Calandra oryzae. - *Rend. R. Accad. Sci. Fis. e Mat. Napoli Ser 3a* , **35**: 244-250.

Ryter A. & Chastellier C. de - 1983 - Phagocyte-pathogenic microbe interactions - *Int. Rev. Cytol.* , **85**: 287-327.

Scheinert W. - 1933 - Symbiose und Embryonalentwicklung bei Russelkäfern. - *Z. Morphol. Ökol.Tiere* , **27**: 76-198.

Schneider H. - 1956 - Morphologische und experimentelle Untersuchungen über die Endosymbiose der Korn und Reiskäfer (Calandra granaria und C. oryzae). - *Z. Morphol. Ökol.Tiere* , **44**: 555-625.

Schwemmler W., Duthoit J.L., Kuhl G. & Vago C. - 1973 - Experimental dissociation of endosymbiosis and nutrition with a synthetic medium of aposymbiontic individuals of Euscelis plebejus - *Z. Morph. Tiere* , **74**: 297-322.

Singh S.B. & Musgrave A.J. - 1966 - Some studies on the chromatin and cell wall of the mycetomal microorganisms of Sitophilus granarius. - *J. Cell Sci.* , **1**: 175-180.

Sneider T.W. - 1986 - DNA methylation : overview and prospectives - *In "Biological methylation and drug design. Experimental and clinical roles of S-adenosylmethionine" - Humana Press- Clifton, New-Jersey* , : 113-125.

Tarsia-in-Curia I. - 1933 - Nuovo osservazioni sull'organo simbiotico di Calandra oryzae (L.). - *Arch. Zool. Ital.* , **18**: 247-264.

Tiegs O.W. & Murray F.V. - 1938 - The embryonic development of Calandra oryzae. - *Q. J. Microsc. Sci.* , **80**: 160-273.

Valovage W.D. & Brooks M.A. - 1979 - Uric acid quantities in the fat body of normal and aposymbiotic German cockroaches,Blatella germanica - *Ann. Entomol. Soc. Am.* , **72**, 5 : 687-689.

Wicker C. - 1980 - Influence of sex,developmental time and food on ß-N-acétyl- glucosaminidase activity in the rice weevil Sitophilus oryzae L. - *Experientia* , **36**: 1059-1060.

Wicker C. - 1983 - Differential vitamin and choline requirements of symbiotic and aposymbiotic S. oryzae (Coleoptera : Curculionidae). - *Comp. Biochem. Physiol.* , **76A**: 177-182.

Wicker C. - 1984a - Etudes d'intéractions nutritionnelles et enzymatiques entre Sitophilus oryzae (Coléoptère Curculionide) et ses bactéries symbiotiques intracellulaires. - *Thèse de Doctorat d'état es-sciences naturelles INSA-UCB Lyon I* , : 1-195.

Wicker C. - 1984b - Comparative acid phosphatase activity in symbiotic and aposymbiotic Sitophilus oryzae (Coleoptera : Curculionidae). - *Endocyt. C. Res.* , **1**: 61-68.

Wicker C. & Nardon P. - 1980 - Rôle des symbiotes et du génotype dans la régulation de l'activité de la 2-N-acétylglucosaminidase chez le Coléoptère Curculionide Sitophilus oryzae L. - *Bull. Soc. Zool. Fr.* , **105**: 191-198.

Wicker C. & Nardon P. - 1982 - Development responses of symbiotic and aposymbiotic weevils Sitophilus oryzae L. (Coleoptera, Curculionidae) to a diet supplemented with aromatic amino-acids. - *J. Insect Physiol.* , **28**: 1021-1024.

Wicker C. & Nardon P. - 1983 - Differential vitamin requirements of symbiotic and aposymbiotic weevils, Sitophilus oryzae - *In "Endocytobiology" De Gruyter and Co (Ed.), Berlin* , **2**: 733-738.

Wicker C., Guillaud J. & Bonnot G. - 1985 - Comparative composition of free, peptide and protein amino acids in symbiotic and aposymbiotic Sitophilus oryzae (Coleoptera, Curculionidae) - *Insect Biochem.* , **15**, 4 : 537-541.

Yadava R.P.S. & Musgrave A.J. - 1972 - Phospholipid patterns of two symbiote harbouring weevils : the rice weevil, Sitophilus oryzae L. and the corn weevil Sitophilus zeamais. - *Comp. Biochem. Physiol.* , **42B**: 197-200.

SIGNALS IN THE PARAMECIUM BURSARIA - CHLORELLA SP. - ASSOCIATION

W. Reisser
FB Biologie, Lahnberge, D-3550 Marburg - Federal Republic of Germany

INTRODUCTION

Endosymbiotic associations of ciliates and coccoid chlorophyceae (*Chlorella* sp.) are frequently observed in freshwater habitats. They show different levels of organizational complexity, ranging from temporary associations with algal partners being digested to permanently stable units where algae are protected from host lytic enzymes by enclosure in special perialgal vacuoles. A stable system may be regarded as an entirely new organism on its own, showing many of the characteristics of a functional unit. It thus offers an excellent possibility to study mechanisms of coordination between partners of different genome types which result from a network of signal exchanges between ciliate and algae.

By far the best studied system among stable ciliate-algae associations is the unit of *Paramecium bursaria* with *Chlorella* sp., the so-called green *Paramecium* (for a general review and associated papers on ciliate-algae associations see Reisser, 1986). In this paper, effects of different modes of signal exchange on the ultrastructural (cell-to-cell recognition), physiological (coordination of partner growth rates), and behavioural (photobehaviour) level in the green *Paramecium* will be discussed with special reference to cell-to-cell recognition systems.

MODES OF SIGNAL EXCHANGE

The cell-to-cell recognition system

Stable ciliate-algae associations such as the green *Paramecium* are characterized by the formation of two vacuole types, i.e., digestive and perialgal vacuoles (Figure 1). Both vacuole types differ in membrane structure and in life history (for details see Meier *et al.*, 1984 ; Reisser, 1986). Digestive vacuoles are highly dynamic structures. During cyclosis

NATO ASI Series, Vol. H17
Cell to Cell Signals in Plant, Animal and
Microbial Symbiosis. Edited by S. Scannerini et al.

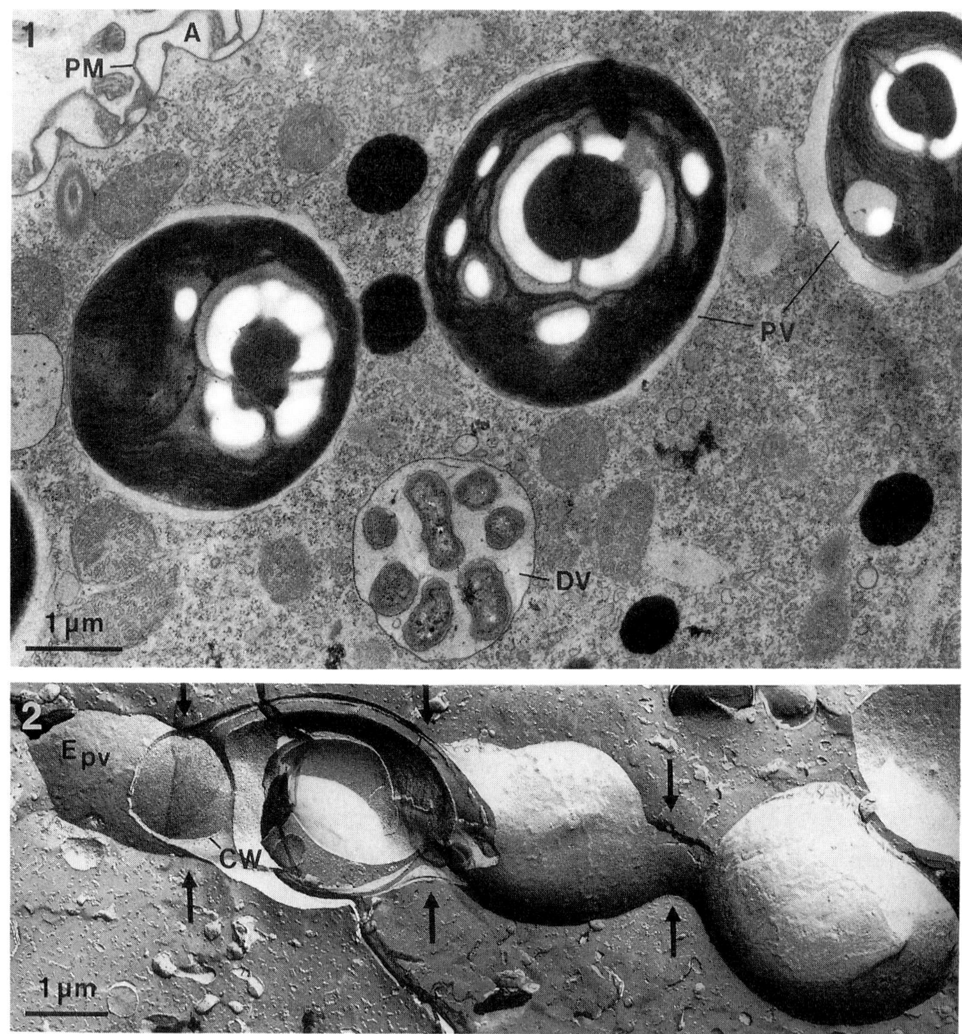

Figure 1 : Perialgal (PV) and digestive (DV) vacuoles in Paramecium bursaria. PVs enclose symbiotic chlorellae. Note peripheral location of PVs and nearby DV with bacteria. PM : plasma membrane, A : alveolus, bar : 1μm (by courtesy of R. Meier, Heidelberg).

Figure 2 : Division stage of the perialgal vacuole (PV) of Paramecium bursaria. After division of the algal mother cell into four autospores the PV invaginates around them (arrows). Note close contact of the PV membrane (Epv : E-face) to the cell wall (CW) of autospores. Bar : 1μm (by permission from Meier et al., 1984).

they migrate through the cell, fusing with each other and with other vesicles, as well as pinching off vesicles, with consequent changes in both size and membrane composition (Meier et al., 1984 ; Allen, 1984). Perialgal vacuoles are more static structures because they are separated from cyclosis events since they do not fuse with either primary or secondary lysosomes, thus protecting enclosed algae from being digested. The formation of perialgal vacuoles results from a cell-to-cell recognition process between ciliate and suitable algae. That process takes place when algal mother cells in perialgal vacuoles divide into autospores which are separated from each other and then immediately enclosed in new perialgal vacuoles (Figure 2), and also when ciliates take up algae from the surrounding medium. In the latter case, algae not suitable for symbiosis formation are sequestered into digestive vacuoles whereas suitable algae are enclosed in perialgal vacuoles. Mechanisms of cell-to-cell recognition and underlying signal exchange between partners are far from being understood in detail, but the available data allow some key points to be outlined (Figure 3). Signals of algae triggering the assembly of special perialgal vacuole membrane material are localized at the algal cell wall surface : algae whose surfaces have been altered by cell wall degrading enzymes or have been covered by specifically binding antibodies or lectins (Reisser et al., 1982) are not enclosed in perialgal but in digestive vacuoles. As could be shown by coupling tests with anionized and cationized ferritin, algae of the genus *Chlorella* have a negative surface charge. Measurements of electrophoretic mobility showed that quantitative differences in charge cannot have any signalling quality since they are randomly distributed among algae, whether or not they trigger the assembly of perialgal vacuole membrane material (Reisser, unpublished results).

Measurements of lectin binding to surfaces of signalling and non-signalling chlorellae after treatment with different cell wall degrading enzymes show that the composition of surface components varies tremendously (Table 1). Recent studies (Table 2) on binding of different lectins to surfaces of signalling and non-signalling chlorellae suggest that several carbohydrate chains of surface components including glucosyl, mannosyl, galactosyl, and fucosyl residues may participate in the formation of a special signalling pattern. That special signalling pattern was shown on chlorellae isolated from *Paramecium bursaria*, on another symbiotic *Chlorella* isolated from *Spongilla* sp. (211-40c) and on a freeliving *Chlorella*

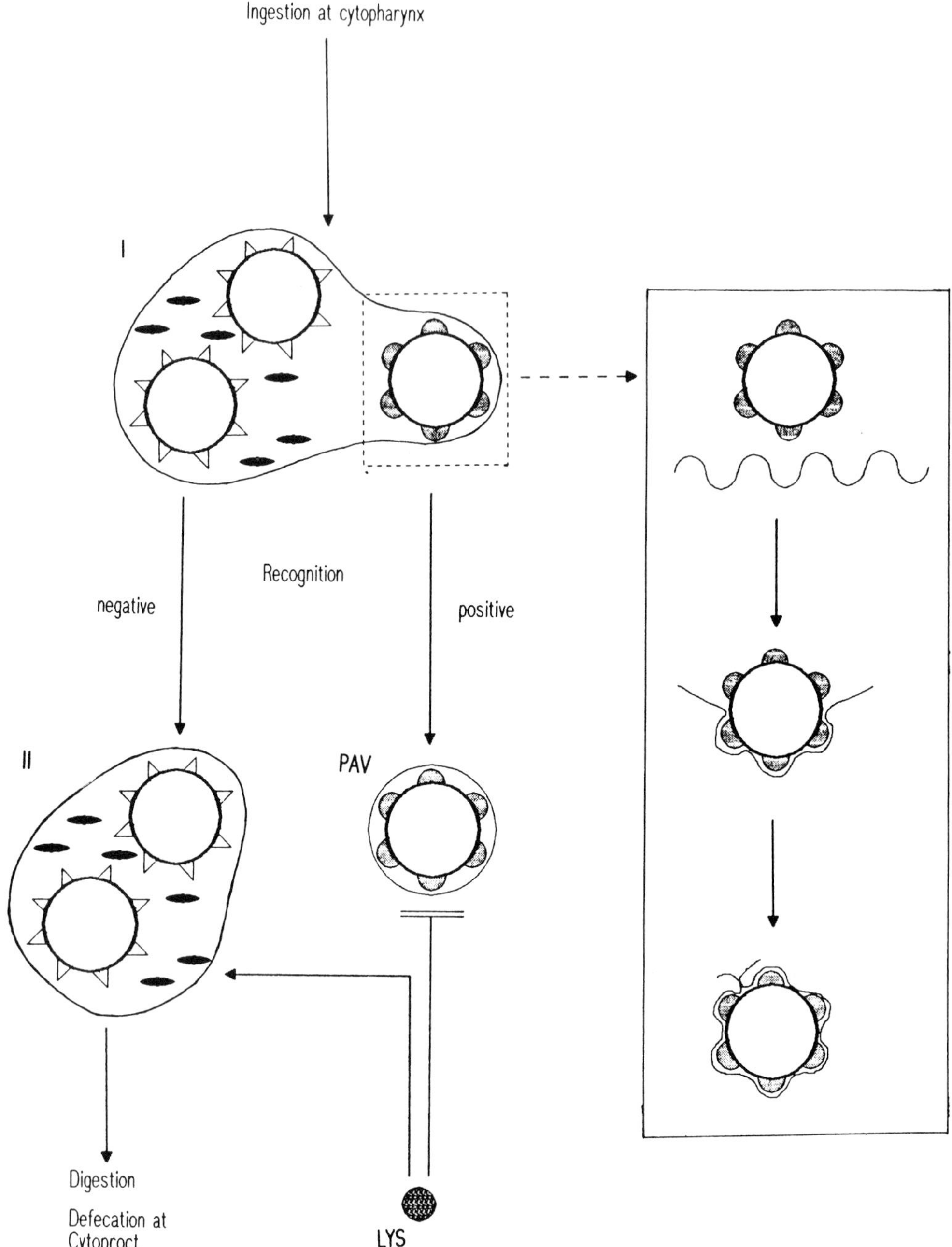

Figure 3 : Schematic drawing illustrating possible mechanisms of cell-to-cell recognition and perialgal vacuole formation in the *Paramecium bursaria* - *Chlorella* sp. - association. I : vacuole being formed at cytopharynx or digestive vacuole with algae (circles) suitable and not suitable for symbiosis formation and bacteria (disks) ; II : digestive vacuole ; PAV : perialgal vacuole ; Lys : lysosome.

Table 1 : Binding of lectin (concanavalin A) to cell wall surfaces of a symbiotic and of a freeliving *Chlorella* sp. before and after treatment with cell wall degrading enzymes.

After incubation[c] with	Incubation time (h)	Lectin[a] bound (%) to symbiotic *C.* of *Paramecium bursaria*[b]	Lectin[a] bound (%) to freeliving *C. fusca* var. *vacuolata* (211-8b)[b]
Control	2	100	100
Cellulase	1	116	161
	2	118	161
Pectinase	1	83	136
	2	78	127

[a]binding was measured photometrically with Fluoresceinisothiocyanate (FITC) labelled lectin as % of bound lectin (+/- 5%) per unit area of algal cell wall surface compared to control ;

[b]for methods of isolating and culturing of algae see Reisser, 1984, numbers in parentheses are identification numbers of Sammlung von Algenkulturen Göttingen (Schlösser, 1982) ;

[c]for protocol of treatment with enzymes see Reisser *et al.*, 1982 ; *C* : *Chlorella*.

lobophora (211-30, see Table 2). Accordingly both 211-40c and 211-30 trigger the formation of perialgal vacuoles in *Paramecium bursaria*, forming so-called artificial symbiotic systems (Niess *et al.*, 1982 ; Reisser, 1987). Other chlorellae not showing that pattern are enclosed in digestive vacuoles. Obviously the special signalling pattern exists only a short time of the alga's life cycle, i.e., only on autospores : when chlorellae isolated from *Paramecium bursaria* are offered again to ciliates, only autospores are taken up into perialgal vacuoles but mother cells are digested. Accordingly, antibodies to autospores are different from those to mother cells (Reisser *et al.*, 1982), and a study of surface components, by measurement of lectin binding, of algae isolated from green *Paramecium* showed that autospores bind about 200% more Concanavalin A per unit surface area than mother cells do (Reisser, unpublished results).

Table 2 : Binding of different lectins to the cell wall surfaces of symbiotic and freeliving Chlorella sp.

Chlorella sp.[a] isolated from	Amount of lectin bound to cell wall surface (rel. units)[b] lectin A	lectin B	lectin C
CILIATA :			
Paramecium bursaria+	100	100	100
Climacostomum virens	115	116	142
Euplotes daidaleos	107	130	115
Stentor polymorphus	129	143	120
PORIFERA :			
Spongilla sp.+	107	108	102
freeliving Chlorella sp.			
C. vulgaris (211-11b)	103	192	197
C. lobophora (211-30)+	96	107	109
C. fusca var. vacuolata (211-8b)	74	122	162

[a]for methods of isolating and culturing of algae see Reisser (1984), numbers in parentheses are identification numbers of Sammlung von Algenkulturen Göttingen (Schlösser, 1982) ;

[b]binding was measured photometrically with Fluoresceinisothiocyanate (FITC) labelled lectins, numbers give relative units of bound lectins (+/-5%) per unit area of algal cell wall surface, lectin A : Concanavalin A (binds to glucosyl and mannosyl chains), lectin B : Arachis hypogaea agglutinin (binds to galactosyl chains), lectin C : Lotus tetragonolobus agglutinin (binds to fucosyl chains) ; C : Chlorella ; + alga triggers the formation of a perialgal vacuole, other algae are digested in digestive vacuoles (see also Niess et al., 1982 ; Reisser, 1987).

Excretion of carbohydrates, which is a characteristic physiological feature of symbiotic chlorellae (Reisser, 1984), most probably does not have any signalling function for the formation of perialgal vacuoles since the latter are also formed around non-excreting chloreallae such as Chlorella lobophora.

Little is known about the receptors of algal signalling surface patterns in ciliates. Receptors must be part of presumptive perialgal vacuole membranes (i.e. cytopharynx or digestive vacuole membranes, or parts of them) and probably of vesicles which form those membranes by assembly and fusion with each other. In the genus Paramecium, perialgal vacuoles are only formed by Paramecium bursaria, so the ability to form special receptors recognizing algal signalling surface patterns is a prerequisite of symbiosis formation. Species such as Paramecium caudatum or Paramecium tetraurelia obviously cannot "understand" special algal signals and hence digest those chlorellae which in Paramecium bursaria trigger the formation of perialgal vacuoles (Reisser, 1986).

Probably the formation of a perialgal vacuole requires an algal surface-ciliate membrane contact which occurs during division of algae into autospores and when ingested algae come into contact with the cytopharynx or the digestive vacuole membrane (I, Figure 3). In both cases the enclosure of algae may be achieved according to a "zipper" mechanism proposed for phagocytotic processes in macrophages (Griffin et al., 1976). That zipper mechanisms does not work in nonsymbiotic ciliates such as Paramecium caudatum where algae cannot escape from digestion. Thus it may be speculated that the ability to form perialgal vacuoles, i.e. to respond to a special signalling pattern, relies on the synthesis of special vesicles which introduce the appropriate receptor components into the membranes of the cytopharynx, digestive vacuoles or growing perialgal vacuoles. Thus the synthesis of those vesicles which shall be tentatively called PAVOV (perialgal vacuole organizing vesicle) would be a special feature characterizing only those ciliates forming perialgal vacuoles, and are thus absent in Paramecium caudatum or Paramecium tetraurelia.

Perialgal vacuole membranes obviously keep their protective character, although the enclosed alga loses its positive signalling surface pattern during its development into a mother cell which - if offered to the Paramecium via ingestion - does not trigger the formation of a new perialgal vacuole (Reisser et al., 1982).

Primary and secondary lysosomes are often observed near perialgal vacuoles (Figure 1) but fusion does not occur. The basic principle endowing the perialgal vacuole membrane with its protective character is still a matter of speculation. As was shown by freeze-fracture studies (Meier et al., 1984), there are some similarities of the intramembraneous particle distributions between perialgal vacuole and old digestive vacuole membranes, which usually also do not fuse with other vesicles. Since there is only a very small distance of about 50nm (Meier et al., 1984) between algal cell wall surface and surrounding vacuole membrane, it is also conceivable that the charge of surface groups such as glycoproteins interacts with opposing vacuole membrane components. Thus, fluidity of those membrane components could be reduced, and contact between the lipid bilayers of perialgal vacuole and lysosomal membranes, and hence vesicle fusion, avoided.

Coordination of partner growth rates

The evolutionary success of stable ciliate-algae associations is based on a well organized cell-to-cell recognition system. It also relies on a plethora of signal exchanges between partners at the physiological level (Figure 4) which have been discussed extensively elsewhere (Reisser, 1986) and shall be mentioned here only briefly. Obviously one of the essential prerequisites for maintaining a permanent association is the coordination of partner growth rates, which guarantees the stability of the system even under conditions which selectively stress only one partner. Special regulating substances or genetic mechanisms have not been found in ciliate-algae associations. Space available for algae is probably not a limiting factor. In *Paramecium bursaria*, chlorellae occupy only about 5 - 8% of the ciliate volume (Reisser, 1986). Algae once enclosed in perialgal vacuoles are protected from host digestion but can successfully populate the host only when their specific ecological requirements from their habitat such as light, CO_2, and supply of nitrogenous and other organic compounds is satisfied. Such physiological factors might be called "signals" since algae respond by changes in growth rate. Thus, an increase in light intensity leads to an increase in the symbiotic algal population size. Consequently, higher amounts of oxygen and excreted sugars are available to the ciliate, and its division rate then increases until a new constant ratio between partners is established. When the light intensity is decreased, production and excretion of photosynthetic products by algae also decreases and both number of algae and host division rate drop. Only a minority of different *Chlorella* spp. taken up in perialgal vacuoles can persist in the ciliate, e.g. by their ability to exploit offered nitrogen compounds etc., other less suitable species are lost from the system by "dilution" (ecological regulation (Reisser, 1986 ; Reisser, 1987)).

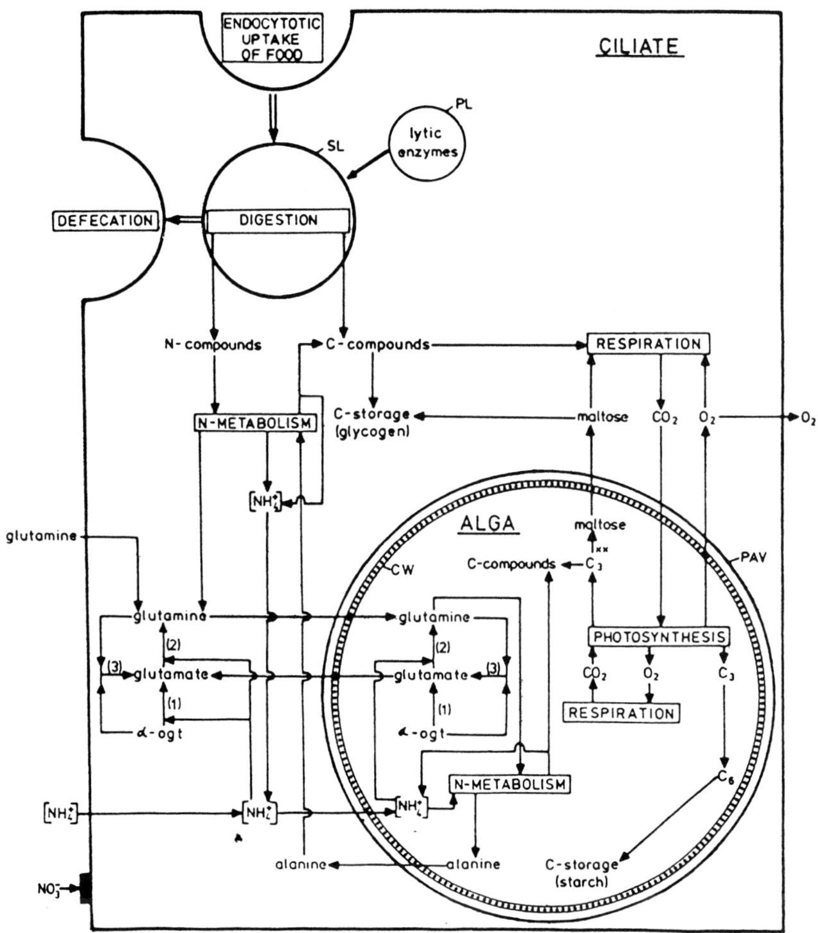

Figure 4 : Schematic drawing of physiological relationships (C- and N-metabolism) of the Paramecium bursaria - Chlorella sp.- association fed with bacteria and grown in the light above the photosynthetic compensation point. ** in darkness maltose is synthesized from starch. CW : cell wall of symbiotic Chlorella ; PAV : perialgal vacuole ; PL : primary lysosome ; SL : secondary lysosome ; ⟹: cyclosis ; (1) NADH-dependent glutamate synthase ; α-ogt : α-oxo-glutarate (by permission from Reisser, 1986).

Photobehaviour

One of the most conspicuous fields of signal exchanges between ciliate and algae is involved in photobehaviour of green ciliates (for a review and associated papers see Niess et al., 1982 , Reisser and Häder , 1984 , Reisser, 1986). Green paramecia gather in a light spot (so-called photoaccumulation) thus showing a qualitatively new behaviour which is not observed with algafree ciliates or the isolated algae. Photoaccumulation is not an oxygen-dependent chemotaxis but results from a series of photophobic step-down reactions each of which requires a cascade of signal exchanges between algae and ciliate (Figure 5). The receptor of the light

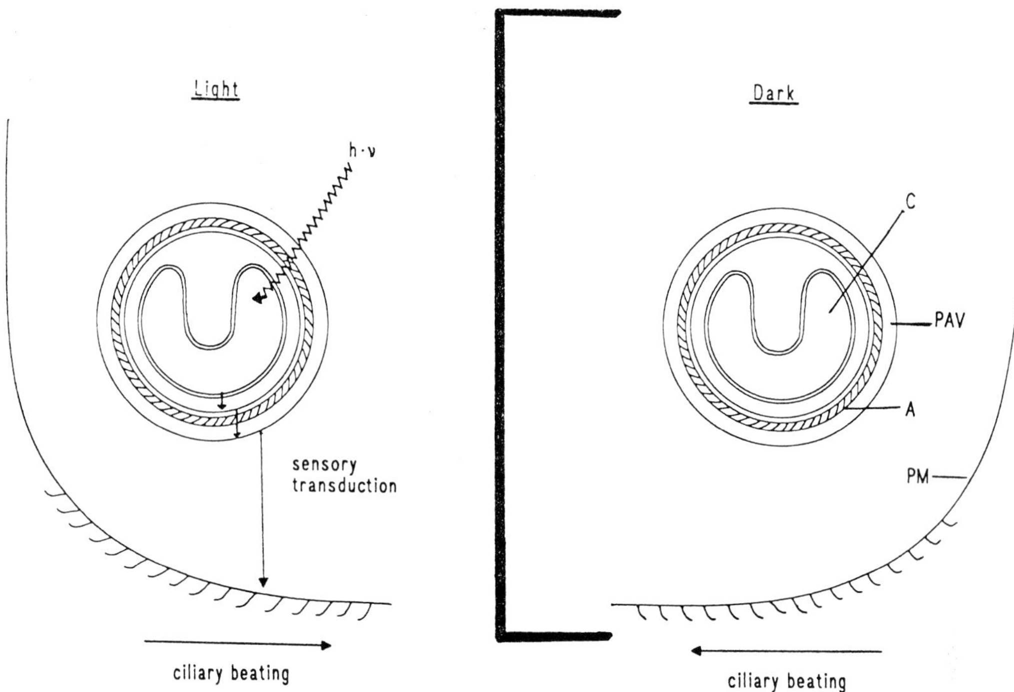

Figure 5 : Step-down photophobic reaction of green Paramecium bursaria. In order to trigger the reversal of ciliary beating (PM : plasma membrane of ciliate) by a light stimulus the according signal must pass inner and outer chloroplast (C) membranes, cytoplasm and plasmalemma of algae, algal cell wall (A), perialgal vacuolar space (PAV), membrane of perialgal vacuole and cytoplasm of ciliate.

stimulus is the chlorophyll of the algae, the effector site is the ciliate plasma membrane where a reversal of ciliary beating is induced. The transduction chain is still unknown. It is not membrane-bound since there is no connection between chloroplast membranes, algal plasmalemma, perialgal vacuole membrane and host plasma membrane. Special transmitter substances have not been found so far. The lower threshold for photoaccumulation of the green Paramecium is about $2.11 Wm^{-2}$, i.e., below its photosynthetic compensation point, and about 50 algae per ciliate. Hence, it may be speculated that signal transduction is related to some chemical compounds or effects resulting from the photosynthetic activity of the whole algal population.

CONCLUSIONS

Endocytobiotic systems may be best understood in ecological terms, i.e., the bigger partner forms an ecological niche in which the smaller and genetically different partners lives.

Thus the study of signal exchange between partners offers a powerful tool to a general concept of symbiosis formation. Obviously the evolutionary success of stable endocytobiotic units depends on the formation of an effective network of signal exchange. It is reasonable to assume that the effectiveness of an association increases the more the signals of partners are specific, i.e. the better they "understand" each other. Thus, contemporary stable endocytobiotic systems are characterized by a marked specificity of partners. Their network of exchanged signals in some cases is so well organized that it generates a qualitative shift in behaviour : the endocytobiotic association reacts as a unit, i.e., as a new organism.

ACKNOWLEDGEMENTS

Support by a grant of the Deutsche Forschungsgemeinschaft and technical assistance of M. Widowski are gratefully acknowledged.

REFERENCES

Allen, R.D. 1984. Paramecium phagosome membrane : from oral region to cytoproct and back again. J Protozool 31: 1-6.

Griffin, F.M., Griffin, J.A., Silverstein, S.C. 1976. Studies on the mechanism of phagocytosis. II. The interaction of macrophages with anti-immuno-globulin IgG-coated bone marrow derived lymphocytes. J Exptl Med 144: 788-809.

Meier, R., Lefort-Tran, M., Pouphile, M., Reisser, W., Wiessner, W. 1984. Comparative freeze-fracture study of perialgal and digestive vacuoles in Paramecium bursaria. J Cell Sci 71: 121-140.

Niess, D., Reisser, W., Wiessner, W. 1982. Photobehaviour of Paramecium bursaria infected with different symbiotic and aposymbiotic species of Chlorella. Planta 156: 475-480.

Reisser, W. 1984. The taxonomy of green algae endosymbiotic in ciliates and a sponge. Br Phycol J 13: 309-318.

Reisser, W. 1986. Endosymbiotic associations of freshwater protozoa and algae. Progr in Protistol 1: 195-214.

Reisser, W. 1987. Naturally occurring and artificially established associations of ciliates and algae : Models for different steps in the evolution of stable endosymbioses. N Y Acad Sci (in press).

Reisser, W., Häder D.P. 1984. Role of endosymbiotic algae in photokinesis and photophobic responses of ciliates. Photochem Photobiol 39: 673-678.

Reisser, W., Radunz, A., Wiessner, W. 1982. The participation of algal surface structures in the cell recognition process during infection of aposymbiotic Paramecium bursaria with symbiotic chlorellae. Cytobios 33: 39-50.

Schlösser, U.G. 1982. Sammlung von Algenkulturen. List of strains. Ber dtsch Bot Ges 95: 181-276.

NUTRITIONAL INTERACTIONS AS SIGNALS IN THE GREEN HYDRA SYMBIOSIS

A.E. Douglas,
John Innes Institute,
Colney Lane,
Norwich, NR4 7UH.
U.K.

1. Introduction

A wide range of cell-cell interactions are mediated by the release of specific molecules by signalling cells, leading to a defined physiological response of target cells. Our concept of signalling derives largely from animal systems. Often the signalling and target cells are in close apposition and the signal is borne on the surface of the signalling cell or passes directly between the cytoplasmic contents of the two cells via gap junctions. Other interacting animal cells are widely-separated and the signal passes through extracellular fluids (e.g. bloodstream) to the target cell. Binding of the signal to a receptor molecule on the target cell represents 'receipt' of the signal, and this interaction between signal and receptor is conventionally known as 'recognition'.

One may anticipate that many of the cell-cell interactions between the partners of alga-invertebrate symbioses are mediated by molecular signalling. Of particular interest are the interactions underlying the maintenance of the "established symbiosis", which can be defined by two characteristics:

a) persistence: the partners do not separate and one organism does not overgrow its partner(s).

b) nutritional interactions: a variety of nutrients are translocated in both directions between living cells of the partners.

Signalling in alga-invertebrate symbioses would be expected

NATO ASI Series, Vol. H17
Cell to Cell Signals in Plant, Animal and
Microbial Symbiosis. Edited by S. Scannerini et al.

to differ topologically from other systems because, in most associations, the algal symbionts are located within the cells of their animal hosts (i.e. they are intracellular). Each algal cell is bounded by a membrane of host origin, known as the perialgal membrane. Any released signals between alga and animal host would pass across the perialgal space between the host membrane and algal cell and both receptors and surface-borne signals of the animal partner would be located on the inner face of the perialgal membrane, and not the cell surface. However, the inner face of the perialgal membrane is topologically equivalent to the outer surface of the host cell membrane because algal symbionts are incorporated by phagocytosis (i.e. the perialgal membrane is derived ultimately from the phagosome).

The association between the noncolonial coelenterate, 'hydra', and green algae of the genus Chlorella is a particularly favourable system for the study of cell-cell interactions (Smith, 1987). The Chlorella cells are located exclusively in the digestive cells of the gastrodermal layer of the hydra. Below, the criteria used to define an established symbiosis are described for this association.

a) Persistence and stability of the association

The persistence of the hydra symbiosis depends on the balanced increase in algal and hydra cells. This is reflected in the constancy of the relative biomass of algae and host under a given set of environmental conditions, as measured by a variety of indices (e.g. number or total volume of algal cells per digestive cell, per animal or per unit host protein). Further, the relative biomass varies consistently with environmental conditions, of which feeding and light regimes have been studied most intensively. A major factor contributing to the constancy of the relative biomass is the regulation of algal growth and division rates so that they closely match those of the hydra; digestion or expulsion of algal cells is rarely observed. Algal growth occurs throughout the cell cycle of the hydra cell, but the cells divide only at the time of host cell division (McAuley, 1982). However, most symbiotic Chlorella cells undergo two successive mitotic divisions to form four daughter cells, and therefore only a small proportion of algal cells divide at host cell division under conditions where a constant number of algae per digestive cell is maintained.

b) Release of photosynthate

The algal symbionts are photosynthetically active and release at least 60-70% of photosynthetically-fixed carbon to the hydra tissues (Douglas, 1987). In algal cells freshly isolated from the association, the released photosynthate is almost exclusively in the form of one compound, the disaccharide maltose, and the release of maltose exhibits a strong pH dependence, with maximal release (2 fmol maltose $cell^{-1}h^{-1}$) at pH 4 and very low rates of release (0.05 fmol $cell^{-1}h^{-1}$) at pH 7. It is widely assumed that the cells also release maltose in the intact association, and that the pH of the immediate environment of algal cells (i.e. perialgal pH) is low, at about 5 units (inferred from data of Mews & Smith, 1982). However, the perialgal pH in the hydra symbiosis remains to be determined directly.

The thesis of this paper is that these two characteristics of an established association are closely linked. It is proposed that the perialgal pH (largely determined by the host) acts as a signal which induces maltose release by the algal symbionts, resulting in the control of algal growth and division. Further, evidence that maltose release by the algal cells may represent a signal which influences the nitrogen metabolism of the host (and hence nitrogen status of the algal cells) will be described.

2. The experimental associations

The first and most important line of evidence that maltose release from the algal cells and regulation of algal growth and division are linked came from studies of the experimental associations constructed between aposymbiotic hydra and heterologous *Chlorella* (Mews & Smith, 1982). A wide range of *Chlorella* strains are phagocytosed by hydra digestive cells and many are retained for prolonged periods, depending upon the strains of hydra and Chlorella, and the culture conditions. However, only those Chlorella strains which exhibit maltose release induced by low pH in culture can form a stable association that persists indefinitely. All these Chlorella strains have been derived from symbioses, particularly with ciliate protozoans such as Paramecium bursaria, and not from the free-living condition.

Most of the heterologous algae in association with hydra release a substantial proportion (>50%) of their photosynthetic carbon to the host tissues and, when freshly isolated from the association, release maltose over several hours at a linear rate (2-3 fmol $cell^{-1}h^{-1}$ at pH 4.5), which is similar to or greater than the native symbionts. A few strains, notably NC64A and UTEX-130, release maltose at a ten-fold lower rate of 0.1-0.3 fmol $cell^{-1}h^{-1}$ at pH 4-5. These algae maintain a stable association with hydra, with the relative biomass of the algal and host partners approaching or comparable to values for associations with high maltose releasing algae (Table 1). However, the hydra containing these algae exhibit the unusual characteristic of regularly expelling coherent pellets, composed principally of algal cells, via the enteron and mouth. Further, a high percentage (up to 20%) of the algae are located in the portion of the digestive cell apical to the host nucleus, whereas in all associations with algae which release maltose at high rates, less than 10% of the algal cells are in an apical position. It is very likely that these algae are destined for expulsion.

These characteristics suggest that strain NC64A is not amenable to regulation by control of its growth and division rates, the usual mode of regulation in alga-invertebrate symbiosis (see above). In support of this interpretation, twice as many cells of strain NC64A as of strains which release maltose at high rates are found in a dividing condition when the hydra cells are stimulated to divide by feeding; cells of strain NC64A also divide when the division of digestive cells is inhibited by starvation (Table 1). The diversion of energy and nutrients into algal biomass which is subsequently lost from

Table 1 Characteristics of the fully established association of the native symbiosis in green hydra and experimental associations with *Chlorella* strains which release maltose at high and low rates (Data from Mews & Smith, 1982; Douglas & Smith, 1984; McAuley, 1986a). The hydra were maintained at 20°C with 12hL:1hD and fed thrice weekly.

Chlorella	maltose release rate at pH 4.5 fmol $cell^{-1}h^{-1}$	mean total algal volume (mm^3) per mg protein	specific growth constant of hydra (d^{-1})	Mitotic index (%) of algal cells 24h after feeding	Mitotic index (%) of algal cells 4d after feeding	% algal cells in apical region of digestive cells
native symbionts	1.19	2.16	0.080	2.5	0.4	7.7
strain 3N8/13-1	2.16	1.69	0.087	3.1	0.7	7.2
strain NC64A	0.17	1.88	0.030	6.0	3.4	16.7

the association probably accounts for the very low growth rate of hydra containing strain NC64A. (The low maltose release rate of strain NC64A is not sufficient to explain the poor growth of the hydra, since aposymbiotic hydra grow at comparable rates to hydra containing native symbionts under standard culture conditions (Douglas & Smith, 1983)).

It is reasonable to conclude that the expulsion of strain NC64A is correlated with, and directly linked to, the low maltose release rate of this alga at low pH. The growth response of symbiotic Chlorella in acidic media is consistent with this contention (Douglas & Smith, 1984; Douglas & Huss, 1986). More than 90% of the cells of all symbiotic Chlorella strains tested were viable when maintained in culture media at pH 2.5-7 for 7 days, and two patterns of pH dependence of growth were observed. Strains which release maltose at rates of one to several fmol $cell^{-1}h^{-1}$ at below pH 5.0 did not grow below this value, but those strains (e.g. NC64A) which release little or undetectable maltose at any pH grew in media of pH as low as 4 units. The sensitivity of high maltose releasing strains of Chlorella (e.g. 3N8/13-1) to low pH was interpreted as a reflection of their release of maltose at low pH. As shown in Figure 1, the growth of strain 3N8/13-1 declines with decreasing pH in the range 6.0-4.0 units, as the maltose release rate by the algal cell increases.

3. The pH model of regulation

Primarily from the study of experimental associations with Chlorella strains which release little maltose, it has been hypothesized that the pH of the perialgal vacuole plays an important role in the control of growth and division of the native symbionts of hydra (Douglas & Smith, 1984). It is proposed that the perialgal pH may vary between a 'restrictive' low value, at which the algal cells release maltose at a substantial rate, and a 'permissive' high value at which growth and division proceeds. The strains which release little maltose and whose growth is relatively insensitive to low pH cannot be regulated entirely by this means and expulsion contributes to

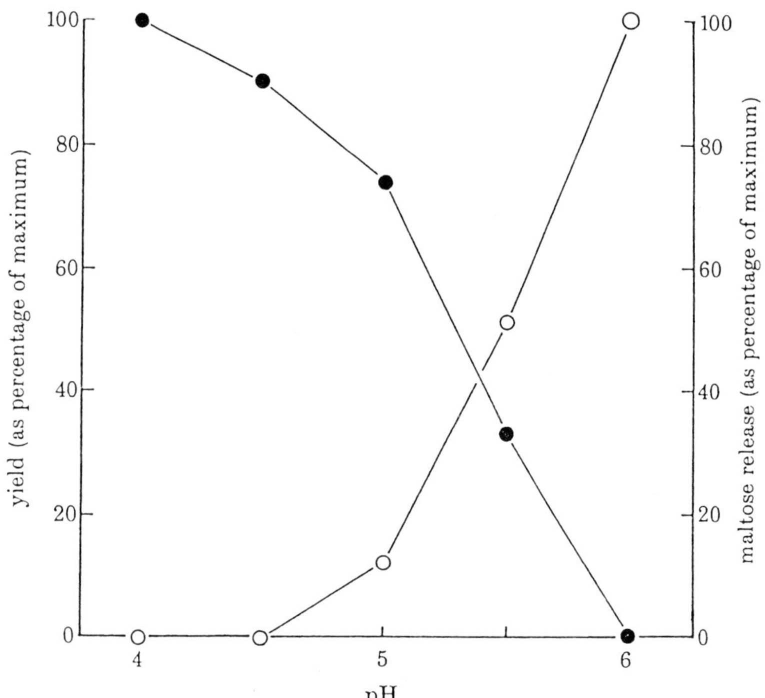

Figure 1 Variation with pH of maltose release (●) and growth (O) of *Chlorella* sp. strain 3N8/13-1 in culture. Maltose release is expressed as % of rate at pH 4 and yield of cells after incubation for 7 days as % of value at pH 6.

---

the control of their total biomass.

The proposed acidic conditions in the perialgal vacuole are believed to be generated and maintained largely by the host cell, perhaps by means of a H^+-pumping ATPase in the vacuolar membrane (see Figure 2). Further, it has been suggested that the magnitude of proton flux across the vacuole membrane (and hence perialgal pH) varies in parallel with the acid load on the host cell. Thus, conditions which increase the acid load (e.g. high CO_2 due to increased host respiration or low algal photosynthesis) would result in a decrease in perialgal pH and tendency towards lower rates of increase in algal biomass; conditions which decrease acid load would have the reverse effect. The response of the relative biomass of alga and host to changes in culture conditions are consistent with this scheme (Douglas & Smith, 1984).

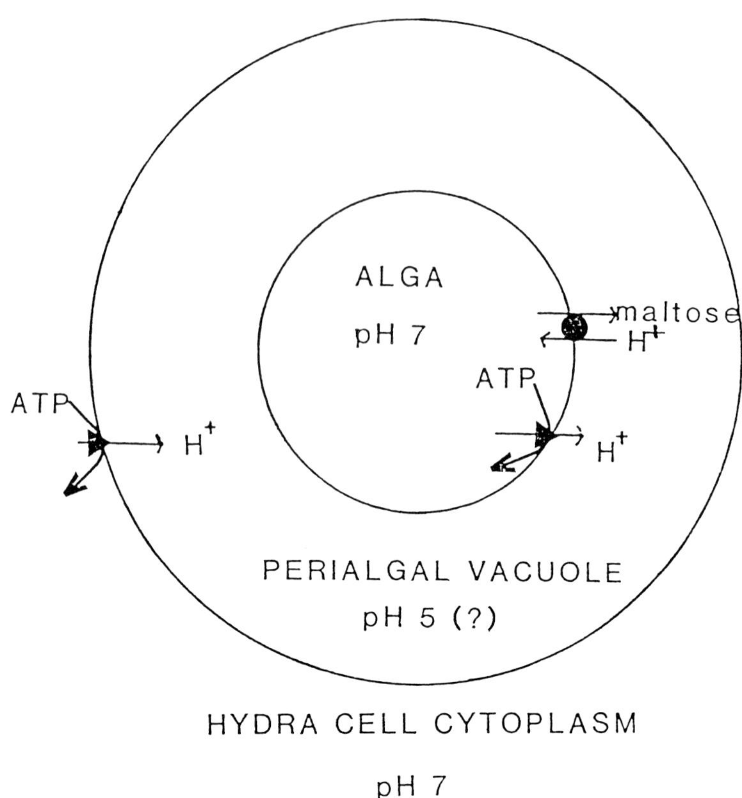

Figure 2 pH profile in the hydra-*Chlorella* symbiosis. A perialgal pH of 5 units has been inferred from ^{14}C tracer studies; and the electrochemical gradients between the vacuole and cytoplasm of the hydra cell (pH 7) and algal symbiont (pH 7) are maintained by H^+-translocating ATPases in the vacuolar membrane and algal plasmamembrane. The pH dependence of maltose release suggests that release is directly linked to H^+ flux across the algal plasmamembrane. It may be coupled directly to H^+ efflux (i.e. maltose/H^+ symport) or alternatively coupled to the passive influx of H^+ (i.e. maltose/H^+ antiport), as shown here, with the electrochemical gradient of protons across the plasmamembrane as the ultimate source of energy.

According to this model the active transport of protons into the perialgal vacuole, and hence low perialgal pH, represent a signal to which symbiotic *Chlorella* cells respond by the release of maltose. As a consequence, the cells retain sufficient carbon and energy to fuel only low rates of growth and division, comparable to those of the host. Shifts in the relative biomass of the partners would result from transient changes in the perialgal pH.

4. Nutritional status of the algal cells in symbiosis

An important factor influencing the ability of symbiotic algal cells to respond to the 'permissive' pH, as proposed in the pH model above, is the status of the cells with respect to nutrients other than carbon. In culture, strains of symbiotic *Chlorella* can grow autotrophically on inorganic nitrogen sources at rates comparable to non-symbiotic *Chlorella* and have no unusual nutritional requirements (Douglas & Huss, 1986). No consistent differences in the nutritional requirements or capabilities of symbiotic *Chlorella* which release maltose at high and low rates have been identified. In the intact symbiosis, the algal cells, of necessity, derive all their nutrients from the surrounding host cell and tracer studies indicate that ^{3}H in the food (*Artemia*) of hydra is translocated to the algal cells (Thorington & Margulis, 1981).

Various lines of evidence suggest that the nutritional status of the algal cells is of critical importance in determining the capacity of the algal cells to divide (McAuley, 1985). In one experiment, hydra were fed on the standard diet of live *Artemia* or 'extracted' *Artemia*, in which the protein content was reduced by 50%. The mitotic index of the hydra cells was increased from 0.6% to 1.8-2.1% in response to both types of food, a result that is consistent with the suggestion of David & Campbell (1972) that hydra cells may divide in response to stretching when the enteron is filled. By contrast, the mitotic index of the algal cells in hydra fed on extracted *Artemia* was 0.74%, half the value (1.58%) for hydra fed on live *Artemia*. These data suggest that the division of the algal cells depends on nutrients derived from the host's food. Although these results could be considered as consistent with a generalized enhancement of the nutritional status of the algal cells, permitting division (McAuley, 1985), McAuley favours the view that the food includes one specific component required for division that he terms a 'division factor'.

The nature of the putative division factor remains unknown, but it has been argued that it is a limiting nutrient and not comparable to the mitosis-promoting factors identified in animal cells. McAuley (1986b & unpublished) has evidence that one or more amino acids may be the major nitrogen source for growth in the symbiosis, and that the algal cells are nitrogen-limited.

The concept of a division factor as a signal from host to the algal symbionts which controls the timing of algal division has been considered as an alternative hypothesis to the pH model (McAuley, 1985; Smith & Douglas, 1987). However, the two ideas are not mutually exclusive. Perialgal pH and the availability of a limiting nutrient may play complementary roles in determining the equilibrium between algal growth and division. For example, Smith (1987) has suggested that the production of the division factor by the host or its receipt by the algal symbionts may be pH-dependent.

5. Signals from the algal symbionts to the host

A valuable approach to investigate signals which may pass from the algal symbionts to the host cell is to compare the characteristics of symbiotic and aposymbiotic hydra of the same strain; any difference is likely to arise from the presence of algal symbionts in the former. One intriguing difference is the higher activity of the ammonia-assimilating enzyme, glutamine synthetase, in the host tissue of symbiotic animals than in aposymbionts (Rees, 1986). Rees has interpreted this finding as evidence for a lower level of free ammonia in symbiotic hydra than aposymbionts, a view supported by the substantially lower rates of release (presumably by diffusion) of ammonia from symbiotic than aposymbiotic hydra (Rees, 1986). Symbiotic *Chlorella* do not assimilate ammonium in media of pH less than 5 units and therefore they probably cannot utilize ammonium in the intact association.

The low concentration of free ammonia in symbiotic hydra may be important in maintaining the stable association and sustained release of maltose from the algal symbionts (Rees,

1986 (Figure 3). Free ammonia would be expected to diffuse into the perialgal space and, under acidic conditions, become protonated to form ammonium. As membranes are virtually impermeable to the protonated base, the perialgal space would act as an 'acid trap' for its accumulation. This would result in a rise in the perialgal pH, with two consequences: firstly, the rate of maltose release would drop and the algal cells would retain a higher proportion of photosynthetically-fixed carbon to support growth; and secondly the algal symbionts may be able to utilize the ammonium as a nitrogen source for growth. The means by which algal growth and division are believed to be controlled (carbon and/or nitrogen limitation) would be circumvented. By this model, high levels of free intracellular ammonia are incompatible with an established symbiosis, by the criteria of regulated algal growth and release of maltose (see Introduction).

Insight into the mechanisms underlying the enhanced assimilation of ammonium by symbiotic hydra comes from the associations with heterologous algae. Hydra containing algae which release maltose at high rates (e.g. strain 3N8/13-1 have levels of glutamine synthetase comparable to or somewhat higher than those with native symbionts, but in those with low maltose releasing algae, the glutamine synthetase activity is similar to values obtained for aposymbionts. It appears that maltose release, or an unknown character correlated with maltose release, is the signal which alters the nitrogen metabolism of the animal. There remain the problems of the mechanisms by which receipt of sugars by the host is translated into a change in nitrogen metabolism and, secondly, the ultimate fate(s) of the amide group assimilated into glutamine by glutamine synthetase.

Concluding remarks

This contribution has described several interlinked signals between the partners in the hydra-*Chlorella* symbiosis that may mediate the persistence of the stable association in which the algal symbionts release substantial amounts of maltose. The proposed signals from host to algae are (a) proton flux into

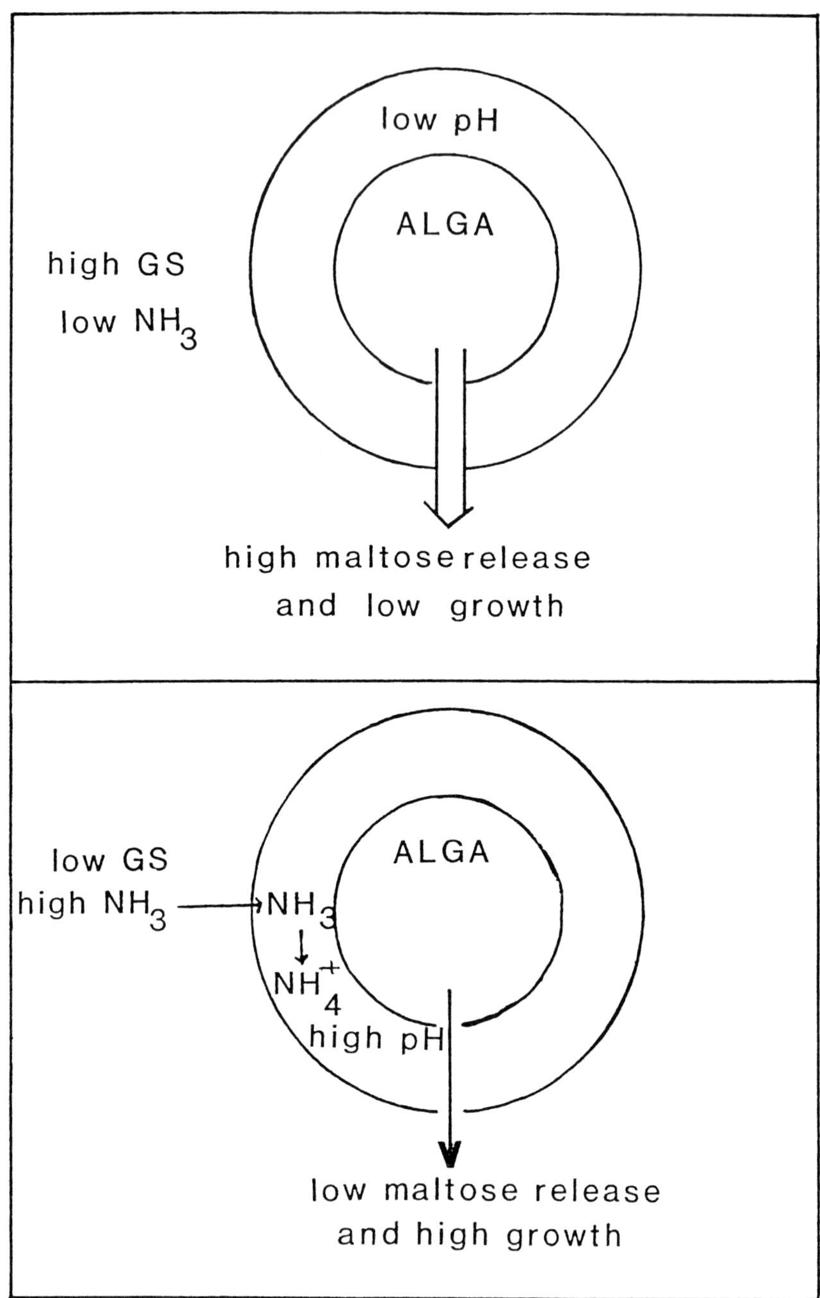

Figure 3 Role of ammonia in the maintenance of the stable hydra symbiosis. a. condition in established association with low levels of free ammonia and high glutamine synthetase (GS) activity in the host fraction. b. hypothetical condition in association between hydra and high-maltose releasing alga, with high free ammonia. See text for details.

the perialgal space and (b) limiting nitrogenous nutrients. Maltose may be a signal passing from the algal cells to the host. Specifically, it has been argued that maltose release is induced by low perialgal pH, and that the loss of algal carbon and/or nitrogen limitation is important in the control of growth and division of the algal cells. The low perialgal pH and nitrogen-limited status of the algal symbionts are sustained by low free ammonia concentration in the host cytoplasm, which in turn results from altered nitrogen metabolism, notably enhanced glutamine synthetase activity. The algal-derived signal mediating this change may be the release of sugars.

The signalling between the hydra and *Chlorella* cells which regulates the growth and division of algal symbionts differs from the classical concept of signalling (see Introduction) in two important respects. Firstly, the proposed signals (perialgal pH, maltose and possibly amino acids) are intrinsically linked to the nutritional relationship between the partners. This contrasts with the various signals which control growth and division in animal systems (e.g. conventional hormones, growth factors), the sole function of which is the change in proliferation rate of target cells. The second difference relates to the interactions between the signal and target cell. Although further studies are required to establish in detail the nature of the interactions in the hydra symbiosis, it is not necessary to invoke specific receptors which bind the signals. In other words, the interactions between signal and target cells mediating the maintenance of the stable symbiosis do not conform to the conventional interpretation of signalling and recognition.

References

David, C.N. & Campbell, R.D. (1972). Cell cycle kinetics and development of *Hydra viridis*. I. Epithelial cells. *J. Cell Sci*. 11, 557-568.

Douglas, A.E. (1987). Alga-invertebrate symbiosis. *Ann. Phytochem. Soc. Europe* 29 (in press).

Douglas, A.E. & Huss, V.A.R. (1986). On the characteristics and taxonomic position of symbiotic *Chlorella*. *Arch. Microbiol*. 145, 80-84.

Douglas, A.E. & Smith, D.C. (1983). The cost of symbionts to the host in the green hydra symbiosis. In _Endocytobiology, Endosymbiosis & Cell Biology_ (ed. W. Schwemmler & H.E.A. Schenk) pp. 631-648. Walter de Gruyter, Berlin.

Douglas, A.E. & Smith, D.C. (1984). The green hydra symbiosis. VIII. Mechanisms in symbiont regulation. _Proc. R. Soc. Lond. B_ 221, 291-319.

McAuley,P.J. (1982). Temporal relationships of host cell and algal mitosis in the green hydra symbiosis. _J. Cell Sci._ 58, 423-431.

McAuley,P.J. (1985). The cell cycle of symbiotic _Chlorella_. I. The relationship between host feeding and algal cell growth and division. _J. Cell Sci._ 77, 225-239.

McAuley,P.J. (1986a). Glucose uptake by symbiotic _Chlorella_ in the green hydra symbiosis. _Planta_ 168, 523-529.

McAuley, P.J. (1986b). Uptake of amino acids by cultured and freshly isolated symbiotic _Chlorella_. _New Phytol_. 104, 415-427.

Mews, L. & Smith, D.C. (1982). The green hydra symbiosis. VI. What is the role of maltose transfer from alga to animal? _Proc. R. Soc. Lond. B_ 216, 397-413.

Rees, V.A.R. (1986). The green hydra symbiosis and ammonium. I. The role of the host in ammonium assimilation and its possible regulatory significance. _Proc. R. Soc. Lond. B_ 119, 219-234.

Smith, D.C. (1987). Regulation and change in symbiosis. _Ann. Bot_. (in press).

Smith, D.C. & Douglas, A.E. (1987). _The Biology of Symbiosis_. Edward Arnold.

Thorington, G. & Margulis, L. (1981). _Hydra viridis_: transfer of metabolites between _Hydra_ and symbiotic algae. _Biol. Bull_. 160, 175-188.

THE ESTABLISHMENT OF ALGAL/HYDRA SYMBIOSES - A CASE OF RECOGNITION OR PREADAPTATION ?

M. Rahat and V. Reich - Department of Zoology,
The Hebrew University of Jerusalem,
91904 Jerusalem, Israel

The various symbioses we find in nature are supposed to be host/symbiont specific as a result of cell-to-cell signals, and recognition of the symbionts by the host (See this volume).

In the algal/hydra symbioses, host/symbiont specificity is apparently expressed by the restricted number of participating species. To date we know only of a few species within two genera of unicellular green algae that can live and reproduce inside the cells of hydra, and not all species of hydra can host these algae (Rahat & Reich, 1986b).

The early studies of host/symbiont specificity in the algal/hydra symbiosis were carried out by injection of free-living or freshly isolated symbiotic algae into aposymbiotic *Hydra viridis*. Results showed that the latter algae were "accepted" by the hydra and for the free-living algae it has been reported that they disappear from the hydra within 1-2 days (Park et al, 1967; Pardy & Muscatine, 1973; Muscatine et al, 1975; Jolley & Smith, 1980). The obvious conclusion was that some characteristics present in symbiotic algae are missing in free-living algae which are therefore not recognized by the host and thus cannot form symbioses (Muscatine et al, 1975; McAuley & Smith, 1982a). These characteristics were supposed to be the excretion of maltose (Cernichiari et al, 1969), specific

NATO ASI Series, Vol. H17
Cell to Cell Signals in Plant, Animal and
Microbial Symbiosis. Edited by S. Scannerini et al.

antigenic determinants in the algal cell wall (Pool, 1979; Meints & Pardy, 1980), specific cell surface charges (McNeil et al, 1981), and the ability of the phagocytosed algae to prevent lysosome-phagosome fusion (O`brien, 1982; Hohman et al, 1982).

With regard to the same characteristics however, it has been reported that some symbiotic algae release "no detectable" amounts of maltose (Mews & Smith, 1982; Douglas & Smith, 1984), that "Algae would be taken up by digestive cells regardless of their antigenic composition" (Pool, 1979), and McNeil & Smith (1982) concluded that "There was no evidence for specific recognition of symbiotic algae during phagocytosis" (see also Rahat & Reich, 1987).

Nevertheless, self/non-self recognition is an established fact even in invertebrates (Coombe et al, 1984), and phagocytosis of non-self hydra cells by epithelial cells of H. attenuata and H. oligactis has recently been reported and shown to be mediated by cell surface molecules (Bosh & David, 1986).

One must assume however, that algae and hydra lived separately as non-symbionts before they joined to form the present day symbioses (Rahat, 1985b). Are the ascribed cell-to-cell signals and recognition essential for the initial formation of the specific algal/hydra symbioses or are they the result of later host/symbiont coevolution ?

"New" algal/hydra symbioses

In a study of initiation of "new" algal/hydra symbioses we used larvae of the brine shrimp Artemia sp. as a vector to infect hydra with various algae and showed that H. viridis forms stable symbioses with a variety of free-living chlorellae. The latter could grow in axenic cultures *in vitro*, in media enriched with organic nutrients, even at very sparse inocculations. Under these conditions no growth has been obtained with chlorellae that could not live as endosymbionts in hydra (Rahat & Reich, 1984; 1985a). Free-living algae phagocytosed by digestive cells of a hydra probably encounter a similar nutrient-rich environment inside the cell vacuole. It is within the vacuole that the preadapted ability of the algae to live and reproduce, or to change the intravacuolar environment according to their needs, determines their future as endosymbionts.

The ability of organisms to affect their environment has been reported for the open sea (Lucas, 1961). We can similarly conceive of an effect of phagocytosed algae on the contents of the phagosome, preventing its fusion with lysosomes (Hohman et al, 1982).

We described recently a new algal/hydra symbiosis: a *Chlorococcum*-like endosymbiotic algae living in the digestive cells of the Japanese brown *H. magnipapillata* (Rahat and Reich, 1985b). At variance with "native" symbiotic Chlorellae that cannot be grown outside their host, we could isolate the symbiotic chlorococci from *H. magnipapillata* and grow them *in vitro* in nutrient enriched media. Also at variance with symbiotic chlorellae, chlorococci were lost from the hydra if grown in the dark, apparently because these algae do not get all their required nutrients from the hosting cell. When green *H. magnipapillata* were starved in the light, the algae overgrew their host and the hydra disintegrated (Table 1). As compared with the *Chlorella/H. viridis* symbiosis which survives starvation in the light for many months and remains green in the dark (Rahat & Reich, 1980), the Chlorococcum/H. magnipapillata symbiosis is apparently "young" and mutual nutritional coevolution of the cosymbionts has not yet been completed.

Table 1: Host/symbiont interaction in H. magnipapillata.

Condition			Algae	Hydra
Feeding -	continuous	light	Abundant	Green
" -	"	Darkness	Disappear	Aposymbiotic
Starvation -	"	light	Overgrow	Disintegrate

When aposymbiotic H. viridis were infected with freshly isolated symbiotic chlorellae many algae were retained in the digestive cells and the hydra attained their full complement of symbionts within one week (Muscatine et al, 1975). However, following infection of H. viridis with free-living chlorellae or infection of H. magnipapillata with symbiotic chlorococci cultured in vitro, relatively few algae are initially retained, and several weeks might pass before "greening" of the hydra is completed.

It has been shown that as in any population, physiological variation and mutations occur also in algal populations (Hochberg et al, 1972). Accordingly we conclude from our data that when free-living algae are phagocytosed by hydra only those algae survive that are preadapted to live and reproduce in the new endocellular environment. Similarly, in a population of hydra infected with free-living algae not all will "green", and some will remain aposymbiotic.

We infected aposymbiotic _H. viridis_ and _H. magnipapillata_, and non-symbiotic African, Australian and European strains of _H. attenuata_ and _H. vulgaris_ with symbiotic and free-living _Chlorella_ and _Chlorococcum_ species (Rahat and Reich, 1986b). With the latter algae we could obtain green _H. attenuata_ and _H. vulgaris_ and symbiotic chlorococci persisted in cells of _H. viridis_ for more then six days (Table 2).

As we have hydra that host some symbiotic algae only and symbiotic algae that can live in other hydra only, we may conclude that the ability to "live together" requires mutual preadaptation.

When we infected aposymbiotic _H. viridis_ with more then one algal species chimeric double infections were obtained. Ensuing interalgal competition determined which alga remained as the sole endosymbiont. In such competitions "native" symbiotic algae succeeded over free-living species (Rahat, 1985a). It has similarly been reported that established "native" symbiotic chlorellae in _H. viridis_ preserve their endocellular "territory" against foreign invading algae (McAuley & Smith, 1982b).

Colonization, competition, succession and territorialism are well known ecological phenomena in abiotic habitats. The occurrence of the above in endocellular symbiosis shows the latter to be an ecological phenomenon (Rahat and Reich, 1986b). Obviously, identical problems are involved in the colonization of biotic-intracellular and abiotic "open" habitats.

Table 2. Infection, persistance of algae and formation of stable symbioses in brown and green hydra with free-living and symbiotic *Chlorella* and *Chlorococcum*.

Infected hydra(1)	Infecting algae: Symbiotic Chlorella(2)	Free living Chlorella(3)	Symbiotic Chlorococcum(4)	Free living Chlorococcum(5)
H. magnipapillata (Jap)	-	-	++	-
H. attenuata (Aus)	-	-	+	-
H. vulgaris (Afr)	-	-	++	-
H. vulgaris (Eur)	-	-	+-	-
H. vulgaris (")	-	-	-	-
H. oligactis (")	-	-	+-	-
H. viridis (Sws)	++	-	+-	-

1. Geographic origins of the various hydra is shown in brackets: Jap = Japan, Aus = Australia, Afr = South Africa, Eur = Europe. Sws = Swiss. 2. Eight free-living strains of *in vitro* cultured *Chlorella* that form stable symbioses with Swiss aposymbiotic *H. viridis* (Rahat & Reich, 1985a), and symbiotic chlorellae freshly isolated from three strains of this hydra. 3. Nine free living strains of *Chlorella* that do not persist in aposymbiotic Swiss *H. viridis* (Ibid). 4. *In vitro* cultured *Chlorococcum* sp. isolated from *H. magnipapillata*. 5. Eight species of free-living *Chlorococcum* obtained from the Cambridge Algal Collection. (-) Infecting algae disappear within three days. (+-) Some algae persist in cells of hydra for more then six days. (+) Symbiosis, algae reproduce and are passed on to buds, hydra pale to the naked eye. (++) Symbiosis, hydra green to the naked eye.

When we establish and study new symbioses we should distinguish between the ability of one species to host or be hosted by the other, and the probability of that symbiosis to compete and survive in nature under given conditions. Algae-hosting hydra that would survive during prolonged starvation (Rahat & Reich, 1980), might not survive if conditions change. Native symbiotic chlorellae in _H. viridis_ outgrow their host in the presence of certain ions (Muscatine & Neckelmann, 1981; Neckelman and Muscatine 1983), and as shown in Table 1, the outcome of _Chlorococcum/H. magnipapillata_ interactions differs in the light and in the dark and depends on the availability of prey.

The ability to "live together" represents intrinsic preadaptations of the potential cosymbionts while ecological fitness of the formed symbiosis depends on the surrounding environment and competing biota.

The reason we do not find green _H. attenuata_ or _H. vulgaris_ in nature might be due to a "failure to meet" with the Japanese _Chlorococcum_ or the "right" chlorellae (Rahat & Reich, 1986b). The "cost of symbiosis" to the host (Douglas & Smith, 1983), under given conditions might also impair the survival of "new" symbioses in nature.

Intraorganismic ecology of symbiotic algae

The various strains and species of chlorellae and _Chlorococcum_ are respectively hosted in _H. viridis_ and _H. magnipapillata_, but only in the endodermal cells of the body and tentacles. The question of host/symbiont specificity might thus be asked also at the cellular level: why is it that only certain cells host algae while other cells remain algae free?

The green H. magnipapillata has been collected in nature, and several mutants of this species were obtained by inbreeding in the laboratory (Sugiyama & Fujisawa, 1977; Sugiyama, 1983). We infected eight such mutants with chlorococci isolated from the green H. magnipapillata: Reg-16; L-4; Ms-1; Sf-1; Maxi-1, Nem-3; Mini-4 and Mh-1. The latter four mutants formed stable symbioses with this alga but differed from each other in the distribution pattern of the symbiotic chlorococci (Fig, 1). In Maxi-1 most algae were in the hypostome with very few in other parts of the hydra. Nem-3 hosted relatively less algae, most of them in the hypostome and in cells somewhat below the tentacles. Mini-4 were completely green and the algae were almost evenly distributed in the polyps. The multiheaded Mh-1 mutants were also green but with few algae only in the hypostome and at the junction of the "heads". In budding and regenerating mutants respectively, corresponding changes in number of algae/host-cell could be followed, e.g. in Maxi-1 and Nem-3 the number of algae increasing in differentiating hypostome cells.

From the pattern of algal distribution found in the mutants of H. magnipapillata, the different number of algae hosted in different endodermal cells, and the changes in number of algae/host-cell corresponding to cell differentiation in budding and regenerating hydra, we must assume an algal/host-cell specificity based on intraorganismic and intracellular nutritional ecology. The intracellular environment of algae-hosting cells obviously differs from that of "nonsymbiotic" cells even in the same hydra and whatever the differences are, they are qualitative as well as quantitative. The different mutants of H. magnipapillata promise to be of special value for a comparative study of algae-hosting cells of hydra.

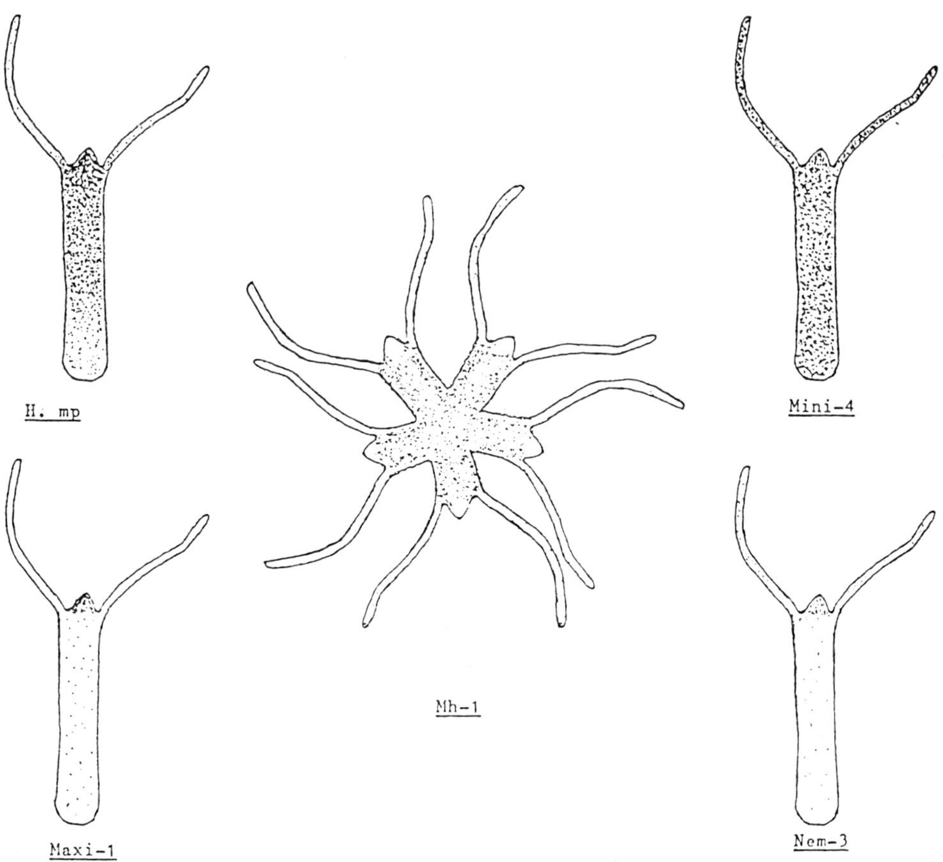

Figure 1: Distribution pattern of symbiotic chlorococci in mutants of H. magnipapillata.

One way to study the intracellular environment would be the use of the various symbiotic chlorellae and Chlorococcum as biochemical probes (Shubert, 1984). Any factor required by the respective algae in axenic culture *in vitro* in defined nutrient media, obviously shows the availability to the algae of that factor, or of its equivalent, inside the host cell vacuole (Rahat & Reich, 1985a; 1986b).

From the above data we conclude that the establishment of algal/hydra symbioses, and of other symbioses as well, and host/symbiont specificity, are the result of chance meetings of preadapted potential cosymbionts. Principles of ecology function also in intraorganismic habitats and determine the consequent formation of stable symbioses. Cell/cell signals and host/symbiont recognition obviously are the result of symbiotic coevolution.

In the exchange of scientific information, terms are used to convey ideas facts and conclusions. We suggest that the term "recognition" should not be used as related to the formation of symbioses and host/symbiont specificity.

To summarise the observed data we propose the "Test tube hypothesis": algae ingested into the vacuole of a hydra cell are subjected to a selection similar to that occurring in any abiotic habitat in nature or in a medium in a test tube. Algae that are preadapted to live and reproduce in the given environment or medium will survive. If their rate of reproduction equals or surpasses that of the hosting cell they will remain as successful colonizers, exposed to mutual selective coevolution toward the formation of a stable symbiosis.

REFERENCES

Bosh,T.C.G., David, C.N. 1986. Immunocompetence in Hydra: Epithelial cells recognize self-nonself and react against it. J. Exp. Zool. 238:225-234.

Cernichiari, E., Muscatine, L., Smith, D.C. 1969. Maltose excretion by the symbiotic algae of Hydra viridis. Proc. R. Soc. Lond. B. 173:557-576.

Coombe, D.R., Ey, P.L., Jenkin, C.R. 1984. Self-nonself recognition in invertebrates. Quart. Rev. Biol. 59:231-255.

Douglas, A., Smith, D.C. 1983. The cost of symbiont to their host in green hydra. In: Endocytobiology II: Eds. W. Schwemmler & H.E.A. Schenk; W. de Gruyter, Berlin, New York. pp. 631-648.

Douglas, A., Smith, D.C. 1984. The green hydra symbiosis. VIII. Mechanisms in symbiont regulation. Proc. R. Soc. Lond. B 221:291-319.

Hochberg, A., Pimstein, R., Rahat, M. 1972. Properties of an ethionine-resistant (ER) mutant of Ochromonas danica. J. Protozool. 19:66-69.

Hohman, T.C., McNeil, P.L., Muscatine ,L. 1982. Phagosome-lysosome fusion inhibited by algal symbiont of Hydra viridis J. Cell Biol. 94:56-63.

Jolley, E., Smith, D.C. 1980. The green hydra symbiosis. II. The biology of the establishment of the association. Proc. R. Soc. Lond. B 207:311-333.

Lucas, C.E. 1961. On the significance of external metabolites in ecology. Symp. Soc. Exp. Biol. 15:190-206.

McAuley, P.J., Smith, D.C. 1982a. The green hydra symbiosis. V. Stages in the intracellular recognition of algal symbionts by digestive cells. Proc. R. Soc. Lond. B. 216:7-23.

McAuley, P.J., Smith D.C. 1982b. The green hydra symbiosis. VII. Conservation of the host cell habitat by the symbiotic algae. Proc. R. Soc. Lond. B. 216:415-426.

McNeil, P.L., Hohman, T., Muscatine, L. 1981. Mechanisms of nutritive endocytosis. II. The effect of charged agents on phagocytic recognition of digestive cells. J. Cell Sci. 52:243-269.

McNeil,P.L., Smith, D.C. 1982. The green hydra symbiosis. IV. Entry of symbionts into digestive cells. Proc. R. Soc. Lond. B 216:1-6.

Meints, R.H., Pardy, R.L. 1980. Quantitative demonstration of cell surface involvment in a plant-animal symbiosis: Lectin inhibition of reassociation. J. Cell Sci. 43:239-251.

Mews, L., Smith, D.C. 1982. The green hydra suymbiosis. VI. What is the role of maltose transfer from algae to animal? Proc. R. Soc. Lond. B 216:397-413.

Muscatine, L., Neckelman, N. 1981. Regulation of numbers of algae in the <u>Hydra /Chlorella</u> symbiosis. Ber. Deutsch. Bot. Ges. BD. 94:571-582.

Muscatine, L., Cook, C.B., Pardy, R.L., Pool,R.R. 1975. Uptake recognition and maintenance of symbiotic chlorellae by <u>Hydra viridis</u>. Symp. Soc. exp. Biol., 29:175-203.

Neckelman, N., Muscatine, L. 1983. Regulatory mechanisms maintaining the <u>Hydra-Chlorella</u> symbiosis. Proc. R. Soc. Lond. B. 219:193-210.

O'Brien, T.L. 1982. Inhibition of vacuolar membrane fusion by intracellular symbiotic algae in _Hydra viridis_ (Florida strain). J. Exp. Zool. 223:211-218.

Pardy, R.L., Muscatine, L. 1973. Recognition of symbiotic algae by _H. viridis_. A quantitative study of the uptake of living algae by aposymbiotic _H. viridis_ . Biol. Bull. 145:565-579.

Park, H.D., Greenblatt,C.L., Mattern, C.F.T., Merril, C.R. 1967. Some relationships between _Chlorohydra_, its endosymbionts and some other chlorophyllous forms. J. Exp. Zool. 164:141-162.

Pool, R.R. 1979. The role of antigenic determinants in the recognition of potential algal symbionts by cells of _Chlorohydra_. J. Cell Sci. 35:367-379.

Rahat, M. 1985a. Competition between chlorellae in chimeric infections of _Hydra viridis_: the evolution of a stable symbiosis. J. Cell Sci. 77:87-92.

Rahat, M. 1985b. Origin and evolution of symbiosis in green hydra. Arch. Sc. Geneve. 38:385-399.

Rahat, M., Reich, V. 1980. Survival budding and osmotrophic nutrition in _Hydra viridis_. In: Developmental and Cellular Biology of Coelenterates. Eds. P. Tardent & R. Tardent; Elsevier, pp. 465-469.

Rahat, M., Reich, V. 1984. Intracellular infection of aposymbiotic _Hydra viridis_ by a free-living _Chlorella_ sp.: Initiation of a stable symbiosis. J. Cell Sci. 65:265-277.

Rahat, M., Reich, V. 1985a. Correlations between characteristics of some free-living Chlorella sp. and their ability to form stable symbioses with Hydra viridis. J. Cell Sci. 74:257-266.

Rahat, M., Reich, V. 1985b. A new alga/hydra symbiosis: Hydra magnipapillata of the 'nonsymbiotic' Vulgaris group hosts a Chlorococcum-like alga. Symbiosis 1:177-184.

Rahat, M., Reich, V. 1986a. Algal/hydra symbiosis: a topic in ecology. In: Environmental quality and ecosystem stability. vol.III. Ed. Z. Dubinsky; Bar Ilan Univ. Press, Israel. pp.331-337.

Rahat, M., Reich, V. 1986b. Algal endosymbiosis in brown hydra: Host/symbiont specificity. J. Cell Sci. 86:273-286.

Rahat, M., Reich, V. 1987. A reevaluation of alleged requirements for the establishment of algal/hydra symbioses and host/symbiont specificity: preadaptation and vacuolar ecology. Endocyt.C.Res. 4:13-23.

Shubert, L.E. (ed). 1984. Algae as ecological indicators. Academic Press.

Sugiyama, T. 1983. Isolating hydra mutants by sexual inbreeding.In: Hydra: Research Methods. Ed. H.M. Lenhoff; Plenum Press, N.Y. pp. 211-221.

Sugiyama, T., Fujisawa, T. 1977. Genetic analysis of developmental mechanisms in hydra I. Sexual reproduction of Hydra magnipapillata and isolation of mutants. Develop. Growth and Differ. 19:187-200.

FACTORS PRODUCED BY SYMBIOTIC MARINE INVERTEBRATES WHICH AFFECT TRANSLOCATION BETWEEN THE SYMBIONTS

Rosalind Hinde, School of Biological Sciences,
The University of Sydney, N.S.W. 2006 Australia

Introduction

The translocation of nutrients between the partners is known to be a vital feature of the association in most symbioses involving close contact between the cells of the partners, and is probably important in all such associations. There is abundant evidence that this is a well-controlled process. Work so far suggests that, at least in marine invertebrates with algal symbionts, translocation is controlled largely by the animal host. In spite of much work in this area, the processes and signals involved are not well understood. This paper will examine evidence that even in the well-studied symbioses between invertebrates and dinoflagellates ("zooxanthellae") the control of translocation is not as simple as suggested by the literature, and will discuss some directions for future research and some potentially useful models.

In symbioses between zooxanthellae and invertebrates about 20-60% of the total photosynthetically fixed carbon is translocated to the host's tissues during photosynthesis (Muscatine, 1967; von Holt & von Holt, 1968a,b; Muscatine & Cernichiari, 1969; Trench, 1971a,b; Muscatine et al., 1972; Hoegh-Guldberg, 1981; Hoegh-Guldberg & Sutton, in preparation; and reviews of Smith, 1974; Trench, 1979 & Hinde, 1983). The actual proportion translocated varies between associations. Only a few, specific compounds are translocated to the host. The most important one is glycerol. Neutral amino acids (often alanine) are also translocated, and glucose and organic acids may be exported from the algae as well, although not all of these compounds are mobile in every association. If the zooxanthellae are isolated and washed clean of host debris, they leak only about 2-5% of their photosynthetic products into the suspending medium. But if homogenized host tissue is added to the washed, isolated zooxanthellae, they begin again to

NATO ASI Series, Vol. H17
Cell to Cell Signals in Plant, Animal and
Microbial Symbiosis. Edited by S. Scannerini et al.

release products of photosynthesis in amounts and types which are very similar to those translocated in the intact association. Homogenates are only effective during (and perhaps briefly after) a period of photosynthesis; if a homogenate of P. versipora is added to a suspension of its zooxanthellae 5 to 10 minutes after they have been removed from a medium containing $^{14}CO_2$, no more ^{14}C-labelled material is released into the medium than if they are suspended in seawater (Hoegh-Guldberg,1981). In similar experiments with Anthopleura elegantissima the rate of release into host homogenate added after a "pulse" of $^{14}CO_2$ is less than half that when host homogenate and $^{14}CO_2$ are added to the algae together (Trench, 1971b) and the export of glycerol decreases, while more alanine, glucose, lipids, and perhaps glycollate are released (Trench, 1971b; Muscatine et al., 1972). This appears to be because the mobile compounds are available for translocation only for a short period after they are formed. Host release factors may be produced only by cells which contain zooxanthellae; aposymbiotic individuals of the anemone Anthopleura elegantissima do not stimulate translocation by zooxanthellae from A. elegantissima, although symbiotic individuals do have host factor activity (Trench, 1971b). Boiling the host homogenate for 10 minutes or more usually abolishes its activity (Muscatine, 1967; Trench, 1971b; Hoegh-Guldberg, 1981; Hoegh-Guldberg & Sutton, in preparation) although this is not the case with Agaricia agaricites (Muscatine et al., 1972). These observations have led to the model that the host produces some specific substance or substances ("host release factor") which stimulates translocation, that it is formed in the host cells in the presence of the zooxanthellae and that it is a protein. These factors are not species-specific, since zooxanthellae from one species of host can be stimulated to release photosynthates into a homogenate of another host species, even where the two hosts are very different. For example, host homogenate from the coral Pocillopora damicornis stimulates release by zooxanthellae from the clam Tridacna crocea (Muscatine, 1967) and the host homogenate of the coral Plesiastrea versipora is effective in stimulating translocation by algae from Zoanthus robustus (Hoegh-Guldberg, 1981; Hoegh-Guldberg & Sutton, in preparation; Hinde & O'Brien, unpublished data).

The control of translocation has also been investigated in other associations between algae and invertebrates. In hydra (a freshwater cnidarian which contains symbiotic Chlorella) pH seems to play a major role in stimulating translocation; maltose is the main mobile compound (as in other associations involving Chlorella) (Cernichiari et al., 1969). Isolated symbiotic Chlorella translocate little or no maltose into the medium at pH 7; maximal rates of translocation are seen at pH 4 (Smith & Douglas, 1987). There have been reports that host homogenates of Hydra tissue have no effect on translocation by isolated Chlorella symbionts (Mews, 1980), that they inhibit it (Cernichiari et al., 1969) and that they stimulate it (but only if gastrodermal tissue alone is used to make the homogenate) (Yu & Dietrich, 1977). In a comparison of cultured symbiotic Chlorella in the stationary and exponential phases of growth, Mews (1980) found no evidence that restriction of growth could stimulate release of photosynthates from Chlorella. The rate of translocation appeared to depend mainly on the availability of excess photosynthate and there was no evidence that the animal could regulate it. In the association between the marine flatworm Convoluta roscoffensis and the prasinophyte alga Tetraselmis convolutae up to 50% of the products of photosynthesis are translocated to the animal's tissues as amino acids (mainly alanine) (Muscatine et al., 1974; Boyle & Smith, 1975). While isolated T. convolutae leaked about 5% of their freshly fixed carbon to an inorganic medium, isolated symbionts in crushed worm tissues leaked 5-16%, suggesting that the worms may produce some sort of release factor (Muscatine et al., 1974). The sacoglossan mollusc Elysia viridis, which retains functional chloroplasts from its food plant, Codium fragile, also shows host release factor activity (Gallop, 1974). The active material is soluble and heat labile. Homogenates of the anemone Anemonia sulcata (which has zooxanthellae) stimulate transport of photosynthetic products from Codium chloroplasts, and homogenates of E. viridis stimulate translocation by zooxanthellae from A. sulcata (Gallop, 1974). For the prokaryotic alga Prochloron, which is an extracellular symbiont of tropical ascidians, release rates of between 15% and 20% have been reported (Alberte et al., 1986), although other authors have obtained lower values (Fisher

& Trench, 1980; Griffiths & Thinh, 1983). Homogenates of foraminiferans with symbiotic diatoms contain factors which affect the formation of the cell walls of the diatoms and stimulate the release of large amounts of photosynthetic products (Lee *et al.*, 1984).

Short-term experiments using carbon-14 as a tracer suggest that a maximum of about 60% of the total products of photosynthesis is translocated to the host. Yet in hosts with zooxanthellae as symbionts, carbon budgets (calculated, using a number of assumptions, according to the model of Muscatine *et al.*, 1983) indicate that up to 98.6% of the total carbon fixed in photosynthesis is translocated to the host (estimates for the coral *Stylophora pistillata* growing in well-lit areas, Muscatine *et al.*, 1983). If the model and assumptions are correct, some other form of translocation must be occurring in addition to the movement of soluble metabolites. There is evidence that fatty acids and triglycerides play important roles in translocation in symbioses involving zooxanthellae (Battey & Patton, 1984).

Host Release Factors in Three Temperate Associations Between Invertebrates and Zooxanthellae

The scleractinian coral *Plesiastrea versipora*, the zoanthid *Zoanthus robustus* and the nudibranch *Pteraeolidia ianthina* occur in Port Jackson, New South Wales. Each has zooxanthellae which have the characteristics of *Symboiodinium microadriaticum*, as defined by Taylor (1973). However, Blank & Trench (1985) have provided evidence that zooxanthellae from various host species should be regarded as belonging to different species. The algae found in *P. versipora*, *Z. robustus* and *Pa. ianthina* have not been studied in enough detail for us to state that they belong to different species, but it seems very likely that they do. *Pa. ianthina* is also interesting because it appears to acquire its symbionts by feeding on a cnidarian, which is the primary host of the zooxanthellae (Rudman, 1982).

In each of these symbioses the algae photosynthesize normally while in the animals, and translocate about 25-50% of their total fixed carbon to the animal host (Hoegh-Guldberg & Hinde, 1986; Hoegh-Guldberg & Sutton, in preparation). In each case glycerol is the major product translocated; the minor constituents of the translocated material are characteristic of each species (Hoegh-Guldberg & Hinde, 1986; Hoegh-Guldberg & Sutton, in preparation). Zooxanthellae from these three hosts stop translocating products of photosynthesis when they are isolated from their hosts (Hoegh-Guldberg, 1981; Hoegh-Guldberg & Sutton, in preparation; Hinde, unpublished). There the similarities between them end.

Plesiastrea versipora is "typical" of associations involving zooxanthellae. In other words, addition of homogenized host tissue causes the isolated zooxanthellae to begin to translocate glycerol, leucine and small amounts of other compounds into the medium; in half-hour incubations 25-55% of the total photosynthetic products are translocated (Hoegh-Guldberg, 1981; Hoegh-Guldberg & Sutton, in preparation; Hinde, unpublished). Thus the effects of the host homogenate on the algae are similar to the effects of being in the host cell, and it is therefore postulated that the activity of the homogenate is similar to the process which controls translocation in the intact host. Regular sampling over two years shows that the algae photosynthesize and release organic compounds into host homogenate at about the same rates in all seasons (Hinde, unpublished).

The zooxanthellae in Zoanthus robustus release about 5% of their photosynthetic products into the medium when suspended in seawater; when suspended in homogenized tissue of Z. robustus they show a significant but very slight increase in release (to about 7%) (Hoegh-Guldberg, 1981; Hoegh-Guldberg & Sutton, in preparation; Hinde & O'Brien, unpublished). They are, however, stimulated to release products of photosynthesis by the addition of a homogenate of Plesiastrea versipora (Hoegh-Guldberg, 1981; Hoegh-Guldberg & Sutton, in preparation; Hinde & O'Brien, unpublished). Again, regular sampling over two years has shown that these results are reproducible and are consistent throughout the year (Hinde & O'Brien, unpublished).

Isolated zooxanthellae from Pteraeolidia ianthina, like those of Zoanthus robustus, show a slight but significant stimulation of release by homogenates of Pa. ianthina (about 2% in seawater, and 2-11% in host homogenate) (Hoegh-Guldberg, 1981; Hoegh-Guldberg & Sutton, in preparation). But, unlike those of Z. robustus, the zooxanthellae of the nudibranch show no response at all to homogenates of the tissue of the coral Plesiastrea versipora (Hoegh-Guldberg, 1981; Hoegh-Guldberg & Sutton, in preparation).

There are a number of possible explanations for these results. (1) Z. robustus and Pa. ianthina may produce host release factors similar to those of P. versipora but these may be inactivated during separation of the animal and algal tissue. This seems unlikely, as diluting host homogenate from P. versipora with that from the zoanthid does not affect its activity (Hoegh-Guldberg, 1981). Since the zooxanthellae from Pa. ianthina do not respond to coral host homogenate, it is unlikely that Pa. ianthina produces such compounds. (2) Z. robustus and Pa. ianthina may control translocation by a mechanism which is destroyed by the physical disruption of the host cell. Such mechanisms might include physical interactions between the membrane of the perialgal vacuole and the plasmalemma of the alga, or the maintenance of electrical, pH or concentration gradients. (3) Translocation in this association may be due to restriction of algal growth, leading to the excretion of excess photosynthate (see below). Restriction of growth might be controlled by the host or be a fortuitous result of symbiosis.

Results with these three associations thus show that the control of translocation in symbioses between invertebrates and zooxanthellae is more complex than previous work suggests, since it is not the same in every association. At the start of this work, we had no reason to suppose that these three associations would differ in the physiology of translocation. Yet, although in each of them the zooxanthellae translocate large amounts of photosynthates in the intact association, and this translocation occurs mainly during photosynthesis and mainly as glycerol (Hoegh-Guldberg, 1981; Hoegh-Guldberg & Sutton, in preparation; Hinde, unpublished), each apparently has a different mechanism of control. To summarize, P. versipora produces a host release factor, and its algae are competent to respond to it. The algae from Z. robustus do not respond to homogenates of their own host, but are competent to respond to *P. versipora* homogenates; this response is a "typical" zooxanthellar reaction to host factor. The algae from *Pa. ianthina* respond neither to their own host's tissue nor to *P. versipora* homogenates. These results suggest that both properties of the algae and properties of the host are involved in the mechanism of translocation.

Attempts to isolate the host release factor from *Plesiastrea versipora* have resulted in partial purification of the active factor, and in evidence that it has two or more components; at least one of these has a molecular weight greater than 8,000 daltons, and one has M.W. less than 8,000 d. (Hinde, unpublished data). Isolation of the active compounds will allow study of the mechanism by which they induce translocation, and of the reasons that some zooxanthellae respond and others do not.

Host Release Factor Activity: Some Models

There are several possible ways in which a soluble host release factor might stimulate translocation. Firstly, it may act as a signal which specifically affects permeability or particular transport systems in the plasmalemma of the algal cells. Secondly, it may trigger a change in the metabolism of the algae so that the mobile compounds are synthesized in much greater amounts than usual, and in greater amounts than can be assimilated; this would make the excess available for translocation, either by diffusion or by the ordinary mechanisms of transport of the relevant compounds. Thirdly, it may affect the availability of one or more inorganic nutrients (such as N or P) to the algae, thus affecting their growth rate and ability to assimilate the organic carbon they produce during photosynthesis.

As pointed out by Hoegh-Guldberg & Sutton (in preparation), cell membranes are freely permeable to glycerol, and it is unlikely that any specific transport mechanism is involved in transferring it from the zooxanthellae to host cells. So it seems probable that the host release factor may not affect glycerol transport as such, but may instead act to increase the rate of its synthesis (see below). Neutral amino acids are the other important mobile compounds in associations between invertebrates and zooxanthellae. Unlike glycerol, they are actively transported in and out of cells (including zooxanthellae - Deane & O'Brien, 1981). It is possible that host release factors specifically affect the amino acid transport systems in the plasmalemma of the algae. Carroll & Blanquet (1984a,b) have shown that zooxanthellae from the scyphozoan Cassiopea xamachana take up alanine by active transport, and that the host produces a factor which inhibits this uptake. They suggest that this factor regulates the translocation of alanine from the host to the algae in the intact association. The inhibitory factor differs from the host release factor (which induces release of glycerol and alanine from the zooxanthellae) in that the effect of the inhibitory factor is not immediately reversed by washing the algae; this, with kinetic evidence, led Carroll & Blanquet (1984) to suggest that the inhibitory factor was not identical with the host release factor. However, the two acting together could well regulate the net translocation of alanine between the partners in response to environmental conditions such as the availability of nutrients and light for the algae and the availability of food for the host.

If stimulation of translocation of a particular compound occurs as a result of its increased mobility (that is, of increased permeability to the compound, or of the induction or stimulation of specific transport systems), and if the symbiotic algal cells normally maintain a pool of the mobile substance, they might respond to increased translocation by producing more of the mobile compound, so tending to maintain the optimal pool size. In this case, any increase in the amount of the mobile compound synthesized would be a secondary effect of the host release factor. On the other hand, if host release factor directly increases the rate at which the mobile compound is synthesized, and the extra material cannot be assimilated or stored it would have to be excreted (released) from the cells. In this case no direct effect of host factor on permeability or transport need be postulated. Since most cells are freely permeable to glycerol and it is not likely to be actively transported, the second type of model seems likely to be able to explain glycerol transport better than the first. Another piece of evidence which may support this idea is the failure of host homogenates from P. versipora to stimulate release when they are added to photosynthesizing zooxanthellae after the end of a pulse of $^{14}CO_2$ (in most experiments the algae have been exposed to host homogenate and $NaH^{14}CO_3$ simultaneously). If the host release factor were stimulating translocation from a reasonably large pool of glycerol in the algae some ^{14}C-glycerol (formed during the pulse) should be released when host homogenate is added. If, on the other hand, host release factor increases the proportion of the early products of photosynthesis used to synthesize glycerol, addition of host homogenate after the end of a period of photosynthesis would come too late to stimulate formation of the "excess" glycerol which is normally available for release in the presence of host release factor. The labelled early products of the Calvin cycle would already have been used to synthesize other compounds and labelled glycerol would not be released. However, it is not possible to distinguish between the models with tests of this kind without knowing the pool size and rate of turnover of glycerol, since if the pool of glycerol is small enough and turns over fast enough, labelled glycerol would only be available for a short time after its formation in any case. Data on pool sizes and rates of turnover are not currently available for any association. These models could be tested more effectively by examining the action of purified host release factor on isolated zooxanthellae under controlled conditions.

Several authors have remarked that zooxanthellae growing in their hosts appear to be nutrient limited, in spite of the known ability of isolated or cultured zooxanthellae to take up nitrate, ammonia, phosphate and the sorts of organic nitrogen compounds which would be readily available in the host (Muscatine, 1973; Summons & Osmond, 1981). When the growth of planktonic algae is limited by the availability of nutrients (at least in the cases of nitrogen, phosphorus and sulphur) they often start to excrete carbon-rich compounds (particularly organic acids, low molecular weight carbohydrates and polysaccharides) (Jensen, 1984). Changes in light, temperature and other environmental conditions can also increase excretion. Excretion is minimal during the exponential phase of growth (Jensen, 1984). Some planktonic algae excrete glycerol into the water (Jensen, 1984). Thus if the host limits the availability of one or more nutrients to the algae, they may respond with a flow of carbon-rich compounds. Such a mechanism is less likely to be able to account for the net export of nitrogen-rich amino acids which occurs from isolated zooxanthellae in host homogenates and which is generally assumed to occur in the host also. Zooxanthellae start to release photosynthetic products within 2 minutes after addition of host homogenate (Muscatine *et al.*, 1972). Unless zooxanthellae lack reserves of these nutrients, it is hard to see how this model could account for such rapid release by isolated zooxanthellae. As we have seen, while about 20-60% of the total fixed carbon is translocated as low molecular weight compounds, the carbon budget model suggests that more than 98% may be translocated altogether in associations involving zooxanthellae (Muscatine *et al.*, 1983). If this estimate is correct, what metabolites could account for this translocation (over and above that of soluble compounds), and how might their movement be controlled? Translocation of neutral lipids is probably important, and may account for most of the translocation in this category, at least in tropical species (Patton *et al.*, 1977; Blanquet *et al.*, 1979; Crossland *et al.*, 1980), and fatty acids may also be translocated (Blanquet *et al.*,1979). In associations between tropical invertebrates and zooxanthellae, the extrusion of droplets of neutral lipids by the zooxanthellae apparently plays an important role in translocation of lipids (Crossland *et al.*, 1980; Kellogg & Patton, 1983; Patton & Burris, 1983), although such droplets were not observed in the giant clam *Tridacna sp.* or the hydrozoan coral *Millepora exesa* (Patton & Burris, 1983). After a period of photosynthesis with $NaH^{14}CO_3$ in the light, the radioactivity in these lipids was mainly in the fatty acyl moieties rather than in the glycerol fraction of the triglycerides (Patton *et al.*, 1983), suggesting that the mobile soluble compounds and the mobile lipids are synthesized by different pathways. Battey & Patton (1984) suggest that the water soluble compounds and the lipids which are translocated to the host are both useful to it, but in different ways. The water soluble compounds appear to be used to support respiration, particularly in the cells which contain the algae, while the lipids act as energy stores, particularly in cells remote from the gastrodermis. The relative importance of each of these translocation processes to the metabolism of the host is yet to be determined (Battey & Patton, 1984), as is the role of lipid translocation in temperate associations, whose lipids are more diverse and less saturated than those of tropical cnidarians (Patton et al., 1977). There has been no investigation of the signals or other control mechanisms which might be involved in the extensive translocation of lipids in tropical hosts of zooxanthellae. So far there has been no study of the possible translocation of other large molecules (polysaccharides, proteins, nucleic acids etc.) and almost no work on the uptake of organic molecules by the zooxanthellae ("reverse translocation"), except for amino acid tranpsort (e.g. Deane & O'Brien, 1981; Carroll & Blanquet, 1984a,b). Models involving physical contact between membranes, or gradients of charge, pH or concentration have not been considered in any detail. The roles of Ca^{2+}, cyclic AMP and other signal systems have not been explored.

Conclusion

Detailed examination of apparently simple models has led to the uncovering of great complexity in the control of translocation between algae and invertebrates, even among the associations between dinoflagellates and marine invertebrates which were once thought to be so uniform. Except in those associations which produce a detectable host release factor, we know nothing of the signals which control translocation from symbiotic algae to their invertebrate hosts. There is still a great deal to be learnt about host release factors and about other cellular signals involved in translocation.

Acknowledgements

I am most grateful to I.O. Hoegh-Guldberg and D.C. Sutton for permission to quote their unpublished work so extensively. Much of it was done at the Roche Research Institute of Marine Pharmacology in Dee Why, Sydney, and the rest in my laboratory. It gives me much pleasure to thank E.J. O'Brien for her excellent research assistance. My own work has been supported by the Australian Research Grants Scheme and by the University of Sydney.

References

ALBERTE, R.S., L. CHENG & R.A. LEWIN (1986) Photosynthetic characteristics of Prochloron sp./ascidian symbioses I. Light and temperature responses of the algal symbiont of Lissoclinum patella. Mar. Biol. 90: 575-587.

BATTEY, J.F. & J.S. PATTON (1984) A reevaluation of the role of glycerol in carbon translocation in zooxanthellae-coelenterate symbiosis. Mar. Biol., 79:27-38.

BLANK, R.J. & R.K. TRENCH (1985) Speciation and symbiotic dinoflagellates. Science, 229:656-658.

BLANQUET, R.S., J.C. NEVENZEL & A.A. BENSON (1979) Acetate incorporation into the lipids of the anemone Anthopleura elegantissima and its associated zooxanthellae. Mar. Biol., 54:185-194.

BOYLE, J.E. & D.C. SMITH (1975) Biochemical interactions between the symbionts of Convoluta roscoffensis. Proc. Roy. Soc. Lond. B, 189:121-135.

CARROLL, S. & R.S. BLANQUET (1984a) Alanine uptake by isolated zooxanthellae of the mangrove jellyfish, Cassiopea xamachana. I. Transport mechanisms and utilization. Biol. Bull. 166:409-418.

CARROLL, S. & R.S. BLANQUET (1984b) Alanine uptake by isolated zooxanthellae of the mangrove jellyfish, Cassiopea xamachana. II. Inhibition by host homogenate fraction. Biol. Bull., 166:419-426.

CERNICHIARI, E., L. MUSCATINE & D.C. SMITH (1969) Maltose excretion by the symbiotic algae of Hydra viridis. Proc. Roy. Soc. Lond. B, 173:557-576.

CROSSLAND, C.J., D.J. BARNES & M.A. BOROWITZKA (1980) Diurnal lipid and mucus production in the staghorn coral Acropora acuminata. Mar. Biol. 60: 81-90.

DEANE, E.M. & R.W. O'BRIEN (1981) Uptake of sulphate, taurine, cysteine and methionine by symbiotic and free-living dinoflagellates. Arch. Microbiol., 128:311-319.

FISHER, C.R. & R.K. TRENCH (1980) In vitro carbon fixation by Prochloron sp. isolated from Diplosoma virens. Biol. Bull., 159:636-648.

GALLOP, A.M. (1974) Evidence for the presence of a 'factor' in Elysia viridis which stimulates photosynthetic release from its symbiotic chloroplasts. New Phytol., 73:1111-1117.

GRIFFITHS, D.J. & L. THINH (1983) Transfer of photosynthetically fixed carbon between the prokaryotic green alga Prochloron and its ascidian host. Aust. J. mar. freshwat. Res., 34:431-440.

HINDE, R. (1983) Host release factors in symbioses between algae and invertebrates. pp709-726 in H.E.A. Schenk & W. Schwemmler (eds) Endocytobiology II - Intracellular space as oligogenetic system. Walter de Gruyter & Co., Berlin & N.Y.

HOEGH-GULDBERG, I.O. (1981). The translocation of newly fixed carbon from zooxanthellae in Plesiastrea versipora, Pteraeolidia ianthina and a zoanthid. Unpublished thesis, University of Sydney.

HOEGH-GULDBERG, I.O. & R. HINDE (1986) Studies on a nudibranch that contains zooxanthellae. I. Photosynthesis, respiration and the translocation of newly fixed carbon by zooxanthellae in Pteraeolidia ianthina. Proc. Roy. Soc. Lond. B, 228:493-509.

HOEGH-GULDBERG, I.O., R. HINDE & L. MUSCATINE (1986) Studies on a nudibranch that contains zooxanthellae. II. Contribution of zooxanthellae to animal respiration (CZAR) in Pteraeolidia ianthina with high and low densities of zooxanthellae. Proc. Roy. Soc. Lond. B, 511-521.

HOEGH-GULDBERG, I.O. & D.C. SUTTON Tissue extracts of four symbiotic invertebrates from temperate latitudes: effect on the photosynthesis and release of newly fixed carbon by isolated zooxanthellae. In preparation.

JENSEN, A. (1984) Excretion of organic carbon as function of nutrient stress. pp. 61-72 in Holm-Hansen, O., L. Bolis & R. Gilles (eds) Marine Phytoplankton and Productivity (Lecture Notes in Coastal and Estuarine Studies, No. 8). Springer Verlag, Berlin.

KELLOGG, R.B. & J.S. PATTON (1983) Lipid droplets, medium of energy exchange in the symbiotic anemone Condylactis gigantea: a model coral polyp. Mar. Biol., 75:137-149.

LEE, J.J., N.M. SAKS, F. KAPIOTOU, S.H. WILEN & M. SHILO (1984) Effects of host cell extracts on cultures of endosymbiotic diatoms from larger foraminifera. Mar. Biol., 82:113-120.

MEWS, L.K. (1980) The green hydra symbiosis. III. The biotrophic transport of carbohydrate from alga to animal. Proc. Roy. Soc. Lond. B, 209:377-401.

MUSCATINE, L. (1967) Glycerol excretion by symbiotic algae from corals and Tridacna and its control by the host. Science, 156:516-519.

MUSCATINE, L. (1973) Nutrition of corals. pp.77-115 in Jones, O.A. & R. Endean (eds) Biology and Geology of Coral Reefs, 2(1). Academic Press, N.Y. & Lond.

MUSCATINE, L. & E. CERNICHIARI (1969) Assimilation of photosynthetic products of zooxanthellae by a reef coral. Biol. Bull. 137:506-523.

MUSCATINE, L., R.R. POOL & E. CERNICHIARI (1972) Some factors influencing selective release of soluble organic material by zooxanthellae from reef corals. Mar. Biol., 13:298-308.

MUSCATINE, L., J.E. BOYLE & D.C. SMITH (1974) Symbiosis of the acoel flatworm Convoluta roscoffensis with the alga Platymonas convolutae. Proc. Roy. Soc. Lond. B, 187:221-234.

MUSCATINE, L., P.G. FALKOWSKI & Z. DUBINSKY (1983) Carbon budgets in symbiotic associations. pp.649-658 in Schenk, H.E.A. & W. Schwemmler (eds) Endocytobiology II - Intracellular space as oligogenetic system. Walter de Gruyter & Co., Berlin & N.Y.

PATTON, J.S., S. ABRAHAM & A.A. BENSON (1977) Lipogenesis in the intact coral Pocillopora capitata and its isolated zooxanthellae: evidence for a light-driven carbon cycle between symbiont and host. Mar. Biol., 44: 235-247.

PATTON, J.S. & J.E. BURRIS (1983) Lipid synthesis and extrusion by freshly isolated zooxanthellae (symbiotic algae). Mar. Biol., 75:131-136.

PATTON, J.S., J.F. BATTEY, M.W. RIGLER, J.W. PORTER, C.C. BLACK & J.E. BURRIS (1983) A comparison of the metabolism of bicarbonate-^{14}C and acetate 1-^{14}C and the variability of species lipid compositions in reef corals. Mar.Biol., 75:121-130.

RUDMAN, W.B. (1982) The taxonomy and biology of further aeolid and arminacean nudibranch molluscs with symbiotic zooxanthellae. Zool. J. Linn. Soc. Lond., 74:147-196.

SMITH, D.C. (1974) Transport from symbiotic algae and symbiotic chloroplasts to host cells. Symp. Soc. exp. Biol., 28:485-520.

SMITH, D.C. & A.E. DOUGLAS (1987) The biology of symbiosis. Edward Arnold, London.

STEWART, C.L. (1986) An investigation of the carbon budget model for symbiotic corals. Unpublished thesis, University of Sydney.

SUMMONS, R.E. & C.B. OSMOND (1981) Nitrogen assimilation in the symbiotic marine alga Gymnodinium microadriaticum: direct analysis of ^{15}N incorporation by GC-MS methods. Phytochem., 20:575-578.

TAYLOR,D.L. (1973) Algal symbionts of invertebrates. Ann. Rev. Microbiol., 27: 171-187.

TRENCH, R.K. (1971a) The physiology and biochemistry of zooxanthellae symbiotic with marine coelenterates. I. The assimilation of photosynthetic products of zooxanthellae by two marine coelenterates. Proc. Roy. Soc. Lond. B, 177:225-235.

TRENCH, R.K. (1971b) The physiology and biochemistry of zooxanthellae symbiotic with marine coelenterates. II. Liberation of fixed ^{14}C by zooxanthellae _in vitro_. Proc. Roy. Soc. Lond. B, _77_:237-250.

TRENCH, R.K. (1979) The cell biology of plant-animal symbiosis. Ann. Rev. Pl. Physiol., _30_:485-531.

VON HOLT, C. & M. VON HOLT (1968a) Transfer of photosynthetic products from zooxanthellae to coelenterate hosts. Comp. Biochem. Physiol., _24_: 73-81.

VON HOLT, C. & M. VON HOLT (1968b) The secretion of organic compounds by zooxanthellae isolated from various types of _Zoanthus_. Comp. Biochem. Physiol., _24_:83-92.

YU, S.L. & W.E. DIETRICH (1977) Effect of host homogenates on photosynthate excreti0n by zoochlorellae of _Hydra viridis_. Proc. Pennsylvania Acad. Sci., _51_:137-138.

Specificity in dinomastigote-marine invertebrate symbioses: An evaluation of hypotheses of mechanisms involved in producing specificity.

Robert K. Trench
Department of Biological Sciences
and the Marine Science Institute
University of California at Santa Barbara
Santa Barbara, California 93106, U.S.A.

INTRODUCTION

It is now fairly well recognized that the symbioses between microalgae and invertebrates demonstrate specificity (Trench, 1979, 1987). The realization that specificity existed in such associations was not forthcoming until the specific identities of the interacting organisms were resolved, and this depended to a large extent on the isolation, culture and taxonomic analysis of the algae involved. After isolation and culture of the 'zoochlorellae' from the flatworm *Convoluta roscoffensis*, Parke and Manton (1967) identified the algae as the prasinophyte *Tetraselmis* (*Platymonas*) *convolutae* (Hori *et al* 1982), and the 'zooxanthellae' symbiotic with *Convoluta convoluta* were identified as the diatom *Licmophora* sp. by Apelt (1969) after they were isolated and brought into culture. Based on the isolation and culture of symbiotic dinomastigotes, D.L. Taylor (1971, 1974) could distinguish between the amphidinioid and gymnodinioid symbionts. Thus, the symbionts in the flatworm *Amphiscolops langerhansi* were recognized as *Amphidinium klebsii*, while those in the chondrophore *Velella velella* were identified as *Amphidinium chattonii* and the symbionts in the radiolarian *Collozoum inerme* were also identified as *Amphidinium* sp. (but see Holland and Carré, 1974 and Trench and Blank, 1987). The gymnodinioid dinomastigotes that live in association with corals and clams and sea anemones etc. have been regarded as the single species *Symbiodinium*

NATO ASI Series, Vol. H17
Cell to Cell Signals in Plant, Animal and
Microbial Symbiosis. Edited by S. Scannerini et al.

microadriaticum since the establishment of that binomial by Freudenthal (1962) following the analysis in culture of the algae from the Caribbean jellyfish Cassiopeia. More recent studies (Blank and Trench, 1985 a,b; 1986; Trench and Blank, 1987) have established that there are more than one species of Symbiodinium. It is also now known that the symbiont of the foraminifer Orbulina universa is Gymnodinium béii (Spero, 1987). Hence there are now three distinct genera of symbiotic dinomastigotes associated with marine invertebrates, Amphidinium, Gymnodinium and Symbiodinium. Careful taxonomic analysis of the Chlorella complex of algae found in Hydra and Paramecium etc. has now been initiated (Douglas and Huss, 1986), but the taxonomic status of the green algae in Anthopleura xanthogrammica remains unresolved.

There are no reports of which I am aware, that show that under natural circumstances Velella harbours Symbiodinium or that the tridacnid bivalves or corals harbour Amphidinium, and the flatworm Amphiscolops appears to be always symbiotic with Amphidinium. Even though another flatworm, Haplodiscus sp. may simultaneously harbour an Amphidinium sp. and a Symbiodinium sp. (Trench and Winsor, 1987), the evidence available indicates that the same two algal species are always present. Again, although the plastids sequestered by the sea slug Elysia viridis may vary seasonally from being derived from Codium to a mixed population including rhodoplasts, the dominant plastid type found in E. viridis, Tridachia crispata and Placobranchus ianthobapsus are derived from the Siphonales (Trench, 1980). There are many other examples that can be cited, but they all point to the same conclusion, that there is specificity in microalgal-invertebrate symbioses, and that the processes involved in the interactions between the two components of the associations are not random ones.

Given that specificity in the associations between microalgae and invertebrates exists, it is reasonable to turn attention to analyses of the mechanisms involved in determining specificity. In considering such processes, it is necessary to explore several distinct aspects of the biology of symbioses and concepts that have been advanced to account for the observed specificity. Some of these include:

(i) the evolutionary initiation of symbioses between microalgae and invertebrates;

(ii) the mechanisms of transmission of symbionts from one sexual host generation to another;

(iii) the mechanisms of selection of symbionts by hosts that do not "inherit" a population of symbionts.

In the succeeding discussion, I shall attempt to illustrate the basic phenomena under consideration with examples drawn predominantly from the symbioses between marine invertebrates and dinomastigotes. However, since there are several aspects of these phenomena that are also demonstrated by freshwater associations involving Chlorella-like algae, some reference, by way of comparison, will be made to them.

Evolutionary initiation of symbiosis

Since there is no direct evidence available on the evolutionary initiation of symbioses between microalgae and invertebrates, concepts pertaining to this topic are perforce speculative. The evolutionary history of symbiosis in most invertebrate groups is very poorly understood. It is generally accepted by invertebrate paleontologists (e.g. Stanley, 1981) that the appearance of the Scleractinia (stony corals) at the beginning of the Triassic (about 230 million years ago) was coincident with their establishing symbioses with dinomastigotes. The evolutionary history of symbioses with microalgae in other marine invertebrate groups, as well as their freshwater counterparts is much more obscure (Trench, 1987).

The evolutionary history of the microalgae involved in symbiosis is also unclear. It is now known that the spectrum of algal groups that are involved in such associations is quite broad. For example, distinct cyanobacteria occur in sponges and in didemnid ascidians; coccoid red algae, diatoms and green algae are known as symbionts in some benthic foraminifers; diatoms occur in some flatworms; phyletically distinct green algae occur as symbionts in *Convoluta roscoffensis*, *Anthopleura xanthogrammica*, and *Hydra viridis*; and different dinomastigotes (amphidinioid and gymnodinioid) exist in symbioses with different marine invertebrates. This wide phyletic distribution of symbiotic algae, in combination with the wide phyletic distribution of their invertebrate hosts, indicates that microalgal-invertebrate symbiosis was polyphyletic in origin (Trench, 1987). Hence it is difficult to conceive of a common force that would have selected for symbiosis among these diverse groups.

In considering the evolutionary initiation of microalgal-invertebrate symbioses it must also be borne in mind that invertebrates possess the ability to distinguish between self and non-self; hence in the first historical contact between algae and animals, there probably was a distinct tendency towards resistance by the animals to invasion by the algae. In support of this view, it can be pointed out that there are numerous invertebrates that harbour algae while their conspecifics do not (Trench, 1987), and cannot be rendered symbiotic merely by exposure to symbiotic algae. The fact that many attempts to form symbioses were successful, as rendered evident by the existence of such large numbers of symbiotic associations, indicates that the algae were able to circumvent the "defence" mechanisms of the hosts and establish a stable relationship that has persisted over protracted periods.

A popular hypothesis for the historical initiation of microalgal-invertebrate symbiosis has been that the algae were captured by herbivorous zooplankton which were incapable of digesting them. The zooplankton themselves subsequently fell prey to cnidarians (for example), and the algae were then phagocytosed by chance along with food particles (Muscatine 1973; D.L. Taylor, 1973, Epp and Lewis, 1981). This mechanism would of course not pertain to the herbivorous hosts such as the symbiotic bivalve molluscs (*Tridacna*, *Corculum*, *Fragum*), and indeed may not be as facile as suggested. First, as indicated above, non-symbiotic hosts are not rendered symbiotic by experimental exposure to symbiotic algae.

Secondly, the non-symbiotic sea anemone Entacmaea which feeds directly on the symbiotic jellyfish Mastigias in the "marine lakes" of Palau, remains non-symbiotic (Trench, 1987). Thirdly, as will be shown later, phagocytosis of microalgae by the endodermal cells of cnidarians does not necessarily indicate the establishment of a symbiosis (Colley and Trench, 1983).

The phylogeny of symbiotic dinomastigotes and dinomastigotes in general is currently in a state of confusion (cf. F.J.R. Taylor, 1980 and Loeblich, 1984). Continuing in a speculative vein, it is possible that the symbiotic dinomastigotes of the genus Symbiodinium may be evolutionarily derived from a line of phagotrophic dinomastigotes (an example of which is Katodinium (=Gymnodinium) fungiforme, Spero, 1982). This possibility is reasonable in light of the recognition that algae in the genus Symbiodinium may possess a peduncle (Loeblich and Sherley, 1979; Trench and Blank, 1987), an organ used in phagotrophic nutrition. Hence, what are currently regarded as mutualistic symbioses involving algae of the genus Symbiodinium may have had their origins in parasitism or predation. The absence of information on the phylogeny of other symbiotic microalgal taxa does not permit even the wildest speculation on how such associations might have been initiated. A driving force for the evolutionary selection of algal symbiosis in nutrient-poor tropical coral reef environments could have been the increased nutrient supply the algae encounter in their hosts. The same force would hardly apply to the symbioses existing in Anemonia sulcata on the British coast or to Anthopleura on the West coast of the United States, unless these symbioses originated elsewhere, in a tropical environment.

Transmission of algae from parents to offspring

Regardless of how symbiotic associations were historically initiated, it is clear that existing associations are successfully perpetuated from one host generation to another. It is significant that in the majority of cases, both invertebrate hosts and symbiotic algae reproduce clonally (Wulff, 1985). Thus, when hosts reproduce asexually, some proportion of the algal population becomes distributed to each of the progeny. This applies equally well to single-celled hosts such as *Paramecium* as to multicellular hosts such as *Hydra*, corals or ascidians.

The transfer of symbionts from parents to offspring during sexual reproduction of the host is more complex. Some hosts "solve" the problem by avoiding it altogether. That is, they release eggs or larvae without symbionts. The larvae or juvenile stages then acquire the algae from the ambient environment. This method, termed the "open system," is found in flatworms (*Convoluta*, *Amphiscolops*), many cnidarians (*Aiptasia*, *Anthopleura* and several corals) and in tridacnid bivalves (Trench, 1987). The alternate mechanism, termed the "closed system" involves the invasion of the eggs or brooded larvae by the algae and subsequent release of offspring carrying a population of symbionts derived directly from the parent (Trench, 1987). Examples of direct "inheritance" have been reported in *Hydra* (Muscatine and McAuley 1982, cf. Thorington and Margulis, 1980) and in the marine hydroid *Myrionema* and some corals (Trench, 1987). In the case of the hydroids, during development of the egg, cytoplasmic extensions of the endodermal cells carrying algae, invade the egg plasm. The maternally derived cytoplasm eventually disappears, and the released algae eventually become sequestered into individual vacuoles as the zygote undergoes cellular differentiation.

Mechanisms of selection of algal symbionts

It should be apparent from the discussion above that specificity in microalgal-invertebrate symbioses exists, even in the "open systems"; thus under natural circumstances, corals or tridacnids do not harbour amphidinioid dinomastigotes and the flatworm *Amphiscolops langerhansi* does not harbour gymnodinioid dinomastigotes. The sea anemone *Aiptasia tagetes* (an "open system") harbours the same species of *Symbiodinium* in Jamaica, Florida, Barbados and Bermuda (Schoenberg and Trench, 1980a). These observations indicate that there is some process of selection which is manifested during the reestablishment of symbioses leading ultimately to the expressed specificity. There are few known examples wherein a given host simultaneously harbours more than one species of algal symbiont; *Anthopleura xanthogrammica* may harbour a dinomastigote and a green alga, and the flatworm *Haplodiscus* simultaneously harbours an *Amphidinium* and a *Symbiodinium* (Trench and Winsor, 1987). In the case of stony corals, given 230 million years of evolution and the demonstrated specificity of their symbioses with dinomastigotes of the genus *Symbiodinium*, the conceptual framework for the exploration of possible co-speciation and co-evolution is now established.

The specificity demonstrated in microalgal-invertebrate symbioses could be brought about by a series of phenomena which, when viewed as the preference of one algal type over another, can be regarded under the general category of recognition. Since the use of the term recognition has recently been the subject of debate, I shall digress slightly to state that in the historical initiation of symbioses, it is difficult to conceive of recognition playing a role since the potential partners were coming into contact with each other for the first time, with no guarantee of forming a successful symbiosis. Having established an association, it is conceivable that recognition plays an important role in the reestablishment of associations in those instances where the symbionts are not transmitted directly from parent to offspring. In the reformation of such associations, the expressed specificity could be determined at the point of intercellular contact between the two partners, based on interactions between electrically charged surfaces and/or interactions between cell surface ligand-receptor molecules. Alternatively, the initial interactions could be non-selective, the "decision" to accept or reject occurring after the symbionts have been internalized. It is equally possible that certain micro-algae without prior contact with hosts possessed antigenic determinants or other macromolecules on their surfaces which mediated ligand-receptor-like interactions with molecules on the animal cell surface resulting in their phagocytosis and maintenance. Such microalgae could be regarded as having been preadapted for the symbiotic way of life, assuming that other parameters of their biology permitting an existence within the host's cells also existed (see Rahat and Reich, 1987).

Is specificity determined on intercellular contact based on ligand-receptor interactions?

It is clear, in many systems where cells interact, (for example during ontogeny), that molecular characteristics of the surfaces play central roles in the final disposition of cell types. Collectively, these membrane-bound glycopeptides are termed "cell adhesion molecules" (CAM). In many plant systems, lectins, which demonstrate specificity for carbohydrate residues, play important roles in ligand-receptor interactions. Is there evidence that similar molecular interactions determine specificity in microalgal-invertebrate symbioses?

In studies of the Florida "strain" of _Hydra viridis_, using algae freshly isolated from the animals, the results of several studies were interpreted as demonstrating specificity as manifested at first contact between animal and algal cells. In summary, the evidence was that larger numbers of live native algae are phagocytosed by animal cells than are killed symbionts or non-symbiotic _Chlorella_ (Pardy and Muscatine, 1973; Pool and Muscatine, 1980; Reisser _et al_ 1982). Differences in the morphologies of the digestive cell surfaces involved with phagocytosis of particles were correlated with the type of particle, and a particular type ("microvillar") appeared specific to native symbionts (McNeil, 1981; cf. McNeil and Smith, 1982). Symbionts treated with antibodies prepared against their surfaces or with lectins demonstrated reduced phagocytosis by animal cells (Pool, 1979; Meints and Pardy, 1980). These pieces of evidence would be consistent with surface recognition phenomena, but Jolley and Smith (1980), finding that latex spheres could be phagocytosed by Hydra cells as readily as algal symbionts, concluded that there was no recognition of symbionts at first intercellular contact. Muscatine et al (1975) also concluded that the initial phagocytosis was non-specific.

The early studies of Schoenberg and Trench (1980b) demonstrated that some species of Symbiodinium, when provided after maintenance in culture to aposymbiotic Aiptasia were "accepted" and proliferated in the animal, while others, although "accepted" grew slowly and yet others were "rejected." In studies using the aposymbiotic scyphistoma (polyp) stage of the jellyfish Cassiopeia xamachana, the phenomena were dissected in more detail. Colley and Trench (1983, 1985) demonstrated that algae freshly isolated from Cassiopeia (homologous) were phagocytosed by the endodermal cells of the scyphistomae using a mechanism involving pseudopodial extensions with attendant microfilaments (see also Fitt and Trench, 1983). Freshly-isolated, killed algae were phagocytosed almost as readily. The process of phagocytosis demonstrated saturation kinetics and could be competitively inhibited if particles such as carmine were provided simultaneously, indicating that similar processes were invoked to elicit the phagocytotic response of the animal's cells by the freshly-isolated algae and the carmine particles. The rapid phagocytosis of freshly-isolated algae was demonstrable regardless of the source from which the algae were derived. However, in all instances, the same algae after maintenance in culture were phagocytosed at very low rates or not at all (Table 1). Ultrastructural analysis demonstrated that the surfaces of the freshly-isolated algal cells were contaminated with adhering animal membranes (probably the host cell vacuolar membrane) and that this membrane, and not the algal cell surface, made contact with the animal cell surface prior to phagocytosis. Removal of the membrane contaminant on the surface of the freshly-isolated algae with Triton X-100 caused them to be phagocytosed at low rates, similar to the cultured algae. Hence, no algal cell surface-animal cell surface interaction is involved in the process of uptake of freshly-isolated algae. The demonstration that scyphistomae of the non-symbiotic jellyfish Aurelia also rapidly phagocytosed freshly-isolated Symbiodinium spp. (Colley and Trench, 1983) adds additional support to the view that the use of freshly-isolated algae for the study of surface recognition phenomena in such systems is unjustified, unless it is first demonstrated that the algal surfaces are free of contaminants.

Table 1. Characteristics of uptake phenomena of symbiotic dinomastigotes by cells of invertebrate hosts (for details, see Colley and Trench, 1983).

State of algae	Host response
Freshly isolated (homologous)	Rapid phagocytosis, multiple phagocytosis by endodermal cells
Freshly isolated (heterologous)	Less rapid phagocytosis, but depending on source, multiple phagocytosis by endodermal cells
Cultured (homologous)	Low rates of phagocytosis; persist
Cultured (heterologous)	Some species phagocytosed at very low to undetectable rates but persist; other species not phagocytosed

Pool (1979) and Pool and Muscatine (1980) suggested that the prior existence of a foreign symbiotic Chlorella inside Hydra cells resulted in this acquisition of surface antigenic determinants similar to those on the surface of native symbionts, rendering the foreign algae more acceptable to *Hydra* when proffered after growth in *Hydra*.

By contrast, Schoenberg and Trench (1980b) found that growth of heterologous algae in *Aiptasia* did not promote subsequent facilitation of uptake of those algae after their reisolation and maintenance in culture. Again, Colley and Trench (1983) found that algae freshly isolated from *Aiptasia*, when provided to scyphistomae of *Cassiopeia* were not taken up as rapidly as algae freshly isolated from *Cassiopeia* until they had first been grown in *Cassiopeia* and then isolated from the animal (Table 2). This observation, in light of the demonstrated surface membrane contamination, indicates that the scyphistomae were responding to their own membrane which in the second instance, surrounded the algae from *Aiptasia*. Being completely aware of the potential dangers in extrapolating from one experimental system to another, since the antibodies were produced against freshly-isolated algae from *Hydra* (the algae have not been successfully cultured), the observations reported by Pool (1979) and Pool and Muscatine (1980) could also be consistent with surface membrane contamination on the freshly-isolated *Chlorella*.

When *Symbiodinium* derived from different host species (heterologous) were provided to the scyphistomae after having first been maintained in culture, initial uptake of algal cells was low, but the natural symbiont *S. microadriaticum* and some others successfully infected the polyps, while others, e.g. *S. pilosum* did not (Colley and Trench, 1983; Trench, 1987). Since the algae replicate clonally inside the hosts' cells, the initial phagocytosis of large numbers of algae is not a prerequisite for the establishment of a successful symbiosis.

Similarly, when the juvenile stages of the flatworm *Amphiscolops* were exposed to cultures of mixed populations of *Amphidinium* (their natural symbiont) and *Symbiodinium*, the animals always selected *Amphidinium* (Taylor, 1971; Trench and Winsor, 1987). These data are consistent with the interpretation that some process of selection does occur when algal cells and animal cells make contact, but the nature of the

Table 2. Effect of previous growth in *Cassiopeia* on phagocytosis of algae from *Aiptasia pallida* (for details, see Colley and Trench, 1983).

Original host	doner host	host response
Cassiopeia	*Cassiopeia*	rapid uptake, multiple phagocytosis
A. pallida	*A. pallida*	reduced uptake, lower levels of multiple phagocytosis
A. pallida	*Cassiopeia*	rapid uptake, multiple phagocytosis

surface interactions is unknown. Perhaps a detailed analysis of the chemical composition of the cell wall of symbiotic dinomastigotes, paying particular attention to the surface material that possesses the histochemical characteristics of a glycoprotein (Trench *et al* 1981; Trench and Blank, 1987), ought to be conducted, and their possible role in chemical recognition analysed.

Does the release of photosynthetic products by symbiotic algae determine specificity?

Smith (1981), McAuley and Smith (1982) and Douglas and Smith (1984) proposed that maltose release by symbiotic *Chlorella* was an important "signal" in eliciting the phagocytosis and sequestration of symbiotic *Chlorella* by the endodermal cells of *Hydra*. This conclusion was based on the observation that inhibition of photosynthesis and hence the release of maltose resulted in the reduction in the number of algae sequestered. Weis (1981) also postulated a role for maltose excretion in recognition of *Chlorella* by *Paramecium bursaria*.

All symbiotic dinomastigotes (Symbiodinium spp. as well as Amphidinium spp.) analysed to date, release the same group of photosynthetic products (Trench, 1987), namely primarily glycerol, along with some glucose, alanine and some organic acids (see also Muscatine, 1967; Muscatine and Cernichiari, 1969; Trench 1971a,b). It should be clear, from the fact that hosts symbiotic with Amphidinium do not "accept" Symbiodinium and visa versa, that the molecules released are not instrumental in determining specificity or persistence. It is known that when algal cells inside their host's cells are killed by herbicides such as DCMU, they are eliminated (Fitt and Trench, 1983), often after lysosomal attack.

Is specificity determined by processes that occur after phagocytosis?

From the discussion above, it should be clear that the processes by which algae gain entrance into animal cells and the processes that determine whether or not those algae that are phagocytosed will persist, are separate. Pertaining to Hydra, Muscatine et al (1975) concluded that recognition of the appropriate Chlorella symbiont occurred after phagocytosis by the endodermal cells. This conclusion was based on the observation that several "foreign" cells and inert particles could be phagocytosed by the endodermal cells of Hydra, but only a select few algae would persist and form a stable association.

In the case of dinomastigote-invertebrate symbioses, it is also abundantly clear that a given host may phagocytose various algae, but only certain types persist. For example, as indicated above, several species of Symbiodinium are "rejected" by the scyphistomae of Cassiopeia when they are provided as cultured cells, either in the free-swimming or coccoid state. However, when these same algae are provided either in the freshly isolated state, or via the nauplii of Artemia, they are taken into the endodermal cells, but most of them are lost within a few hours or days. Again, although the larvae of *Amphiscolops* do not take up *Symbiodinium* spp. when provided after culture, these same algae may gain entrance to the animals via *Artemia*, but once more, they do not persist and are regurgitated, morphologically intact (Trench and Winsor, 1987). Finally, the endodermal cells of the non-symbiotic jellyfish *Aurelia* may also phagocytose symbiotic algae, but these algae do not persist (Colley and Trench, 1983).

The fate of algae after initial phagocytosis goes beyond sequestration into host vacuoles at the bases of endodermal cells. Several studies of the process of infection of *Hydra* by *Chlorella* have indicated that following the initial phagocytotic uptake of the algae, there is a decline in algal numbers followed by rapid proliferation of the algae (Cook, 1980; Jolley and Smith, 1980; McAuley and Smith, 1982; McNiel and Smith, 1982). A similar phenomenon exists in the case of the infection of the scyphistomae of *Cassiopeia* with *Symbiodinium microadriaticum* (Colley and Trench, 1985).

When freshly-isolated *S. microadriaticum* or some other *Symbiodinium* spp. (e.g. those from *Aiptasia*) are introduced into the scyphistomae of *Cassiopeia*, following the initial uptake of large numbers of algae, the algal population declines to less than 50% of the initial level within a few days, and then increases dramatically (Colley and Trench, 1983). Algae that do not persist (e.g. those derived from *Zoanthus sp.* or *Anthopleura*) show a rapid decline to zero population, though sometimes over a protracted period of several days.

In the case of algae that successfully infect, it was demonstrated that after they were sequestered, the animals' endodermal cells discontinued phagocytotic activity, withdrew from lining the coelenteric cavity and invaded the mesoglea, carrying the algae with them. Within these "wandering amoebocytes" the algae divided rapidly, such that in the released ephyrae (immature jellyfish), the majority of the algae are found in the mesoglea (Colley and Trench, 1985).

The observed decline in algal numbers could be as a result of expulsion or digestion. Using the technique of colloidal gold labelling of lysosomes, Colley and Trench (1985) demonstrated a 4% frequency of lysosomal fusion with the vacuoles containing algae. As this did not approximate the observed 50% decline in algal population, it was concluded that digestion of the algae was not responsible for the decline in the algal population, and the alternative, exocytosis seemed more consistent with the observations.

In the situation where more than one algal type gains access to a potential host, post-phagocytotic processes seem to "sort" the algae out, such that usually only one type persists. Provasoli, Yamasu and Manton (1968) demonstrated that *Tetraselmis* (*Platymonas*) *covolutae* "displaced" other algae that were first taken up by *C. roscoffensis*. In the flatworm *Amphiscolops*, Trench and Winsor (1987) found that when *Symbiodinium* and *Amphidinium* both gained access to the worm, the natural symbiont *Amphidinium* was transported to the peripheral parenchyma, while the unnatural one, *Symbiodinium* remained in the digestive region until regurgitated.

These observations are consistent with the interpretation that there are processes that occur after phagocytosis that determine that some types of cells will persist and others not. The nature of these processes are unknown, but the possibility of ligand-receptor interactions between the surface of the algae and the host's cell vacuolar membrane cannot be ignored. On each division of the algae, after the daughter cells escape from the parent cell wall, a "new" surface is presented to the internal surface of the vacuole. Thus, a process of continuous internal monitoring of the algal population by the host cell can be envisioned.

Summary and conclusions

From the discussion above the concept that microalgal-invertebrate symbioses demonstrate specificity is irrefutable. A given invertebrate host, under natural circumstances, harbours a particular algal symbiont. Although natural associations may be highly circumscribed, laboratory analyses suggest that often the restriction is not precise, and that a given host may be able to initiate an association with more than one type of alga, but the unnatural alga may not persist or may be eliminated in preference for the natural symbiont if the latter is provided after an initial infection with the former. All the observations taken together indicate that there are processes of selection involved in the acquisition of symbionts by hosts, but the mechanisms involved at the cellular and molecular levels are unknown.

If the specificity observed is viewed as an end product of a range of processes that may begin in the broad ecological sphere and culminate at molecular levels, then there are several factors that may be drawn upon to explain the observations. Such factors include (Trench et al 1981).

(1) ecological, behavioural and physiological parameters of the algae and the potential hosts which regulate their distribution in time and space.

In the broad ecological sense, because of the discontinuity between the Atlantic and Indo-Pacific regions (an event that occurred with the last closing of the Panama bridge some 6000 yrs. ago) animals that are restricted to one region would not be exposed to symbiotic algae found in the other. In the specific case of the temperate sea anemone A. elegantissima, the Symbiodinium sp. that it harbours grows best in culture at 15°C, while most Symbiodinium spp. from tropical hosts do not grow below about 21°C. It is therefore most unlikely for physiological reasons that Anthopleura would naturally be exposed to Symbiodinium spp. which occur in the tropical Pacific.

(2) possible surface-bound macromolecular "recognition" modulators which come into play during intercellular contact.

In no instance involving microalgal-invertebrate symbioses has any surface macromolecules that are involved in recognition been isolated and characterized. The conclusion that such phenomena exists is for the most part inferential. It is clear, for example, that the scyphistomae of Cassiopeia and the juveniles of Amphiscolops discriminate between different algae on first contact when those algae are provided from culture (Colley and Trench, 1983; Trench and Winsor, 1987). The mechanisms involved are unknown, but could well involve ligand-receptor interactions. However, this does not represent the only process of selection leading to the expressed specificity, and could be regarded as an initial screening that is not very precise. It must be borne in mind that in the coelenterates at least, the same cells that phagocytose and harbour symbiotic algae are also involved in nutritive phagocytosis of food particles.

(3) events that occur after the initial phagocytosis which determine whether an alga that was engulfed would be sequestered and maintained or eliminated.

Unnatural symbiotic algae are phagocytosed by the cells of symbiotic and nonsymbiotic invertebrates, but these algae generally do not persist, while the natural symbionts always do. Clearly, some process or processes occur after phagocytosis wherein the "decision" to keep and maintain an alga, or to eliminate it, is made. The cellular or molecular interactions which occur within the phagocytotic vesicle are unknown. Similarly, the mechanisms through which an animal cell recognizes defunct individuals in a population of symbiotic algae, resulting in the elimination of such defunct cells, are unknown. The same algal cell surface molecules that may influence recognition on initial phagocytosis may also be involved in recognition processes that occur inside the animals' cell vacuole, which is derived from the plasmalemma.

(4) the influence of selective forces acting on the established association.

Different symbiotic dinomastigote species appear to possess different photoadaptive capabilities (Chang _et al_ 1983). It is therefore possible that in the natural environment, a given species of alga may initiate a symbiosis with a host living in a habitat to which the alga is poorly adapted. Environmental constraints would probably act against the persistence of that species of alga in that host. Depending on the extent to which the host relies on the algae for reduced carbon for its nutritional maintenance, the host may also not persist, unless an appropriate symbiont becomes available.

Taking all the evidence available into consideration, the conclusion that the specificity expressed in microalgal-invertebrate symbioses is an end result of a series of processes is almost inescapable. I believe that the phase of initial contact between algae and potential host may be selective, but is not restrictively specific. Events that occur after phagocytosis, be they based on cellular/molecular interactions between the algal cell surface and the animals' vacuolar membrane or ecological or physiological constraints, ultimately result in one algal type persisting as symbiont rather than any other.

Acknowledgement

Preparation of this manuscript was supported by a NSF grant BSR83-20450.

References

Apelt, G. 1969. Die Symbiose zwischen dem acoelen Trubellar *Convoluta convoluta* und Diatomen der Gattung *Licmophora*. *Mar. Biol.* 3:165-187.

Blank, R.J. and Trench, R.K. 1985a. Speciation and symbiotic dinoflagellates. *Science* 229:656-658.

Blank, R.J. and Trench, R.K. 1985b. *Symbiodinium microadriaticum*: A single species? *5th Int'l. Coral Reef Congr. (Tahiti)* 6:113-117.

Blank, R.J. and Trench, R.K. 1986. Nomenclature of endosymbiotic dinoflagellates. *Taxon* 35:286-294.

Chang, S.S., Prézelin, B.B. and Trench, R.K. 1983. Mechanisms of photoadaptation in three strains of the symbiotic dinoflagellate *Symbiodinium microadriaticum*. *Mar. Biol.* 76:219-229.

Cook, C.B. 1980. Infection of invertebrates with algae. In: *Cellular Interactions in Symbiosis and Parasitism*, C.B. Cook, P.W. Pappas and E.D. Rudolph (eds.). Ohio State University Press, Columbus. pp. 47-74.

Colley, N.J. and Trench, R.K. 1983. Selectivity in phagocytosis and persistence of symbiotic algae by the scyphistoma stage of the jellyfish *Cassiopeia xamachana*. *Proc. R. Soc. Lond.* (B) 219:61-82.

Colley, N.J. and Trench, R.K. 1985. Cellular events in the reestablishment of a symbiosis between a marine dinoflagellate and a coelenterate. Cell Tissue Res. 239:93-103.

Douglas, A.E. and Huss, V.A.R. 1986. On the characteristics and taxonomic position of symbiotic Chlorella. Arch. Microbiol. 145:80-84.

Douglas, A.E. and Smith, D.C. 1984. The green hydra symbiosis. VIII. Mechanisms in symbiont regulation. Proc. R. Soc. Lond. (B) 221:291-319.

Epp, R.W. and Lewis, W.M. Jr. 1981. Photosynthesis in copepods. Science 214:1349-1350.

Fitt, W.K. and Trench, R.K. 1983. Endocytosis of the symbiotic dinoflagellate Symbiodinium microadriaticum Freudenthal by endodermal cells of the scyphistomae of Cassiopeia xamachana and resistance of the algae to host digestion. J. Cell Sci. 64:195-212.

Freudenthal, H.D. 1962. Symbiodinium gen. nov. and Symbiodinium microadriaticum sp. nov., a zooxanthella. Taxonomy, life cycle and morphology. J. Protozool. 9:45-52.

Holland, A. and Carré, D. 1974. Lex Xanthelles des Radiolaires sphaerocollides, des Acanthaires et de Velella velella. Infrastructure-cytochemie-taxonomie. Protistologica 10:573-601.

Hori, T., Norris, R.E. and Chihara, M. 1982. Studies on the ultrastructure and taxonomy of the genus Tetraselmis (Prasinophyceae). I. Subgenus Tetraselmis. Bot. Mag. Tokyo 95:49-61.

Jolley, E. and Smith, D.C. 1980. The green hydra symbiosis. II. The biology of the establishment of the association. Proc. R. Soc. Lond. (B) 207:311-333.

Loeblich, A.R. 1984. Dinoflagellate evolution. In: Dinoflagellates, D.L. Spector (ed.). Academic Press, New York, London, pp. 481-552.

Loeblich, A.R. and Sherley, J.L. 1979. Observations on the theca of the motile phase of free-living and symbiotic Zooxanthella microadriatica (Freudenthal) comb. nov. J. mar. biol. Ass. U.K. 59:195-206.

McAuley, P.J. and Smith, D.C. 1982. The green hydra symbiosis. V. Stages in the intracellular recognition of algal symbionts by digestive cells. Proc. R. Soc. Lond. (B) 216:7-23.

McNeil, P.L. 1981. Mechanisms of nutritive endocytosis. I. Phagocytic versatility and cellular recognition in Chlorohydra digestive cells; a scanning electron microscopic study. J. Cell Sci. 49:311-339.

McNeil, P.L. and Smith, D.C. 1982. The green hydra symbiosis. IV. Entry of symbionts into digestive cells. Proc. R. Soc. Lond. (B) 216:1-6.

Meints, R.H. and Pardy, R.L. 1980. Quantitative demonstration of cell surface involvement in a plant-animal symbiosis: lectin inhibition of reassociation. J. Cell Sci. 52:243-269.

Muscatine, L. 1967. Glycerol excretion by symbiotic algae from corals and Tridacna and its control by the host. Science 156:516-519.

Muscatine, L. 1973. Nutrition of corals. In: Biology and Geology of Coral Reefs, Vol. II. O.A. Jones and

R. Endean (eds.). Academic Press, New York. pp. 77-115.

Muscatine, L. and Cernichiari, E. 1969. Assimilation of photosynthetic products of zooxanthellae by a reef coral. Biol. Bull. 137:506-523.

Muscatine, L., Cook, C.B., Pardy, R.L. and Pool, R.R. 1975. Uptake, recognition and maintenance of symbiotic Chlorella by Hydra viridis. Symp. Soc. Exp. Biol. 29:175-203.

Muscatine, L. and McAuley, P.J. 1982. Transmission of symbiotic algae to eggs of green hydra. Cytobios 33:111-124.

Pardy, R.L. and Muscatine, L. 1973. Recognition of symbiotic algae by Hydra viridis. A quantitative study of the uptake of living algae by aposymbiotic H. viridis. Biol. Bull. 145:565-579.

Parke, M. and Manton, I. 1967. The specific identity of the algal symbiont of Convoluta roscoffensis. J. mar. biol. Ass. U.K. 47:445-464.

Pool, R.R. 1979. The role of algal antigenic determinants in the recognition of algal symbionts by Chlorohydra. J. Cell Sci. 35:367-379.

Pool, R.R. and Muscatine, L. 1980. Phagocytic recognition and the establishment of Hydra viridis-Chlorella symbiosis. Endocytobiology Vol. I. W. Schwemmler and H.E.A. Schenk (eds.). Walter de Gruyter & Co. Berlin, New York. pp. 225-238.

Provasoli, L., Yamasu, T. and Manton, I. 1968. Experiments on the resynthesis of symbiosis in Convoluta rescoffensis with different flagellate cultures. J. mar. biol. Ass. U.K. 48:456-479.

Rahat, M. and Reich, V. 1987. A reevaluation of alleged requirements for the establishment of alga/hydra symbioses and host/symbiont specificity - preadaptation and vacuolar ecology. Symbiosis (in press).

Reisser, W., Radunz, A. and Wiessner, W. 1982. Participation of algal surface structures in the cell recognition process during infection of aposymbiotic Paramecium bursaria with symbiotic Chlorella. Cytobios. 33:39-50.

Schoenberg, D.A. and Trench, R.K. 1980a. Genetic variation in Symbiodinium (=Gymnodinium) microadriaticum Freudenthal, and specificity in its symbiosis with marine invertebrates. I. Isoenzyme and soluble protein patterns of axenic cultures of S. microadriaticum. Proc. R. Soc. Lond. (B) 207:405-427.

Schoenberg, D.A. and Trench, R.K. 1980b. Genetic variation in Symbiodinium (=Gymnodinium) microadriaticum Freudenthal, and specificity in its symbiosis with marine invertebrates. III. Specificity and infectivity of S. microadriaticum. Proc. R. Soc. Lond. (B) 207:445-460.

Smith, D.C. 1981. The role of nutrient exchange in recognition between symbionts. Ber. Deutsch. Bot. Ges. 94:517-528.

Spero, H.J. 1982. Phagotrophy in Gymnodinium fungiforme (Pyrrhophyta); the peduncle as an organelle of ingestion. J. Phycol. 18:356-360.

Spero, H.J. 1987. Symbiosis in the planktonic foraminifer Orbulina universa and the isolation of its symbiotic dinoflagellate Gymnodinium béii, sp. nov. J. Phycol. (in press).

Stanley, G.D. Jr. 1981. Early history of scleractinian corals and its geological consequences. Geol. 9:507-511.

Thorington, G. and Margulis, L. 1980. Transmission of the algae and bacterial symbionts of green hydra through the host sexual cycle. Endocytobiology Vol. I. W. Schwemmler and H.E.A. Schenk (eds.). Walter de gruyter & Co. Berlin and New York. pp. 175-222.

Taylor, D.L. 1971. On the symbiosis between Amphidinium klebsii (Dinophyceae) and Amphiscolops langerhansi (Turbellaria, Acoela). J. mar. biol. Ass. U.K. 51:301-313.

Taylor, D.L. 1973. The cellular interactions of algal-invertebrate symbioses. Adv. Mar. Biol. 11:1-56.

Taylor, D.L. 1974. Symbiotic algae: taxonomy and biological fitness. In: Symbiosis in the Sea, C.B.W. Vernberg, (ed.). University of South Carolina Press, Columbia. pp. 254-262.

Taylor, F.J.R. 1980. On dinoflagellate evolution. BioSystems 13:65-108.

Trench, R.K. 1971a. The physiology and biochemistry of zooxanthellae symbiotic with marine coelenterates. I. The assimilation of photosynthetic products of zooxanthellae by two marine coelenterates. Proc. R. Soc. Lond. (B) 177:225-235.

Trench, R.K. 1971b. The physiology and biochemistry of zooxanthellae symbiotic with marine coelenterates. II. Liberation of fixed ^{14}C by zooxanthellae in vitro. Proc. R. Soc. Lond. (B) 177:237-250.

Trench, R.K. 1979. The cell biology of plant-animal symbioses. Ann. Rev. Plant Physiol. 30:485-532.

Trench, R.K. 1980. Uptake, retention and function of chloroplasts in animal cells. Endocytobiology, Vol. I. W. Schwemmler and H.A.E. Schenk, (eds.). Walter de Gruyter & Co. Berlin, New York. pp. 703-727.

Trench, R.K. 1987. Dinoflagellates in non-parasitic symbioses. In: The Biology of Dinoflagellates. F.J.R. Taylor (ed.). Blackwell, Oxford. pp. 531-570.

Trench, R.J. and Blank, R.J. 1987. Symbiodinium microadriaticum Freudenthal, S. goreauii, sp. nov.; S. kawagutii, sp. nov., and S. pilosum sp. nov.; gymnodinioid dinoflagellate symbionts of marine invertebrates. J. Phycol. (in press).

Trench, R.K., Colley, N.J. and Fitt, W.K. 1981. Recognition phenomena in symbioses between marine invertebrates and zooxanthellae; uptake, sequestration and persistence. Ber. Deutsch. Bot. Ges. 94:529-545.

Trench, R.K. and Winsor, H. 1987. Symbiosis with dinoflagellates in two pelagic flatworms, Amphiscolops sp. and Haplodiscus sp. Symbiosis 3:1-22.

Weis, D.S. 1981. Further tests for the hypothesis that algal maltose excretion and cell surface ligands are signals in the infection of *Paramecium bursaria* by *Chlorella*. *Ber. Deutsch. Bot. Ges.* 94:547-556.

Wulff, J.L. 1985. Clonal organisms and the evolution of mutualism. In: *Population Biology and Evolution of Clonal Organisms*, J.B.C. Jackson, L.W. Buss and R.E. Cook (eds.). Yale University Press, New Haven. pp. 437-466.

APPLICATIONS OF GENETIC ENGINEERING TO "SYMBIONTOLOGY" IN AGRICULTURE

Marco P. Nuti, Maria B. Pasti, Andrea Squartini
Dipartimento di Biotecnologie Agrarie, Università di Padova,
Via Gradenigo 6, 35100 Padova (Italy).

1. INTRODUCTION

Symbiosis represents an ancestral form of cell-to-cell interaction, often offering the partnership unique adaptations to environmental conditions. Agricultural activities have added selective pressure on symbiotic systems, particularly if some managerial practices are considered; these include conventional plant breeding, massive use of fertilizers and pesticides, cropping forest land and marginal areas. As a consequence, symbioses such as mycorrhizae, actinorrhizae, Rhizobiaceae/Leguminosae, *Termitomyces*/termites, lichens, tend to be disrupted or become inappropriate for the new food, feed, and energy demands.

DNA technology is nowadays considered one of the most promising tools to overcome these problems, i.e. to enhance crop yield and improve the quality of food,

NATO ASI Series, Vol. H17
Cell to Cell Signals in Plant, Animal and
Microbial Symbiosis. Edited by S. Scannerini et al.

fodder, and raw materials for agro-industrial processes taking into consideration the ecological impact of agricultural productivity.

Genetic engineering of whole plants is being exploited <u>via</u> the use of T-DNA vectors, protoplast transformation or direct DNA injection (Hooykaas and Schilperoort, 1984; de la Pena et al., 1987; Grimsley et al., 1987).

Other applications of DNA technology in agriculture include the use of genetically manipulated microorganisms and enzymes for processing of agricultural products (Priest, 1984; Nuti et al., 1987), the construction of suitable gene probes for ecological surveys and assessment of field-released, genetically engineered organisms (Dean-Ross, 1986; Nuti, 1987).

As for direct, mutually beneficial, interactions between microbes, plants and animals, genetic engineering has proved a powerful tool to: (a) understand the intimate mechanisms controlling the interactions; and (b) develop new strategies for the exploitation of the technological capabilities of symbiotic systems relevant to agriculture.

Two examples are presented here, the first dealing with the complex sequence of events which leads to the establishment of functional dinitrogen-fixing symbioses between rhizobia and leguminous plants, and the second with symbiotic and associative partnership between microbiota and african higher termites.

2. RHIZOBIUM-LEGUME SYMBIOSES

Symbiotic dinitrogen fixation is a specific interaction between soil bacteria belonging to the Rhizobiaceae

(Rhizobium, Bradyrhizobium and Azorhizobium) and legumes, occuring at the root or stem level. It is of great agronomic importance, and rhizobia can be used directly to inoculate seeds or soil, thus providing the emerging plant with its symbiotic partner. The functional symbiosis requires cooperation between the plant and the bacterium.

During the last decade, gene technology has allowed a rapid expansion of the identification of the sym(biotic) genes that are involved in the interaction (fig. 1). DNA technology has also been applied to the analysis of gene expression of legume symbiotic genes (Verma et al., 1985) or to study mutant alleles of host plant genes which block nodule formation and functioning (Gresshof et al., 1987; Nap et al., 1987). As has been highlighted by others in this volume (Brewin et al., Dazzo et al., Okker et al., Torrey) the steps leading to root nodule formation are many and complex. However, functional nodules will emerge only when bacterium and plant are able to exchange prompt, effective signals. In contrast to tumorogenic and pathogenic responses (Rogowski et al., 1987), the establishment of a symbiosis requires positive action by the plant, i.e. recognition signal(s); in this there is an implicit concept of specificity. In other words, the plant must first discriminate between potentially symbiotic and pathogenic/ opportunistic bacteria, then between specific and non-specific partners, and eventually between dinitrogen fixing and Fix endosymbionts. On the other hand, rhizobia must be able to transmit the signals needed to avoid exclusion at each of the steps of the overall infection process.

The presence of the exopolysaccharides or β-glucans could elicit the activation of compounds related to the pre-infection situation, like the flavonoids whose structure is close to that of phytoalexins, the latter being known as molecules induced in the presence of pathogens. Rhizobia have apparently evolved the ability to convert this into a highly specific signal for the induction of the promoter of a gene central for the control of nodulation, i.e. nod D. This gene can further activate the transcription of other nodulation genes nod ABC, nod FE and nod M (Fig. 2). The appearance of gene products from the above operons could be scattered in time, and the order could affect the whole process. Among the consequences, modification of surface components could occur, leading to the binding of specific plant receptors. The reaction might be mediated by lectins, although the plant could play a more active role by forming "modifying substances" whose nature remains to be elucidated. As a result, the bacteria will adhere to the exterior of the root hairs.

The specificity of the induction of hair curling could result from the positive responses to interrogative signal(s) aimed at verifying which nodulation gene is expressed at that moment. This implies that different rhizobial species, sharing the same nod operons, would be recognized as different according to the timing of nod gene expression. As soon as the answer has been given, hac genes are allowed to be expressed and the root hair will undergo Shepherd's crook formation. At this stage the bacteria are still some distance from the cortex, but cell division is induced in the latter, leading to the formation of nodule promordia. The existence of a nodule organogenesis-inducing principle (noip) acting as a diffusable signal can be postulated. The nature of this signal remains to be identified (Dénarié and Truchet, 1979).

The correct onset of nodulation appears to be dependent on the prompt induction of bacterial genes controlling infection thread formation, after root hairs have curled. Hac genes then need to be switched off to allow subsequent steps to occur; should their expression persist, diffuse delayed nodulation would be observed. The presence of a light-dependent plant signal inhibiting hair curling has been reported (Knight et al.,1986), and this might be a role

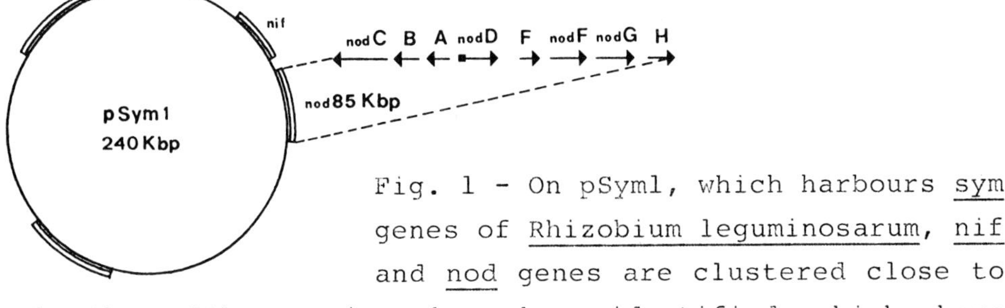

Fig. 1 - On pSyml, which harbours sym genes of Rhizobium leguminosarum, nif and nod genes are clustered close to each other. Other regions have been identified, which share DNA homology with pTi plasmid of Agrobacterium tumefaciens.

for one of the early nodulins. Two nodulation genes are involved in the control of infection thread formation, i.e. nod FE; their activity is also required to allow the stop signal of hair curling to be effective. At the same time they appear to be related to the synthesis of lipopolysaccharides.

The model described above is like a biochemical or genetic "conversation" between the bacterium and the plant. Its significance is to ensure that only the true, specific and effective symbionts will proceed in their mutual interaction.

The plant can reject the prokaryiote at any step if a correct biochemical "answer" is not given. The specificity of the "dialogue" keeps the plant safe from a generalized invasion of its roots, while selecting for a potential benefit in terms of nitrogen intake. All the reactions described above occur within a limited period of time, and the timing of the nodulation process for inoculated seeds of Pisum sativum can be as follows: after 3-4 days from sowing the infection will start; nodule formation will begin after six days, while the bacteria will be released in the cortical cells nine to ten days after sowing. Leghemoglobin will appear after 12 days, and nitrogenase activity will start around the fourteenth day.

A tight biochemical interdependance is also needed for the

Fig. 2

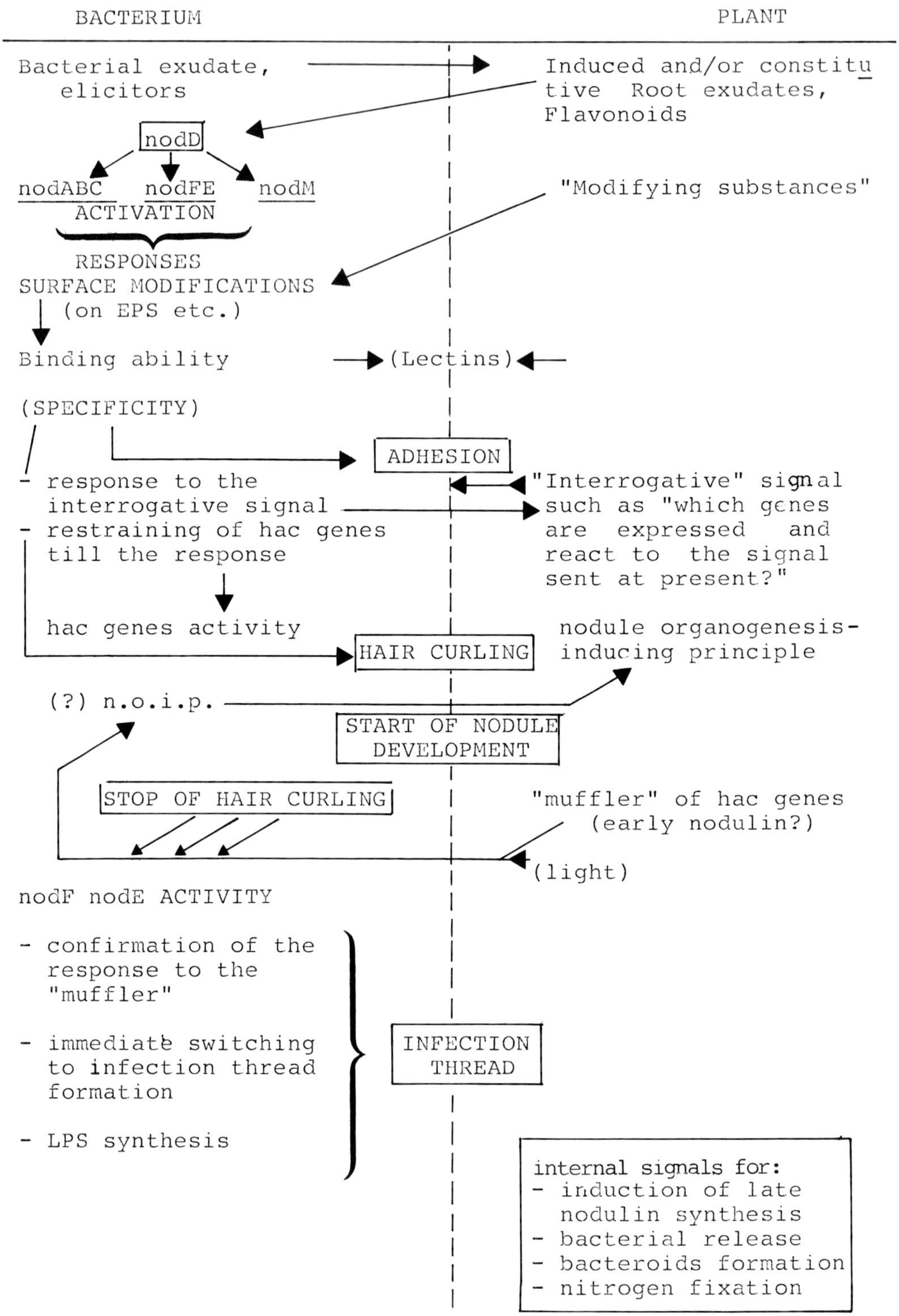

BACTERIUM | PLANT

Bacterial exudate, elicitors

Induced and/or constitutive Root exudates, Flavonoids

nodD

nodABC nodFE nodM

ACTIVATION

"Modifying substances"

RESPONSES

SURFACE MODIFICATIONS (on EPS etc.)

Binding ability

(Lectins)

(SPECIFICITY)

ADHESION

- response to the interrogative signal
- restraining of hac genes till the response

"Interrogative" signal such as "which genes are expressed and react to the signal sent at present?"

hac genes activity

HAIR CURLING

nodule organogenesis-inducing principle

(?) n.o.i.p.

START OF NODULE DEVELOPMENT

STOP OF HAIR CURLING

"muffler" of hac genes (early nodulin?)

(light)

nodF nodE ACTIVITY

- confirmation of the response to the "muffler"
- immediate switching to infection thread formation
- LPS synthesis

INFECTION THREAD

internal signals for:
- induction of late nodulin synthesis
- bacterial release
- bacteroids formation
- nitrogen fixation

metabolic aspects of the symbiosis. For instance, in soybean the infected cells will support the pathways of ammonia assimilation and conversion to intermediates suitable for transport. Xanthine is exported to the neighbouring uninfected cells, where it undergoes further enzymatic reactions leading to ureide nitrogen being conveyed to the transport system.

An intensive effort has been devoted to genetic improvement of Rhizobiaceae, and genetically engineered rhizobia have recently been produced. Genes for hydrogen uptake (hup) have been cloned from Bradyrhizobium japonicom (Haugland et al., 1984) and R.leguminosarum (Tichy et al., 1985).

Genes for the resistance to pesticides, like 2,4-D have been transferred from Alcaligenes eutrophus to R.trifolii and R.leguminosarum (Nuti et al., 1984). Genes have been modified in R.meliloti, leading to an increased N_2-fixation (F. Cannon, pers. comm., 1986). Molecular cloning of genes involved in carbon metabolism in R.meliloti and B.japonicum, like the dicarboxylate transport genes (dct) which control the carbon influx to the bacteroids, has been reported recently (O'Gara et al., 1987).

Genes encoding glutamine syntethase, which regulates ammonia assimilation in R.leguminosarum, have been isolated and characterized (Filser et al. 1986).

3. SYMBIOSIS MICROBIOTA/HIGHER TERMITES

In the forest and forest-savannah of West Africa and South America, higher termites (Termitidae), comprising four fifths of the described species or Isoptera, are among the most abundant insects. Despite their importance in nutrient cycling, they have received little attention so far, except for demonstrations that microbiota occur in the hindgut, and that the association possibly plays a significant role in digestion (Breznak, 1982; O'Brian and Slaytor, 1982). Among the microorganisms associated with higher termites, the fungus Termitomyces has been reported to be able to enter

symbiotic relationships with species of Macrotermes (Thomas, 1983). Actinobacteria have also been isolated from these termites and shown to possess a fairly high cellulolytic activity (Pasti and Belli, 1985).
Saprophytic microorganisms have been isolated from the hindgut of M.michaelsenii, and shown to belong to different genera, including the ubiquitous Pseudomonas, Escherichia and Bacillus (Osore and Okech, 1983).
The diet of Termitidae ranges from wood to leaves, grass, humus, detritus, and herbivorous dung. Its main component is represented by ligno-cellulose, and on passage through the termite gut this substrate undergoes degradation; in the case of cellulosic component degradation can account for 65-99% of the initial substrate. Lignocellulose degradation is likely to occur during the construction of nests, as well as during the nest "life", although evidence for that by in situ assays remains to be obtained. The following model might contribute to a better understanding of the role of microbiota in the "life" of M.michaelsenii nests (Fig. 3).
These termites grow the fungus Termitomyces on a structure called "fungus comb" which derives from chewed, undigested plant fragments. Ligno-cellulolytic actinobacteria could act as starters for the bioconversion of the undigested material into a compost which, after further enrichment with faecal pellets, would then support the growth of Termitomyces mycelium. The formation of "white nodules" (or mycotêtes), which are represented by fairly compact fasciculate hyphae with distal blastoconidia (Fig. 4), is now possible. They might well represent a suitable enzyme source for termites, enhancing ligno-cellulose digestion. The invasion of the "comb" by mycelial strands would be followed by the formation of the edible mushroom. In our opinion, three main factors can contribute to sporocarp production by Termitomyces: the presence of an invading mycelium, mechanically transported by the termite workers; the presence of a composted substrate having C/N lower than the original undigested ligno-cellulosic biomass; and the

appropriate environmental conditions in this particular area of the nest (pCO_2, T°C and humidity).

However a number of points remain to be elucitated, including: (a) the recognition pattern leading to the specific symbiotic relationship between *Termitomyces* sp. and *Macrotermes* sp.; (b) the functional role of microbiota during ligno-cellulose decomposition; and (c) the mechanism controlling the transfer or the origin of microbial components during the formation of a new nest. Gene technology could represent a useful approach to a better understanding of this microbe/animal interaction in several ways.

Ligninolytic- or cellulolytic-minus mutants of *Termitomyces* and Actinobacteria could be isolated by site-directed mutagenesis, which might help in deciphering signal exchanges and the role of partners in substrate decomposition; genetically marked strains might help in tracing the origin of microbiota during the foundation of new termite nests.

Another application of genetic engineering is the cloning of genes from termite-associated microorganisms, involved in ligno-cellulose degradation. So far, one endoglucanase has been successfully cloned from *Streptomyces* associated with the hind-gut of *Reticulitermes* (Coppolecchia et al., 1986; Mastromei et al., 1987).

Fig. 3 - Interactions between a higher termite, *Macrotermes michaelsenii*, and microbiota associated with it i.e. the symbiotic fungus *Termitomyces* sp. and bioceonotic actinobacteria.

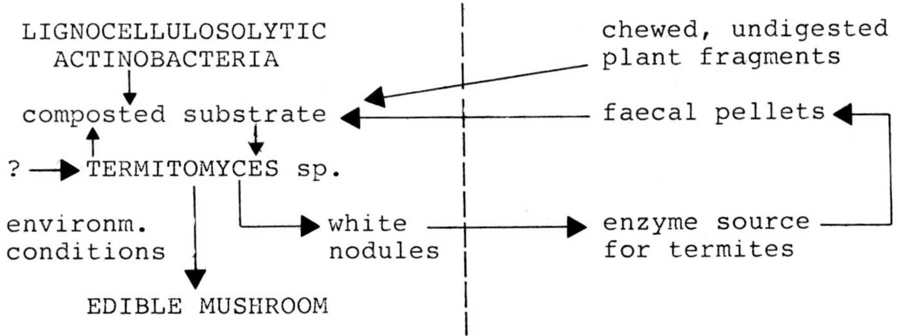

Fig. 4 - Scanning electron micrograph of a "white nodule" growing in the fungus comb (courtesy of prof. R. Locci).

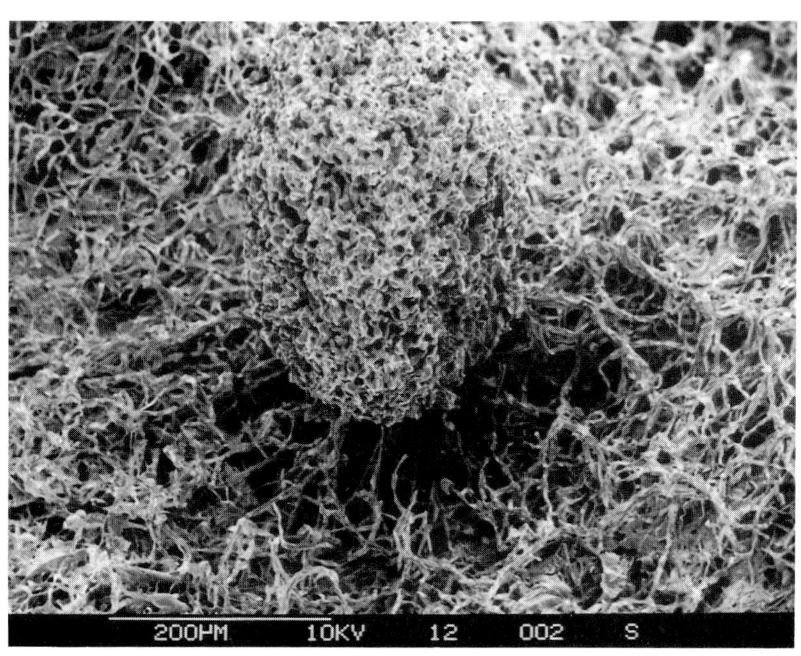

4. REFERENCES

Breznak J.A. (1982) - Intestinal microbiota of termites and other xylophagous insects. Ann. Rev. Microbiol. 36, 323-343.

Coppolecchia R., Dessi M.R., Lepidi A., Mastromei G., Nuti M.P., Polsinelli M. (1986) - Cloning of cellulase genes from Streptomyces A2. In "Genetic, physico-chemical approach for analysis of biological catalysts", Florence (Abstr. vol.).

Dean-Ross D. (1986) - Release of genetically engineered organisms: hazard assessment. Amer. Soc. Microbiol. News 52, 572-575.

De La Pena P.J.J., Schilperoort R.A. (1984) - Transgenic rye plants obtained by injecting DNA into young floral tillers. Nature 325, 274-276.

Dénarié J., Truchet G. (1979) - La symbiose Rhizobium-Légumineuses: roles respectifs des partenaires. Physiol. Veg. 17, 643-667.

Filser M., Moscatelli C., Lamberti A., Vincze E., Guida M., Salzano G., Iaccarino M. (1986) - Characterization and cloning of two R.leguminosarum genes coding for glutamine synthetase activities. J. Gen. Microbiol. 132, 2561-2569.

Gresshof P.M., Olssen J.E. Day D.A., Schuller K.A., Mathews A., Delver A.C., Krotzky A., Price G.D., Carrol B. (1987) - Plant Host genetics of nodulation initiation in soybean. In "Molecular Genetics of Plant-Microbe Interactions" (D.P.S. Verma and N. Brisson Eds.) Martinus Nijhoff Publ., Dordrecht, pp. 85-90.

Grimsley N., Hohn T., Davies J. W., Hohn B. (1987) - Agrobacterium-mediated delivery of infectious maize streak virus into maize plants. Nature 325, 177-179.

Haugland R.A., Cantrell M.A., Beaty J.S., Hanus F.J., Russell S.A, Evans H.J. (1984) - Characterization of R.japonicum hydrogen uptake genes. J. Bacteriol. 159, 1006-1012.

Hooykaas P.J.J., Schilperoort R.A. (1984) - The molecular genetics of crown-gall tumorigensis. Adv. Genetics 22, 209-283.

Knight C..D., Durbin R.D., Langston P.J. (1986) - Role of glutamine synthetase adenylation in the self-protection of Pseudomonas syringae subsp. tabaci from Its-toxin, Tab- toxinine- β-lactam. J. Bacteriol. 166, 224-229.

Mastromei G., Coppolecchia R., Dessi M.R., Nuti M.P., Polsinelli M. (1987) - Cloning of Streptomyces cellulase genes. In "Proc. IV Eur. Cong. on Biotechnol." Amsterdam (Abstr. vol.).

Nap J.P., Moerman M. van Kammen A., Govers F., Gloudemans T., Fraussen H., Bisseling T. (1987) - Early nodulins in root nodule development. In "Molecular Genetics of

Plant-Microbe Interactions" (D.P.S. Verma and N. Brisson Eds.) Martinus Nijhoff Publ. Dordrecht, pp. 14-19.

Nuti M.P., Casella S., Pasti M.B. (1984) - Evaluation of rhizobia genetically engineered for pesticide resistance. In "Advances in Nitrogen Fixation Research" (C. Veger, W.E. Newton Eds.) pp. 268, Nijhoff/ Junk/Pudoc, The Hague.

Nuti M.P. (1987) - Gene expression and environmental quality. In "Proc. IV Int. Symp. Environm. Biogeochem.", in the press.

Nuti M.P., Pasti M.B., Squartini A. (1987) - Impatto delle biotecnologie innovative sul settore agro-alimentare. l Latte XII, 345-351.

O'Brian R.W., Slaytor M. (1982) - Role of microorganisms in the metabolism of termites. Aust. J. Biol. Sci. 35, 239-262.

O'Gara F., Boesten B., O'Regan M., Kiley B., Higisson B., Condon K., Birkenhead K., Manian S. (1987) - Molecular biology of genes involved in carbon metabolism in R.meliloti and B.japonicum. In " Molecular Genetics of Plant-Microbe Interactions" (D.P.S. Verma, N. Brisson Eds.), pp. 295-297, Nijhoff Publ., Dordrecht.

Osore H., Okech M.A. (1983) - The partial purification and some properties of cellulase and β-glucosidase of Termitomyces conidiophores and fruit bodies. J. appl. Biochem. 5, 173-179.

Pasti M.B., Belli M.L. (1985) - Cellulolytic activity of Actinomycetes isolated form termites (Termitidae) gut. FEMS Microbiol. Lett. 26, 107-112.

Priest F.G. (1984) - Strain selection and improvement. In "Extracellular enzymes", Aspects of Microbiol. vol. 9 (J.A. Cole, C.J. Knowles, D. Schlessinger Eds.), Amer. Soc. Microbiol., pp. 55-62.

Rogowsky P., Close T.J., Kado C.J. (1987) - Dual regulation of virulence genes of Agrobacterium plasmid pTiC58. In "Molecular Genetics of Plant-Microbe Interactions"

(D.P.S. Verma and N. Brisson Eds.) pp. 14-19, Martinus Nijhoff Publ., Dordrecht.

Thomas R.J. (1981) - Ecological studies on the symbiosis of *Termitomyces* Heim with Nigerian Macrotermitinae. Ph. D. Thesis, University of London.

Tichy H.V., Richter P., Lotz W. (1985) in "Genetic Engineering of Plants and Microorganisms important for Agriculture" (Magnien E., de Nettancourt D., Eds.) pp. 64-65, Nijhoff/Junk, Dordrecht.

Verma D.P.S., Lee J.S., Katinakis P., Sutton B. (1985) - Nodule specific genes of soybean. In "Analysis of plant genes involved in the Legume-*Rhizobium* symbiosis", pp. 74-84, OECD, Paris.

ACKNOWLEDGEMENTS

Research work supported by CNR, ITALY. Special grant I.P.R.A. Subproject 1, paper n° 1498.

THE APPLICATION OF MONOCLONAL ANTIBODY TECHNOLOGY TO THE STUDY OF CELL-CELL INTERACTIONS

Green, J.R., Jones, J.L. and Callow, J.A.
Department of Plant Biology, University of Birmingham, U.K.

1. INTRODUCTION

Cell signalling and recognition are key components in the processes involved in setting up cell-cell interactions and in particular those involved in symbiotic relationships. The successful outcome of a symbiotic relationship will depend on establishing and maintaining an interface between the organisms involved. Some or all of these processes involve the cell surface and cell surface receptors. In animal systems the plasmamembrane is the main structure involved, whereas plants and bacteria have the added complication of the cell wall. Evidence from a variety of systems has indicated that it is the cell surface molecules, comprising membrane glycoproteins or extracellular polysaccharides which are important in cell interactions and this chapter will focus on this particular aspect.

The identification and characterisation of cell surface molecules has been facilitated in the last ten years by the advent of monoclonal antibody (Mab) techniques (Galfre and Milstein, 1981; Barclay and Williams, 1986). Using Mabs it is possible to analyse complex mixtures of molecules, such as those making up a membrane, without having any previous knowledge about the detailed structure of the components of the mixture. In addition, it is not necessary to have absolutely pure material with which to immunise the animals. After immunising a mouse or rat with the antigens of interest (cells, membranes or glycoproteins/polysaccharides) a large number of antibody secreting hybridomas are generated (see fig. 1) . Antibodies of interest can be screened in binding assays (ELISA, radio-immunoassay) or microscopically (indirect immunofluorescence, immunocytochemistry or E.M. immunogold). Mabs which identify molecules of interest can then be used to study the structure and function of the membrane from which the antigens were derived. The first section of this chapter will discuss some of the key results concerned with cell surfaces which have been obtained using Mabs. The second section will outline some of the questions relevant to symbiotic interactions which can be addressed using Mabs. The final section will discuss the approach we are using to study cell recognition in a model system, between gametes of the brown alga *Fucus* during fertilisation.

NATO ASI Series, Vol. H17
Cell to Cell Signals in Plant, Animal and Microbial Symbiosis. Edited by S. Scannerini et al.

2. ANALYSIS OF CELL SURFACE COMPONENTS USING MONOCLONAL ANTIBODIES

Mabs have been used to study the distribution, localisation, molecular properties and functional associations of membrane components. Much of this work has been performed on cells derived from rats , mice and man, and in particular on the cells involved in the immune system itself. Thus, many of the examples given below will refer to these systems. As yet there is very little information regarding the components of the plant plasmamembrane, although there is a large amount of data on the structure of the plant cell wall.

a) Distribution and localisation.

The distribution of particular antigens on different cell types and tissues within an organism can be investigated using Mabs in binding assays on cells derived from different regions (e.g. brain or thymus), and also by immunocytochemistry and immunohistochemistry. More detailed localisation can be obtained at the ultrastructural level using immunogold labelling of sections and E.M.

Using these techniques with a large number of Mabs to cell surface glycoproteins on animal cell surfaces (e.g. in the rat) it seems that the molecules recognised are distributed in three different ways. Some molecules are expressed on a wide variety of cell types: a good example of this would be the class I major histocompatibility antigens. Some molecules have a more restricted distribution to a few cell types: examples of these include the rat markers Thy-1, W3/13 and MRC-OX-2 which are expressed on lymphoid cells and neuronal cells. Finally there are a small number of molecules which are cell type specific. Molecules restricted to a single cell type are more dificult to find as compared with the other two categories and two examples are the surface immunoglobulins of B-lymphocytes and the T cell receptor for antigen (Barclay and Williams, 1986).

Cell surface antigens can either be localised to particular domains on the membrane or they can be distributed evenly over the whole cell. For example, gut epithelial cells have distinct regions on their surface, the apical and basal regions, which are made up of different membrane components and which carry out separate functions. The plasmamembrane of male gametes usually comprises a number of domains covering flagella/tail regions and body/ head regions. Resting lymphocytes and unfertilised eggs usually display their surface molecules in a random pattern over the cell. In a study of the distribution of the auxin carrier protein in pea stems, a Mab against this molecule was localised in the plasmamembrane of elongated cells in the stem specifically to their basal ends (Jacobs and Gilbert,1983). It therefore seems possible that the general findings for distribution/localisation patterns in animals will also apply to plants.

b) Molecular characterisation.

Mabs have been used extensively to characterise the surface molecules which they recognise. Using Western blotting or immunoprecipitation it is possible to assess the molecular weight of the antigen identified and by use of periodate oxidation and protease treatments it is possible to assess whether the antibody is binding to protein or carbohydrate determinants. The antibodies can also be coupled to sepharose to make affinity columns which can be used to purify antigens of interest. In animal systems most of the antigens identified by Mabs have been integral membrane glycoproteins (one notable exception is the Thy-1 antigen). Initially the primary structure of some of these membrane glycoproteins was elucidated by protein sequencing but this has now been superseded by using molecular biology techniques and obtaining the sequence of DNA corresponding to the antigen in question. Mabs have an important part to play in this latter approach too since the purified antigens can be used to obtain partial sequence data for constructing oligonucleotide probes.

Sequence data on a variety of cell surface molecules has shown that some of these molecules can form superfamilies of related structures. One example of this is the immunoglobulin superfamily (Barclay and Williams, 1986; Hunkapiller and Hood, 1986). Members of this superfamily all share a common sructure called the immunoglobulin homology unit, composed of about 100 amino acid residues and characterised by a centrally placed disulphide bridge that stabilises a series of anti-parallel beta strands into the antibody fold. The variable (V) and constant (C) homology units, defined for the respective portions of the immunoglobulin chains from which they were identified, have similar but distinct three-dimensional structures. Members of this superfamily include a variety of molecules involved in immune function (including immunoglobulins, T-cell receptor, MHC class I and II antigens) and others which function in other capacities (including Thy-1, OX-2 and N-CAM). One point of interest is that many of these structures have functions associated with recognition and this will be further discussed below.

In addition to providing information on the protein components of cell surface molecules, Mabs have also enabled the carbohydrate portion of these molecules to be analysed. For mammalian glycoproteins it seems that antigenic determinants are more often based on protein than carbohydrate probably because mammals have in common a wide range of their carbohydrate structures. Thus immunisation across species does not generally provoke production of anti-carbohydrate antibodies. However there are cases where anti-carbohydrate antibodies are produced and these include immunisations using foetal material and tumour cells (Feizi, 1985). The carbohydrate may be immunogenic in these cases because carbohydrate structures not generally in contact with the adult immune system are expressed on these tissues. Many of the Mabs generated so far against plant membranes and glycoproteins seem to recognise carbohydrate determinants (e.g. *Chlamydomonas* cell wall glycoproteins (Smith et al, 1984)) .

Anti-carbohydrate antibodies might be expected to react with a number of different glycoproteins/glycolipids, since on any one cell, carbohydrate structures are likely to be shared by a number of different glycoproteins. In contrast anti-protein antibodies might be expected to react with a specific molecule, although there are possibilities of cross reactions involving a short segment of sequence in proteins that are otherwise unrelated. Since the use of Mabs to dissect complex antigenic mixtures of glycoproteins depends on the fact that molecule-specific antibodies recognising protein determinats can be obtained, one would usually try to avoid producing antibodies against the carbohydrate portion of the molecule. However it is clearly possible to obtain MAbs to specific sugar sequences which could be used to analyse complex carbohydrates and to probe the function of oligosaccharide side chains. It should also be noted that the determinants recognised by antibodies are small and correspond to about three to six saccharide or amino-acid residues. This means that very detailed analysis of specific epitopes making up a molecule can be analysed.

c) Functional associations.

Monoclonal antibodies can be used to separate cells on the basis of their surface markers using the fluorescence activated cell sorter (FACS) and the functions of these cells can then be assessed in various ways. In addition Mabs can be used to perturb *in vitro* assays, indicating a functional association of the antigen recognised by the antibody (Mason et al, 1983). Thus Mabs have been shown to affect processes such as proliferation and differentiation and more specifically to block recognition processes. Two instances in which recognition/adhesion structures have been investigated are those involving a) N-CAM and b) LFA-3:CD2. Neural cell adhesion molecule (N-CAM) is a brain specific molecule which mediates adhesion of neuronal cells to each other. Experiments using antibodies and purified N-CAM have shown that binding is homopholic (self). It is of interest that this molecule is one of the Ig superfamily (Edelman, 1983).

Interactions between different cell types are usually mediated by different cell surface antigens. Antibodies to LFA-3 (found on a variety of cell types in the mouse) and CD2 (found on mouse lymphocytes) have been used to purify these antigens and to show that they bind to each other, this being an example heterotypic binding (Selvaraj et al 1987).

3. CELL-CELL INTERACTIONS

The key questions which can be approached using Mab technology in relation to cell-cell interactions, in particular symbiotic interactions are as follows:

1. What is the molecular nature of the receptors involved in recognition and adhesion?

The first step in many of the symbioses is recognition and adhesion between the host and symbiont. Mabs coulds be used to identify the molecules involved in these processes. The antibodies would have to be selected on the basis of functional assays designed to select for those that will block recognition/adhesion.

2. What are the structural components of the interface between the two organisms involved in the interaction?

Often a specialised region of membrane, or interface, occurs between the organisms forming the symbiosis (e.g. extrahaustorial membrane, peribacteroid membrane, vacuole) . By preparing Mabs to molecules present at the interfacial region it would be possible to characterise the structures present. Obviously this task would be facilitated if the membranes at the interface could be isolated away from the main bulk of membrane material so as to be able to immunise mice with an enriched fraction. Mabs could be screened in binding assays and by EM Immunogold techniques.

3. How does the interface differ from the normal membrane or surface of each participant?

A panel of Mabs which bound to antigens present in the normal membranes of each symbiont could be used to determine whether particular antigens were either included or excluded from the interfacial region.

4. Which components of the interface have been derived from each organism?

This problem could be addressed by determining whether the Mabs identifying antigens of interest also labelled the Golgi membranes, endoplasmic reticulum and assorted vesicles within the two cell types involved in the symbiosis.

5. How does the interaction develop during a time course?

Indirect immunofluorescence or EM immunogold tequniques with Mabs could be used to examine the development of a specialised structure during a time course of the interaction. In this way it should be possible to see how soon the surfaces of each symbiont alter after initiation of the interaction.

6. What are the functions of the components of the interface?

Molecules within the interfacial region could be involved in recognition processes, suppressing rejection mechanisms and in membrane transport of a variety of nutrients. Identification of functional associations requires the development of assays which can be used to measure or assess a particular function. If this can be achieved then Mabs can be used to perturb these in vitro assays, so that antigens with particular functions can be identified.

4. FERTILISATION IN *FUCUS*

Fertilisation in the marine brown algae of the genus *Fucus* has proved to be an ideal system for studies of cellular recognition in plants. The interaction between sperm and egg is highly species specific and it can be easily observed and assessed in a quantitative bioassay. In addition the gametes are naked protoplasts and can be obtained in large quantities (Callow, 1985) . We are using monoclonal antibody technology to study the molecular basis of fertilisation and recognition in *Fucus*.

a) Fertilisation

The mature *Fucus* plant bears receptacles at the tips of its branches and when fertile these contain many conceptacles. Each conceptacle opens out onto the surface by an ostiole from which sterile filaments protrude. The conceptacles of male plants contain antheridia in which the spermatozoids develop and of female plants, oogonia in which the eggs develop. The *Fucus* eggs are brown, spherical (about 80 μm in diameter) and are enclosed only within a plasmamembrane. In contrast the sperm are pear-shaped cells, biflagellate (the body is about 5 μm long) and bright orange in colour due to carotenoids accumulating in the eyespot region of the chloroplast. The shorter anterior flagellum bears mastigonemes on its surface.

Initially the sperm are attracted to the egg by a pheromone. After sperm make contact with the egg there is species-specific binding followed by fusion of the sperm and egg plasmamembranes (plasmogamy). This is followed by the rapid release of a polyuronide alginic acid by the egg which covers the egg surface with a cell wall and may contribute to a block to polyspermy,

This rapid release of wall material forms the basis of an assay for fertilisation, since it can be visualised with the fluorescent brightener Calcofluor White ST. Eggs to which sperm have been added are incubated in Calcofluor and then examined with a fluorescence microscope. The proportion of fertilised eggs can be measured by counting those which are fluorescing brightly (Callow et al, 1978).

b) The molecular basis of gamete recognition

The initial chemoattraction of sperm to eggs is mediated by an octotriene and is not species-specific. Specificity probably lies therefore at the level of sperm binding and plasmogamy. Using the bioassay for fertilisation it has been possible to establish that pre-treatment of eggs with two glycosidases (α-fucosidase and α-mannosidase) inhibits fertilisation. In addition the lectins ConA and RCA 120 bind strongly to eggs but not to sperm and this inhibits fertilisation. Fucose-binding protein however binds to both gametes causing inhibition of fertilisation. Glycoprotein fractions have been isolated from egg membranes using lectin affinity columns. The glycoproteins bind to sperm and cause species-specific inhibition of fertilisation. In a further study, a polyclonal antiserum against surface antigens of *Fucus* sperm flagella inhibited fertilisation in a species specific manner (summarised in Callow, 1985). This evidence taken as a whole suggests that there are specific molecules present on the surfaces of eggs and sperm which are involved in the recognition and fertilisation process. The interactions also seem to involve carbohydrate determinants which are possibly part of the structure of these molecules. We are now attempting to identify and characterise these surface receptors using Mab techniques. The main objectives of the work are to determine:

Figure 1. Strategy for production of monoclonal antibodies against *Fucus serratus* (F.s.) sperm

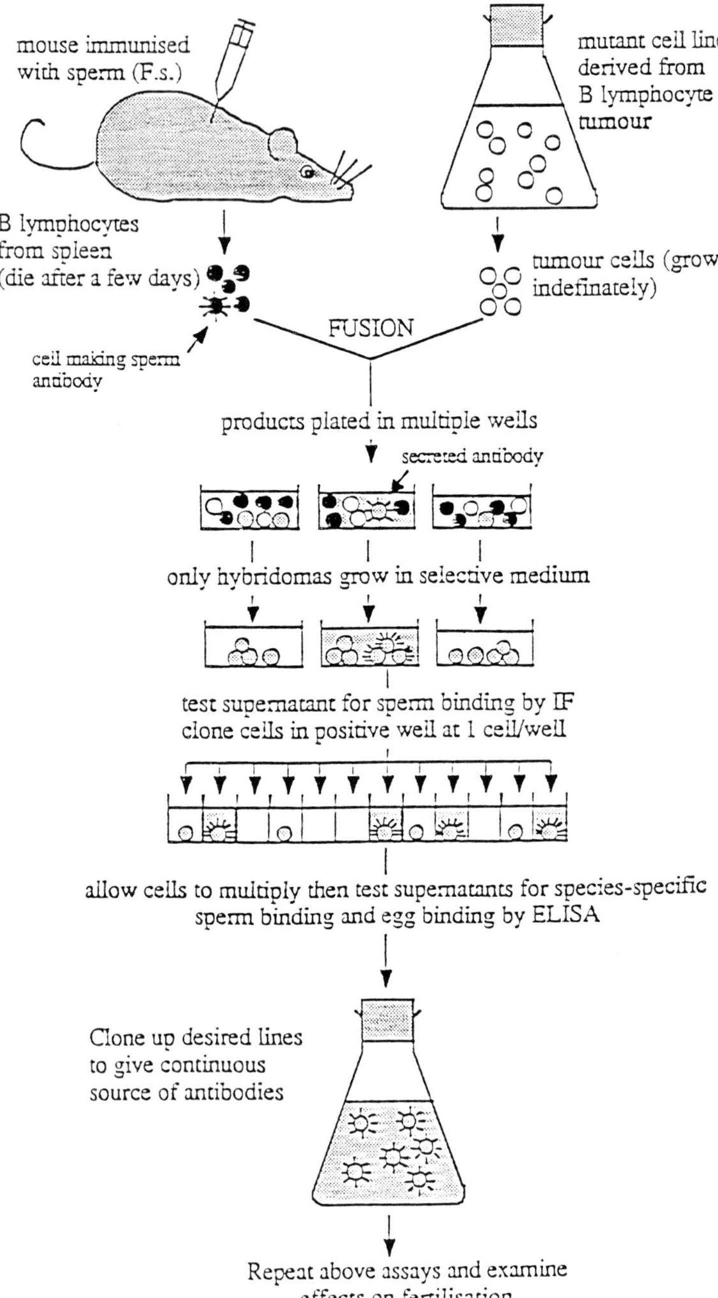

1. The molecular nature of the receptors governing species-specific egg-sperm recognition.
2. The cellular location and distribution of the receptors.
3. How the receptors of one species differ from a related species.

c) Strategies in the production of Mabs against Fucus sperm cells

The strategy we are using in this work is to prepare Mabs to cell surface antigens of the *Fucus* gametes, followed by selection of those antibodies that will block the fertilisation assay. The protocol outlined in fig.1 has been used to generate Mabs against the sperm cell surface. Mice were immunised with intact, live sperm cells from *Fucus serratus*. After fusion of the immune spleen cells with the NSO myeloma cell line and selection in HAT, the tissue culture supernatants from growing hybridomas were tested after 10-14 days. The supernatants were tested for binding in ELISA using fixed sperm cells and fixed egg vesicles and they were also tested using indirect immunofluorescence of fixed sperm cells.

In the initial fusions a large number of antibody-producing cell lines were obtained and in order to reduce the number of antibodies to be tested in the second stage antibodies were selected on two criteria. The first was that Mabs which bound specifically to sperm cells and did not bind to eggs were selected on the basis that previous findings on recognition systems involving two different cell types generally employ different receptors. In addition, the experiments cited above, using lectins to block fertilisation in *Fucus,* suggested a heterophilic interaction. The second was that Mabs which identified antigens with restricted distribution on the surface of the sperm cell were also selected, whether they bound to eggs or not, since these would be of interest with regard to the overall organisation of the sperm cell surface.

A summary of the binding characteristics of some of the Mabs which have been selected on these criteria is illustrated in Table 1 (Jones, Callow and Green, manuscript in preparation). The data show that there are several Mabs (FS 4,9,2,5) which bind to membranes from both gametes whereas the remainder of the Mabs bind to sperm cells only. Observations of binding to sperm cells by immunofluorescence shows that there are several different patterns of labelling by the antibodies. One group (FS4,3,8) binds uniformly to the whole of the sperm , including the body and both flagella. Within this group FS4 differs from the other two antibodies in that it also binds to egg membranes. A second group shows preferential binding to the body of the sperm (FS9,10,1,6,) with weaker labelling of both flagella.. In a third group five antibodies (FS7, 2,5,11,12) bind preferentially to the front end of the sperm cell: these can be further subdivided in that FS 7 and FS2 show some binding to the posterior flagellum, whereas FS5,FS11 and FS12 do not bind to this region.. The results show that

TABLE 1. BINDING CHARACTERISTICS OF MONOCLONAL ANTIBODIES TO FUCUS

Mab	Sperm(IIF)			Egg(ELISA)
	A.F.	Body	P.F.	
FS4	+++	+++	+++	+
FS3	++	++	++	-
FS8	+	+	+	-
FS9	+	+++	+	+
FS10	+	+++	+	-
FS1	+	++	+/-	-
FS6	+/-	+	+/-	-
FS7	+++	++	+	-
FS2	++	+	+/-	+
FS5	++	+	-	+/-
FS11	++	+	-	-
FS12	+	+	-	-

Mab : Monoclonal antibody
IIF : Indirect immunofluorescence
A.F. : Anterior flagellum
P.F. : Posterior flagellum

Binding is recorded on a scale of - (no binding), +/-, +, ++ +++ (strong binding)

there are antigens specifically expressed by one gamete (sperm) and not the other. These antigens will presumably have functions associated particularly with the sperm cell, possibly involved in fertilisation. The surface of the sperm cell seems to be made up of a number of domains: some antigens are uniformly distributed over all these domains whereas some have restricted distribution or are concentrated in a particular domain. These results can be compared with data for mammalian sperm , which show that the surface membrane is organised into a number of structural and functional domains (Holt, 1983; Primakoff and Myles, 1983).The next stage of the work involves biochemical characterisation of these surface antigens and testing the Mabs in the fertilisation assay.

REFERENCES

Barclay, A.N. and Williams, A.F. 1986. The lymphocyte cell surface. In: Weir,D. (ed) Handbook of Immunology vol3 . Blackwells, Oxford.

Callow, J.A. 1985.Sexual recognition and fertilisation in brown algae. J.Cell Sci. (Suppl), Proc. 6th John Innes Symposium:Cell surface in Plant Growth and Development 2: 219-232.

Callow,M.E.,Evans,L.V.,Bolwell,G.P.and Callow,J.A.1978 .Fertilisation in brown algae I. SEM and other observations on *Fucus serratus*.. J.Cell Sci. 32: 45-54.

Feizi, T. 1985. Demonstration by monoclonal antibodies that carbohydrate structures of glycoproteins and glycolipids are onco-developmental antigens. Nature 314: 53-55.

Galfre,G. and Milstein, C. 1981. Preparation of monoclonal antibodies:strategies and procedures. Methods Enzymol. 73: 1-46.

Holt,W.V. 1983. Membrane heterogeneity in the mammalian spermatozoon. Int.Rev.Cytol. 87: 159-194.

Hunkapiller,T. and Hood,L. 1986. The growing immunoglobulin gene superfamily. Nature 323: 15-16.

Jacobs,M. and Gilbert, S.F. 1983. Basal localisation of the presumptive auxin transport carrier in pea stem cells. Science 22: 1297-1300.

Mason,D.W.,Arthur,R.P.,Dallman,M.J.,Green,J.R.,Spickett,G.P. and Thomas,M.L. 1983. Functions of rat T lymphocyte subsets isolated by means of monoclonal antibodies. Immunol. Rev. 74 :57-82.

Primakoff,P. and Myles,D.G. 1983. A map of the guinea pig sperm surface constructed with monoclonal antibodies. Dev.Biol. 98 :417-428.

Selvaraj,P.,Plunkett,M.L.,Dustin,M.,Sanders,M.E.,Shaw,S.and Springer,T.A. 1987. The T lymphocyte glycoprotein CD2 binds the cell surface ligand LFA-3. Nature 326 : 400-403.

Smith,E.,Roberts,K.,Hutchings,A. and Galfre,G. 1984.Monoclonal antibodies to the major structural glycoprotein of the Chlamydomonas cell wall. Planta 161: 330-338.

THE USE OF MONOCLONAL ANTIBODIES TO INVESTIGATE PLANT-MICROBE INTERACTIONS IN PEA ROOT NODULES CONTAINING RHIZOBIUM LEGUMINOSARUM

Brewin[1], N.J., Bradley[1], D.J., Wood[1], E.A., Kannenberg[1], E.L., VandenBosch[1], K.A. and Butcher[2], G.W.

John Innes Institute, Colney Lane, Norwich NR4 7UH[1]
Monoclonal Antibody Centre, AFRC, Babraham, Cambridge CB2 4AT[2]

1. INTRODUCTION

a) The Rhizobium-legume symbiosis

The plant-microbe interactions that characterise the Rhizobium-legume symbiosis may be divided somewhat arbitrarily into those concerned with nutritional interactions (biotrophy) and those concerned with nodule development (morphogenesis). At the most intimate stage in the symbiosis, several thousand Rhizobium bacteroids are sustained within each infected plant cell. The host plant maintains a microaerobic oxygen concentration in the vicinity of the bacteroids, sufficient to permit oxidative respiration without denaturation of the highly oxygen-sensitive nitrogen fixation enzyme system. The host plant cells also regulate the supply of respiratory substrates needed by the bacteroids, and assimilate ammonia produced and excreted as a result of N_2 fixation. However, before these elaborate physiological interactions can take place, another series of Rhizobium-legume interactions must occur in order to bring about the specific infection of legume roots by Rhizobium and the subsequent differentiation of the highly organised legume nodule structure. As our understanding of these morphogenetic processes unfolds, it will be interesting to compare and contrast these symbiotic interactions with those between a plant and a microbial pathogen. Furthermore, the study of the infection processes for both microbial symbionts and microbial pathogens should help to extend our knowledge and understanding of the normal processes that govern plant cell organisation and morphogenesis.

NATO ASI Series, Vol. H17
Cell to Cell Signals in Plant, Animal and Microbial Symbiosis. Edited by S. Scannerini et al.

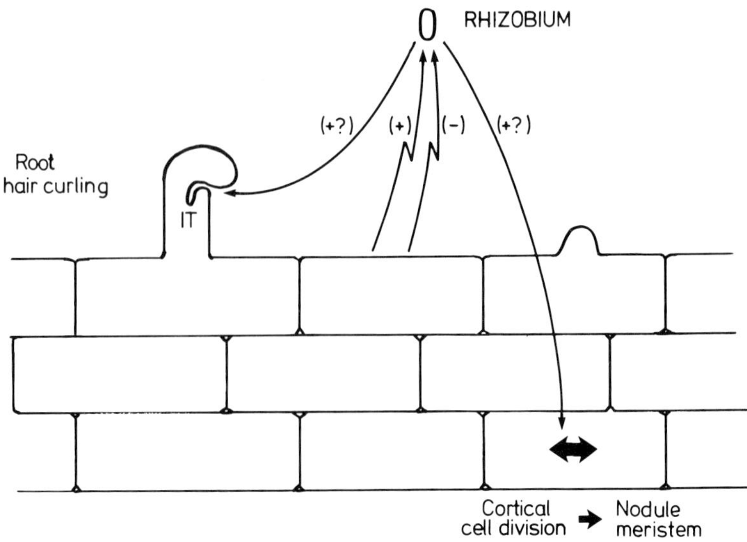

Figure 1. Successive stages in root nodule morphogenesis.
Above. 1: Root exudates act as positive and negative inducers of *Rhizobium* *nod* gene transcription; 2: at the site of emergence of root hairs, cortical cell divisions generate a nodule meristem; 3: root hair curling diverts the point of cell wall growth from the apical tip to the inwardly growing infection thread.
Below. 1: The infection thread traverses the cortex as an intracellular tunnel sheathed by plant cell membrane and plant cell wall; 2: the infection droplet lacks a plant cell wall and bacteria are engulfed by plant cell membrane; 3: within the plant cytoplasm, bacteria divide and differentiate into nitrogen-fixing bacteroids, and at all stages they remain enclosed by a plant-derived peribacteroid membrane; 4: uninfected nodule cells differentiate for nitrogen assimilation and transport.

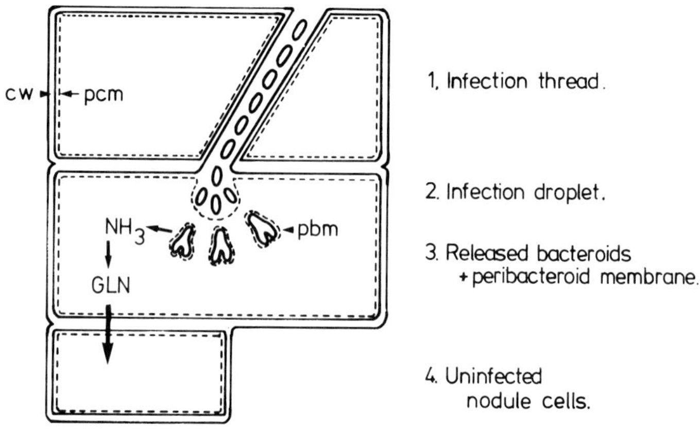

b) Nodule morphogenesis

Stages in the development of the legume root nodule are summarised in Figure 1. It can be seen that the internalisation of Rhizobium by plant cells proceeds in two stages: first, to form the infection thread tube, and second, to form the peribacteroid membrane. In the first stage the plant cell surface is sheathed by plant cell wall and the bacterium by capsular polysaccharide (Finan et al, 1985), whereas in the second stage the plant cell membrane is exposed directly to the bacterium within the infection droplet or the peribacteroid space. In order to investigate these very interesting surface interactions, a combination of biochemistry and cell biology is needed. As an approach to this problem, we have attempted to produce monoclonal antibodies that will help to identify and analyse individual molecular components that are involved in plant-microbe surface interactions within the legume root nodule. Our studies have focussed on antigens of the peribacteroid membrane or bacteroid surface. The immunogens used have been derived from a preparation of membrane-enclosed bacteroids isolated after crude sucrose gradient fractionation of nodule homogenates (Brewin et al, 1985, 1986; Bradley et al, manuscript in preparation). A similar experimental approach involving monoclonal antibodies could well prove useful for the study of other host-symbiont interactions, and therefore in this article we will concentrate on methodological aspects of our research programme.

c) Why choose monoclonal antibodies?

The isolation of monoclonal antibodies (McAb) involves a considerable investment of effort and expense, and therefore their use in preference to conventionally produced antisera must be carefully justified for each individual research application. A major advantage of monoclonal antibodies is that a hybridoma cell line provides an inexhaustible supply of monospecific antibody so that, in effect, the McAb can be used as an off-the-shelf biochemical reagent or specific molecular probe which is itself capable of being further modified, e.g. by conjugation to sepharose, iodine-125, or colloidal gold, in order to diversify the range of applications available. Another advantage of McAbs is that they can be generated in response to relatively small amounts of antigen in complex and otherwise intractable biological systems which need not necessarily be pure or even well characterised (Brewin et al, 1985, 1986): once isolated, the McAb can subsequently be used as a molecular probe for the further molecular dissection of such a system. A third advantage of McAbs exploits their unique antigen specificity which is of crucial importance for immunochemical localisation studies. Furthermore, the use of a number of different McAbs that react with different sites (epitopes) on the same macromolecule make it possible to learn about the molecular architecture within that macromolecule (Thomas et al., 1984). In our case, this may be relevant to the study of plant membrane glycoproteins and to bacterial lipopolysaccharide (LPS) molecules.

2. PRODUCTION OF MONOCLONAL ANTIBODIES

The principles and practice of hybridoma technology are described in many review articles (Goding, 1987; Galfre & Butcher, 1986; Galfre & Milstein, 1981). In order to derive hybrid myelomas a large number of spleen cells from an immunised animal are fused with cells of a suitable myeloma line by the use of polyethylene glycol. Hybrid cell lines are isolated using a form of 'HAT' selection, which involves growth on culture medium supplemented with hypoxanthine, aminopterin and thymidine. The original myeloma line is unable to grow on this medium because it is sensitive to aminopterin as a result of a mutation either in the enzyme hypoxanthine-guanine phosphoribosyl transferase (which is necessary for the synthesis of purines from hypoxanthine) or in the enzyme thymidine kinase (which is necessary for growth on exogenous thymidine). The spleen cells (lymphocytes) are also unable to grow in vitro for more than a few cycles in culture medium, and hence the only survivors of the selection are hybrid myelomas.

The choice of species for the efficient production of hybridomas is confined mainly to mice, rats or hamsters. Cross-species fusions can be successful, but with increasing phylogenetic distance there are increasing problems of karyotype instability and low fusion efficiency. Whenever possible, intra-species fusions should be performed using inbred strains of animal, particularly if there is any intention to produce ascites fluid, because this can only be achieved if there is histocompatibility between the introduced hybridoma line and the host.

A crucial element in the success of a fusion experiment is the choice of screening system used in order to identify cell lines producing the most desirable McAbs. The antigen/antibody binding assay must be simple, sensitive and specific, so that several hundred hybrid cell culture supernatants can be screened per day. It is often useful to combine a simple primary screen, e.g dot immunoassay or ELISA, with a more specific secondary screen, such as immunostaining of western blots, immunogold staining of thin sections of tissue on EM grids, competitive antibody binding assay, or immunofluorescence.

As soon as useful hybridoma cell lines have been identified by the antibody screening tests, they are cloned once or twice by limiting dilution into microtitre dishes. This procedure serves two important functions: firstly it ensures that the antibody-producing cell line is genuinely monoclonal, and secondly it ensures that a relatively stable antibody-producing karyotype is selected as a derivative of the original polyploid hybridoma. After cloning, the cell line can be either frozen in liquid nitrogen or grown in culture bottles to generate large quantities of culture supernatant containing antibody. Alternatively, the hybrid cell line can be injected into the peritoneal cavity of an animal in order to generate an ascites tumour which gives rise to a very high concentration of antibody (20 mg/ml) in the peritoneal fluid. However, it should be remembered that ascites fluid is not truly 'monoclonal' because it also contains other antibodies naturally present in the animal which harbours the ascites tumour.

3. MODIFICATIONS OF THE ANTIGEN

When preparations of plant material are used as immunogens for rats or mice in order to generate a range of hybridoma cell lines, it is frequently found that certain classes of monoclonal antibody specificity predominate among the cell lines isolated, and these all react with one or a few immunodominant antigens, especially carbohydrate groups (Anderson et al, 1984; Brewin et al, 1985, 1986). Modified immunisation or screening strategies are needed in order to diversify the range of McAb specificities obtained, and this is particularly necessary if, for example, tissue-specific antigens are being sought.

a) Immunodepletion

The principle of this system is that the most abundant class (or classes) of antibody raised from a first fusion experiment are conjugated to sepharose and used as an immunoaffinity column to deplete the corresponding antigen(s) from the material to be used as immunogen in a subsequent fusion experiment. A possible disadvantage of this procedure is that certain classes of interesting antigen could be lost during the immunodepletion step as a result of a physical association with the immunodominant antigen whose specific removal is being sought.

In our own studies we found that, using a crude peribacteroid membrane

(PBM) preparation for immunisation of rats, we isolated large numbers of McAb (typified by AFRC MAC 64) that all reacted to the same carbohydrate epitope present on a series of plant membrane proteins. In addition, a large number of McAb were isolated that were typified by AFRC MAC 57 and reacted with crude bacteroid lipopolysaccharide (LPS), which was a major contaminant of these crude peribacteroid membrane preparations. Thus, by passing the crude PBM preparation through sepharose immunoaffinity columns containing MAC 57 and MAC 64, it was possible to deplete the corresponding antigens from the immunogen. As a result, three new classes of McAb antigen specificity were identified in the hybridoma cell lines derived from this myeloma fusion.

b. Deglycosylation

Because most of the antibodies isolated from the first fusion experiments reacted with carbohydrate antigens, we deglycosylated a preparation of peribacteroid material using trifluoromethane sulphonic acid (Edge et al 1978). After this treatment, the PBM material did not react with either MAC 57 or MAC 64, but unfortunately the residual material was very insoluble, perhaps as a result of the removal of solubilising sugar groups from otherwise hydrophobic membrane glycoproteins. Thus, although some faint protein bands could be visualised on SDS gels after silver staining, the rat immunised with deglycosylated PBM material only yielded a single antibody specificity class from the hybridoma lines screened after fusion. In future, periodate treatment might be preferable as a milder form of deglycosylation, although this has also been found to generate some insoluble protein material.

c. Specific screening

Neither immunodepletion nor deglycosylation has proved to be entirely satisfactory as procedures for extending the range of antigen specificities. More success has been achieved by using normal immunogen and putting a lot of effort into the screening systems used to discriminate the antigen specificities of the antibody-producing cell lines. By careful examination of staining patterns on western blots and the protease and periodate sensitivity of antigens, it was possible to pick out four antibody-producing cell lines with novel antigen specificities from among 178 that were similar to MAC 57 or MAC 64.

4. MODIFICATIONS OF THE IMMUNE RESPONSE

An alternative approach to the problem of avoiding immunodominant antigens is to precondition the animal so that it responds differently to the immunogens that are presented. We are currently investigating these techniques as a way to enrich for hybridoma cell lines producing monoclonal antibodies that react with nodule-specific and non-carbohydrate antigens.

a) Tolerisation

Suppression of the immune response can be used as a means to manipulate the pattern of antibody-producing cells within immunised rats and mice (Matthew & Paterson, 1983). One technique is to induce tolerisation with immunosuppressive drugs such as cyclophosphamide, and another is to inject neonatal animals with a preparation of antigen. A subsequent immunisation with a related antigen will then result in the production of antibodies only to those antigens which are 'novel' and for which the animal has not become tolerised. In principle, this is a good way of identifying antigens that are specific to a particular tissue cell type or differentiation state (Golumbeski & Dimond 1986), but problems are often encountered with the immunisation of new-born animals and the mortality rate can be extremely high.

b) Coimmunisation

An alternative to the tolerisation technique has recently been described which appears to be very simple and straightforward (Barclay & Smith, 1986). In this study, an attempt was being made to isolate hybridoma cell lines producing antibodies that react with *Dicytostelium* cell surface antigens that were specific to the differentiated (aggregating) state. This objective was achieved by a primary immunisation of BALB/c mice with living aggregation stage cells, followed by secondary immunisation with a preparation of aggregating cells mixed with polyclonal BALB/c antiserum derived from isogenic mice that had been immunised with undifferentiated *Dicytostelium* cells. By this simple stratagem, approximately 20% of all anti-*Dicytostelium* McAbs obtained from fusion experiments were specific for differentiation antigens, whereas, if the polyclonal antiserum was not included, none of the 100 clones isolated yielded McAb reacting specifically with differentiation antigens.

There is not yet a satisfactory immunological explanation for the success of this coimmunisation scheme, although the phenomenon is well documented. One plausible theory is that of 'antigen deviation', which suggests that the antigen is sequestered away from the most antigenic state as a result of its association with the coimmunised antiserum (McKearn, 1978). An alternative explanation, discussed by Barclay and Smith (1986), is that the coinjected polyclonal serum induces the synthesis of anti-idiotype antibodies which in turn modify the immunological response of the immunised animal.

This method of enriching for McAbs that react with a specific subset of potential antigens has several advantageous features. Firstly, it is much less time-consuming than cascade selection methods which require successive cycles of immunodepletion in order to remove unwanted antigens from a complex mixture. Secondly, it has the potential of permitting various combinations of antisera or McAb to be added to the primary immunisation mixture in order to elicit a precise selective response. Thirdly, this method might help to overcome the problem of common antigens (frequently oligosaccharides) being shared by otherwise dissimilar macromolecules. Because these oligosaccharides are strong immunogens they frequently confound attempts to isolate McAbs that uniquely define a single species of glycoprotein molecule. However, following the coinjection of a reactive antiserum, it should be possible to suppress the immunogenicity of these abundant oligosaccharide side-chains.

5. SUMMARY OF MONOCLONAL ANTIBODIES OBTAINED

Three different immunisation procedures have been used to isolate monoclonal antibodies reacting with antigens in the peribacteroid membrane (Bradley et al, manuscript in preparation). In the first procedure, the antigen used was the unmodified peribacteroid membrane material recovered by osmotic shock treatment of isolated membrane-enclosed bacteroids. In the second procedure this material was immunodepleted of MAC 57 antigen (bacterial LPS) and MAC 64 antigen (plant membrane glycoproteins) because these antigens were known to be highly immunogenic. For the third procedure, the peribacteroid material was treated with trifluoromethane sulphonic acid in order to deglycosylate and enrich for protein antigens.

After fusion and screening of cloned hybridoma lines from individual rats, these three immunisation methods yielded respectively 4, 3 and 1 classes of monoclonal antibody (McAb) with unique antigen specificities (Figure 2). Ultrastructural immunogold localisation studies revealed peribacteroid membrane antigens (identified by MAC 202, 206 or 209), and the novel identification of an infection thread matrix glycoprotein of plant origin (recognised by MAC 204). With the exception of MAC 209, all of these epitopes were sensitive to periodate treatment, and with the exception of MAC 206 they were also sensitive to protease digestion: these data indicate that most of our McAb react with the carbohydrate components of plant glycoproteins. Several of the epitopes recognised by these McAb were found to be more abundant in extracts of nodule tissue than in uninfected roots (MAC 64, 202, 204, 206). In the case of the peribacteroid membrane antigens, this may simply reflect the fact that infected plant cells have about 50 times as much antigen-containing membrane material (peribacteroid and plasma membrane) compared to uninfected plant cells where only the plasma membrane is labelled by immunogold staining. In addition, two McAbs allowed the identification of new bacterial antigens: MAC 203 identified a bacterial LPS antigen expressed upon infection but not in free-living cultures of Rhizobium and MAC 115 identified a bacterial polypeptide (55 Kd) that was present in both free-living and bacteroid forms. There were also some McAbs of broader specificity which cross-reacted with antigens present in both plant and bacterial cytoplasms.

A previous study (Brewin et al, 1986) which focussed on antigens of the bacteroid surface, isolated a number of useful McAb. Some LPS antigens were found to be strain-specific (e.g. MAC 113, MAC 114), some antigens were enriched within the nodule (e.g. MAC 57 for strain 8401, pRL1JI and MAC 203 for strain 3841) and some antigens (especially the O-antigen side chains of pea bacteroids) were diminished in nodule bacteria. In the case of the MAC 203 antigen we have recently shown that the expression of this epitope can be induced in free-living cultures when grown in low oxygen concentrations, especially when glycerol is also present.

Using the McAb that we have already isolated, we can now identify a wide range of components of the bacterial surface; the plant cell plasma membrane, the peribacteroid membrane and the infection thread matrix. These McAb can be used for further studies at the levels of cytochemistry, biochemistry and genetics to increase our understanding of surface interactions between the plant and the bacterium at all stages during the development of the Rhizobium-legume root nodule symbiosis (Bradley et al, 1986).

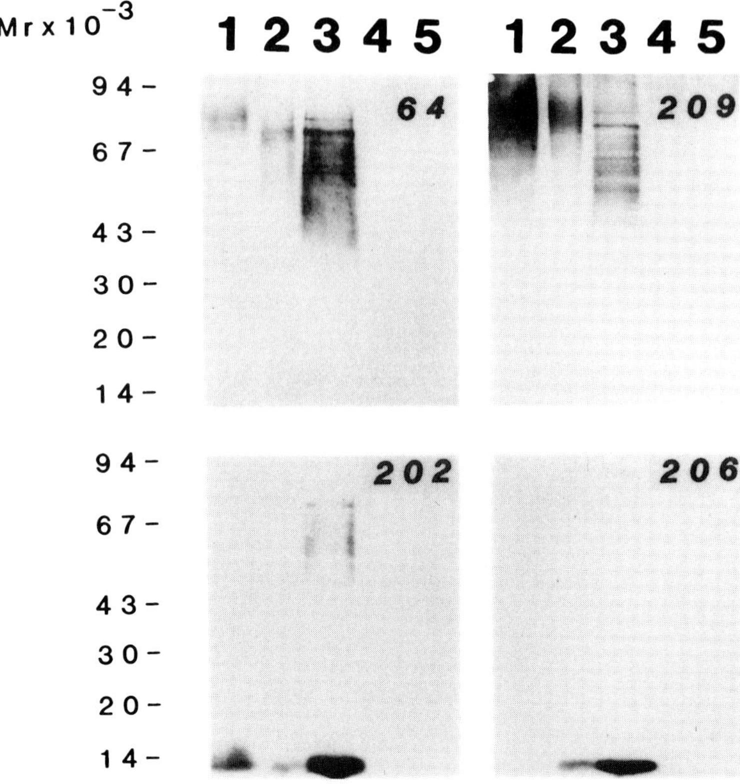

Figure 2. Immunophosphatase staining of Western blots using four McAbs: MAC 64, MAC 209, MAC 202 and MAC 206, showing different patterns of antigen specificity. Gel tracks were loaded with equivalent concentrations of: 1. uninfected pea root homogenate; 2. nodule supernatant; 3. peribacteroid material; 4. bacteroids; 5. free-living bacteria.
Periodate treatment of Western blots destroyed the antibody binding site for MAC 64, MAC 202 and MAC 206. Protease pretreatment before SDS PAGE destroyed the high molecular weight antigens stained by MAC 64, MAC 209 and MAC 202, but not the low molecular weight antigens stained by MAC 202 or MAC 206. When each of these four McAbs were used for immunogold staining of nodule thin sections, the localisation patterns were indistinguishable: all four McAb labelled membrane antigens present in the peribacteroid and plasma membranes.

REFERENCES

Anderson, M. A., Sandrin, M. S., and Clarke, A. E. (1984). A high proportion of hybridomas raised to a plant extract secrete antibody to arabinose or galactose. Plant Physiol. 75, 1013-1016

Barclay, S. L. and Smith, A. M. (1986). Rapid isolation of monoclonal antibodies specific for cell surface differentiation antigens. Proc. Natl. Acad. Sci. 83, 4336-4340.

Bradley, D. J., Butcher, G. W., Galfre, G., Wood, E., Brewin, N. J. (1986). Physical association between the peribacteroid membrane and lipopolysaccharide from the bacteroid outer membrane in Rhizobium-infected pea root nodule cells. J. Cell. Sci. 85, 47-61.

Brewin, N. J., Robertson, J. G., Wood, E. A., Wells, B., Larkins, A. P., Galfre, G. and Butcher, G. W. (1985). Monoclonal antibodies to antigens in the peribacteroid membrane from Rhizobium-induced root nodules of pea cross-react with plasma membranes and Golgi bodies. EMBO J. 4. 605-611.

Brewin, N. J., Wood, E. A., Larkins, A. P., Galfre, G., and Butcher. G. W. (1986). Analysis of lipopolysaccharide from root nodule bacteroids of Rhizobium leguminosarum using monoclonal antibodies. J. Gen. Microbiol. 132, 1959-1968.

Edge, A.S.B., Faltynek, C.R., Hof, L., Reichert, L. E. and Weber, P. (1981). Deglycosylation of glycoproteins by trifluoromethanesulfonic acid. Anal. Biochem. 118, 131-137.

Finan, T. M., Hirsch, A. M., Leigh, J. A., Johansen, E., Kuldau, G. A., Deegan, S., Walker, G. G., Signer, E. R. (1985). Symbiotic mutants of R. meliloti that uncouple plant from bacterial differentiation. Cell 49, 869-877.

Galfre, C., Milstein, C. (1981). Preparation of monoclonal antibodies: strategies and procedures. Methods Enzymol. 73, 3-46.

Galfre, G. and Butcher, G. W. (1986). Making monoclonal antibodies. In: Immunology in Plant Science, Ed. T. L. Wang, Cambridge University Press.

Goding, J. W. (1987). Monoclonal antibodies: principles and practice. Academic Press (London) (Second Edition).

Golumbeski, G. S. and Dimond, R. L. (1986). The use of tolerisation in the production of monoclonal antibodies against minor determinants. Anal. Biochem. 154, 373-381.

Matthew, W. D. and Paterson, P. H. (1983). The production of a monoclonal antibody that blocks the action of a neurite outgrowth-promoting factor. Cold Spring Harbor Symp. 47, 625-631.

McKearn, T. J., Sarmiento, M. and Weiss, A., Stuart, F. P. and Sitch, F. W. (1978). Selective suppression of reactivity to rat histocompatibility antigens by hybridoma antibodies. Current Topics in Microbiol. and Immunol. 81, 61-65.

Thomas, B., Penn, S. E., Butcher, G. W. and Galfre, G. (1984b). Discrimination between the red- and far-red-absorbing forms of phytochrome from Avena sativa I. by monoclonal antibodies. Planta 160, 382-384.

IMMUNOCYTOCHEMICAL STUDIES OF SYMBIOTIC DEVELOPMENT AND METABOLISM IN NITROGEN-FIXING ROOT NODULES

Kathryn A. VandenBosch
John Innes Institute, Colney Lane, Norwich NR4 7UH, UK

1. INTRODUCTION

Immunogold labelling is an excellent tool for addressing developmental questions in symbiotic interactions, such as the identification of specific determinants on cell surfaces, or the localisation of enzymatic components within subcellular compartments. The technique employs specific antibodies as affinity probes for particular macromolecules in biological tissues. Colloidal gold particles conjugated to secondary antibodies or to protein A bind to the primary antibody probes, and thus serve as electron-dense markers of an antigen when the tissue is examined in the electron microscope. The attractive feature about gold labelling is that it combines the specificity of an immunoassay with the high resolution of the electron microscope, therefore providing molecular information within the context of cell ultrastructure. These techniques have been of great value in developing our understanding of the *Rhizobium*-legume symbiosis.

Within legume root nodules, nitrogen-fixing *Rhizobium* bacteroids exist in a delicate balance with the host plant. The plant establishes the conditions of oxygen concentration, and high energy supply that enable the bacteria to fix nitrogen. In addition, the plant cells assimilate ammonia produced by the bacteria and process it for export to the rest of the plant. These symbiotic functions dictate that the nodule is a highly specialised plant organ.

Intricate interactions between the symbionts lead to the establishment of an effective root nodule. Some of the stages of morphological development are illustrated in Figures 1-3. Prior to infection, rhizobia bind to the host root and elicit root hair curling. The bacteria typically invade the root from a region of the curled root hair and become encased in a tunnel of plant cell wall material, the infection thread. Later, the rhizobia are released from the infection thread via endocytosis and are then separated from the host plant cytoplasm only by a plant-derived membrane, the peribacteroid membrane.

NATO ASI Series, Vol. H17
Cell to Cell Signals in Plant, Animal and Microbial Symbiosis. Edited by S. Scannerini et al.

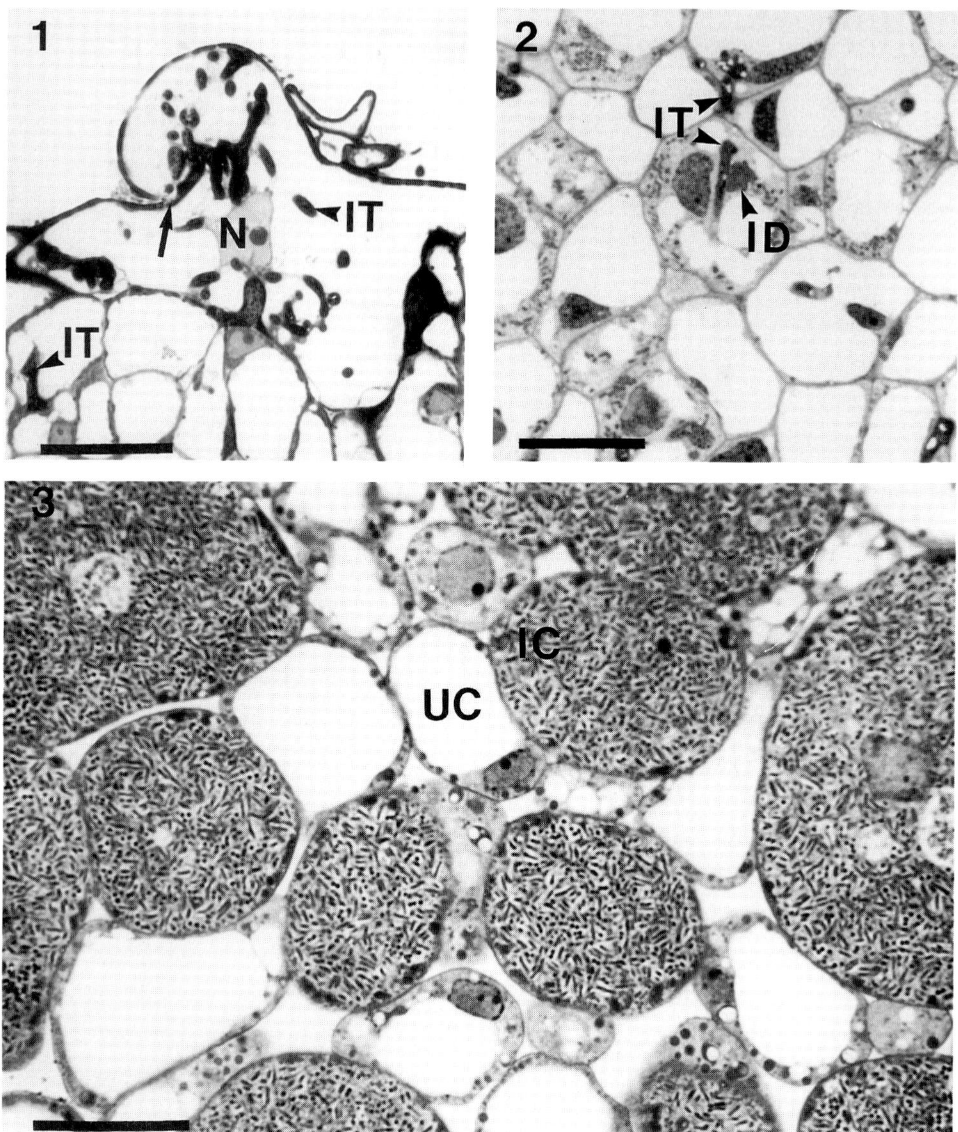

Figures 1-3. Stages in the development of the Rhizobium/legume symbiosis. Fig. 1: An infected root hair of bean (Phaseolus vulgaris). Infection threads ramify from the point of infection (arrow) and have penetrated the root cortex. Bar = 20 μm. Fig. 2: Release of bacteria from infection threads in a young pea nodule (Pisum sativum). Bar = 20 μm. Fig. 3: Infected and uninfected host cells in a mature soybean nodule (Glycine max). Bar = 20 μm. Figure 1 is reproduced courtesy of the Journal of Bacteriology (VandenBosch, K.A. et al., 1985: J. Bacteriol. 162, 950-959). Symbols used in figures: B = Rhizobium bacteroid; C = host plant cytoplasm; IC = infected cell; ID = infection droplet; IT = infection thread; N = plant cell nucleus; PBM = peribacteroid membrane; UC = uninfected cell.

(Some exceptions to this pattern are known from primitive legumes and the non-legume Parasponia, where the nitrogen-fixing bacteria remain within the infection thread. See article by J. G. Torrey, this volume). In the mature nodule, both the rhizobia, now termed 'bacteroids' and the plant cells become morphologically and physiologically specialised. Throughout nodule development, there are thus many opportunities for cell-cell interactions between the symbionts: between colonising bacteria and the root surface, between rhizobia in the infection thread and the young nodule cells, and between nitrogen-fixing bacteroids and the peribacteroid membrane.

Another type of cell-cell interaction takes place between plant cell types within the nodule. More than half of the cells in the nodule typically remain uninfected. In soybean and related species that transport fixed nitrogen in the form of ureides, these interstitial cells become specialised for ureide production, and thus perform a supporting role to the infected cells (Hanks et al, 1983). Frequent plasmodesmata interconnect infected and uninfected cells. The two cell types are non-randomly arranged such that every infected cell borders at least one uninfected cell, suggesting maximum opportunity for communication (Selker and Newcomb, 1985). It is not yet known what factors trigger a cell to differentiate as an infected or an uninfected cell.

The discussion that follows will outline selected immunocytochemical methods for the localisation of cellular constituents by light and electron microscopy. Examples drawn from studies of legume nodules will illustrate how these techniques are advancing our comprehension of symbiotic organisation and development.

2. SPECIMEN PREPARATION FOR IMMUNOLABELLING

Successful application of immunogold labelling techniques requires good preservation of antigenicity as well as ultrastructural detail. Specimen preparation for conventional electron microscopy usually involves fixation in glutaraldehyde, followed by post-fixation in osmium tetroxide, dehydration in ethanol or acetone, and finally embedding in an epoxy resin which is polymerised with heat. This standard procedure preserves ultrastructure beautifully, but may alter the tertiary structure of macromolecules, and hence affect antigenicity adversely.

Most methods of specimen preparation for immunocytochemistry still rely upon a primary fixation in aldehydes, although some people prefer a milder fixation in formaldehyde over glutaraldehyde for this purpose. Post-fixation in osmium tetroxide, however, often completely masks antigenicity and prevents antibody binding. This masking effect is often overcome by treating the sections with periodate before labelling. Periodate treatment may remove sugars from glycoconjugates, and so this procedure is not suitable for all antigens (Brewin *et al*, 1985).

An important advance in immunolabelling came with the development of polar acrylic embedding resins, such as Lowicryl K4M and the London Resins (LR). The polar nature of these resins interferes less with antigenicity than does non-polar epoxy resin. Moreover, because the acrylic resins have very low viscosity and can be polymerised with ultraviolet light, tissue infiltration and embedment can be conducted at low temperatures.

The fixation/embedding schedule currently in use in our laboratory is as follows: small ($< 1\ mm^3$) pieces of nodule tissue are fixed in cacodylate buffered-glutaraldehyde overnight at 4^oC, and then dehydrated in an ethanol series. The first dehydration step is carried out on ice; subsequent steps are carried out at lower temperatures (to -35^oC) as the concentration of ethanol increases. The tissue is then gradually infiltrated with LR White resin at -20^oC. The resin, which contains 0.5% benzoin methyl ether as a catalyst, is polymerised by illumination with UV light. We find that this schedule yields excellent structural preservation and retention of antigenicity in most cases.

3. ELECTRON MICROSCOPIC LOCALISATION OF ANTIGENS

Labelling a single antigen. In order to visualise an antigen within a tissue, labelling is carried out after the specimen has been embedded and sectioned. Thin sections (70-90 nm) are collected on to grids, and the grids are incubated first in a blocking buffer, followed by an antibody that is specific for the antigen of interest. Lastly, the sections are treated with colloidal gold conjugated to Protein A or a secondary antibody. The conjugate binds specifically to the primary antibody probe so that the gold marks the position of the antigen on the section.

Immunogold labelling studies of root nodules fall roughly into two categories: localisation of key nodule proteins using polyclonal antibodies generated against a purified immunogen, and identification of interesting determinants using monoclonal antibodies made against heterogeneous and uncharacterised immunogen. Examples of both types are discussed below.

The enzyme uricase catalyses the penultimate step of ureide synthesis in soybean related legumes. Enzyme assays performed on separated infected and uninfected cells from soybean nodules had suggested that uricase activity occurred in the uninfected cell fraction (Hanks et al, 1983). However, because there was cross contamination of the fractions, it was impossible to determine whether the infected cells also functioned in the final steps of ureide synthesis. A means for localising this enzyme in situ was required in order to resolve this question. Using protein A-gold and rabbit antibodies against the nodule-specific form of uricase, it was determined that uricase is exclusively present in the enlarged peroxisomes of uninfected cells (VandenBosch and Newcomb, 1986). Moreover, the enzyme could be localised in time as well as space. Uricase was first detected in developing peroxisomes coincident with the release of rhizobia from infection threads in adjacent infected cells.

Another example of the shortcomings of tissue fractionation is seen in attempts to localise the oxygen-binding protein leghaemoglobin (Lb) in legume nodules. It had been argued that Lb was present in the nodule cytoplasm, in the peribacteroid space, or in both locations. Robertson et al (1984) used immunogold labelling to demonstrate clearly that Lb in pea nodules was in the cytoplasm of infected cells and not in peribacteroid spaces. Recently, using a similar labelling procedure it has been shown that Lb also occurs in uninfected cells of soybean nodules, but at a lower concentration than in infected cells (VandenBosch and Newcomb, manuscript in preparation).

Brewin and coworkers have used monoclonal antibodies (McAbs) made against peribacteroid membrane components (Brewin et al, 1985) and isolated bacteroids (Brewin et al, 1986) to probe the cell surfaces of the symbionts. The cytological distribution of these antigens has been examined using immunogold labelling. This technique has been essential for the characterisation of the corresponding antigens as components of the plant membrane, the infection thread matrix, or the rhizobial cell surface (see article by N J Brewin, this volume).

Double-labelling techniques

Because colloidal gold particles can be made in a range of diameters (Slot and Geuze, 1985), more than one antigen may be localised in a single experiment using gold probes of different sizes. Double-labelling techniques facilitate correlative studies of antigens expressed in the same tissue. This can be accomplished using two primary antibodies raised in different species, followed by colloidal gold conjugated to the appropriate secondary antibodies. Such a labelling scheme is diagrammed in Figure 4.

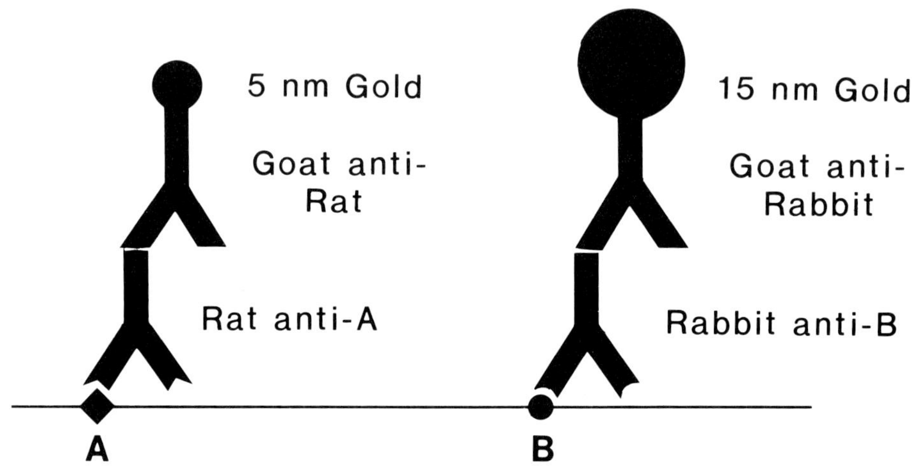

Figure 4. Schematic representation of double labelling of two antigens in a single specimen, using primary antibodies raised in different species.

An example of double labelling is shown in Figure 5. From a group of McAbs raised in rats against peribacteroid material, a McAb (MAC 203) was obtained which recognises a symbiosis-specific form of rhizobial lipopolysaccharide (LPS) (Bradley et al, manuscript in preparation). In order to determine which stages of bacterial development express this form of LPS, thin sections of nodule tissue were incubated first in MAC 203, followed by goat anti-rat immunoglobulin conjugated to 5 nm gold particles. The same sections were then treated with a polyclonal antiserum raised in a rabbit against purified nitrogenase, the enzyme responsible for reducing atmospheric dinitrogen to ammonia. The polyclonal antibodies were visualised with goat anti-rabbit immunoglobulin conjugated to 15 nm gold particles. The results clearly demonstrate that this form of LPS is present in nitrogen-fixing bacteroids but not in undifferentiated bacteria at earlier stages of nodule development.

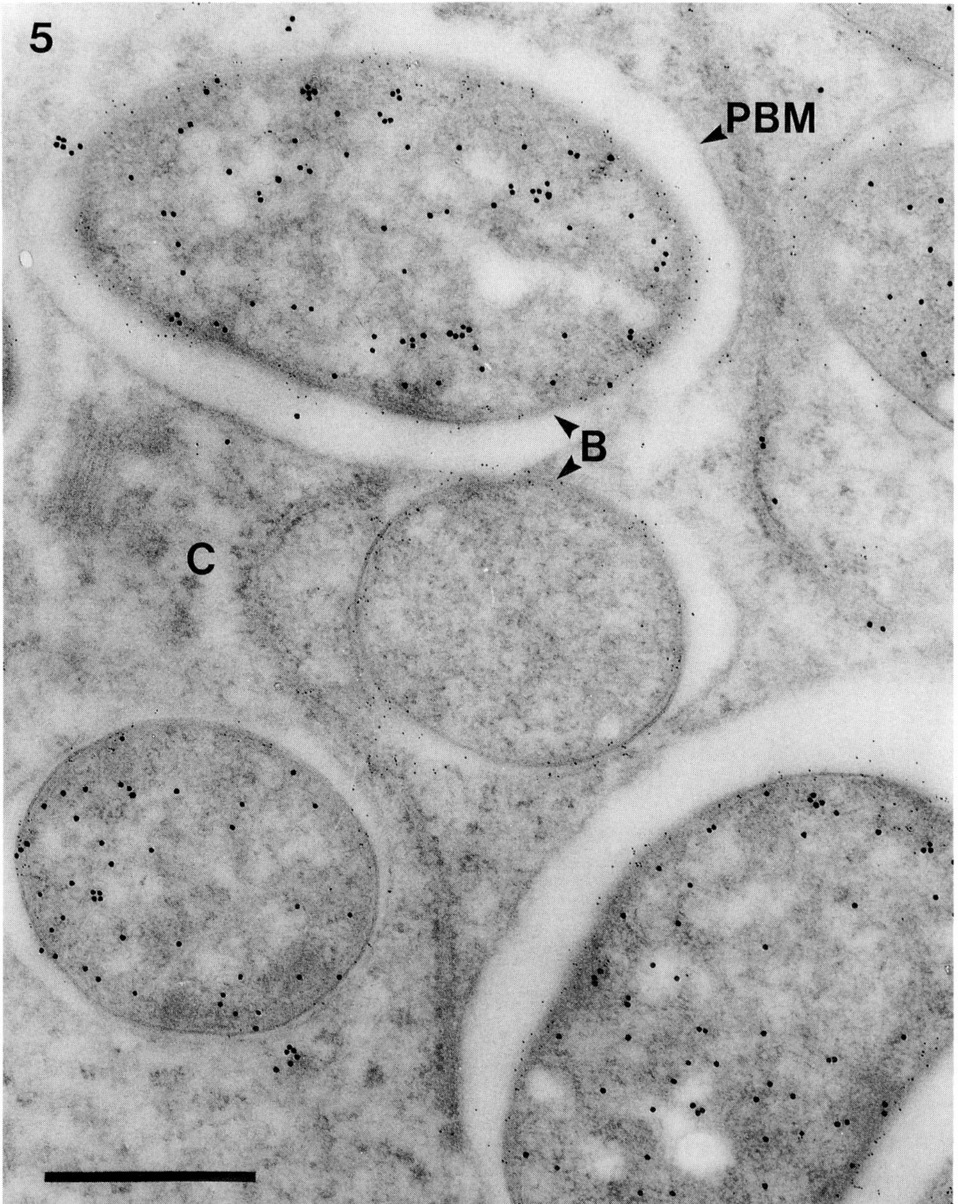

Figure 5. Double labelling of nitrogenase (large gold particles) and a symbiosis-specific surface antigen (small gold particles) in Rhizobium leguminosarum bacteroids. Note that one bacteroid appears to lack nitrogenase. Pea nodules were fixed in glutaraldehyde in cacodylate buffer, dehydrated in ethanol and embedded in LR White acrylic resin. Thin sections were incubated sequentially in the following immunoreagents: i. Rat McAb against bacterial lipopolysaccharide (MAC 203). ii. Goat anti-rat gold (5 nm particles). iii. Rabbit anti-nitrogenase. iv. Goat anti-rabbit gold (15 nm particles). Bar = 0.5 μm.

Other double-labelling techniques are available when both primary antibodies have been raised in the same species. Polyclonal antibodies raised in rabbits can be detected using protein A conjugated to different-sized gold particles (Titus and Becker, 1985). Alternatively, one of the antibodies can be biotinylated, and then detected with a streptavidin-gold probe that binds to the biotin (Bradley et al, 1986).

4. SILVER ENHANCEMENT OF GOLD LABELLING

Light microscopic visualisation of gold labelling is also desirable for the determination of tissue-wide distribution of antigens. However, rarely is the antigen suffiently concentrated for the gold to be resolved with a light microscope. The gold labelling signal may be amplified by silver enhancement (Holgate et al, 1983; Danscher and Nörgaard, 1983). In this procedure, gold labelled sections are incubated in a silver developer containing silver ions and a reducing agent; metallic silver then precipitates onto the gold grains, yielding a dense, black signal visible with a light microscope. Immunogold-silver staining may be performed on semi-thin (0.2-2 um) sections of tissue processed as for electron microscopy (VandenBosch, 1986) thus affording correlative localisation studies using light and electron microscopy.

Figures 6 and 7 depict the light microscopic visualisation of two antigens in sections of pea nodules that have been gold-labelled and silver enhanced. Figure 6 demonstrates the presence of nitrogenase in *Rhizobium* bacteroids and absence of the enzyme from infection threads. Figure 7 illustrates the distribution of a form of bacterial lipopolysaccharide. This form of LPS, which is detected by the McAb MAC 57, is present in both free-living and bacterial forms of *Rhizobium leguminosarum* (Brewin et al, 1986). In Figure 7, it can be seen that the MAC 57 antigen is present in bacteria in infection droplets and released bacteria. In addition, it can be discerned that, unlike nitrogenase, LPS is a surface component of the bacteria.

Immunogold-silver staining has several advantages over other immunoassays for light microscopy. It is more sensitive than immunoperoxidase staining. It does not require a special microscope, as does immunofluorescence, and silver-staining is more permanent than immunofluorescence. Furthermore, it avoids the problem of autofluorescence which is common in plant tissues.

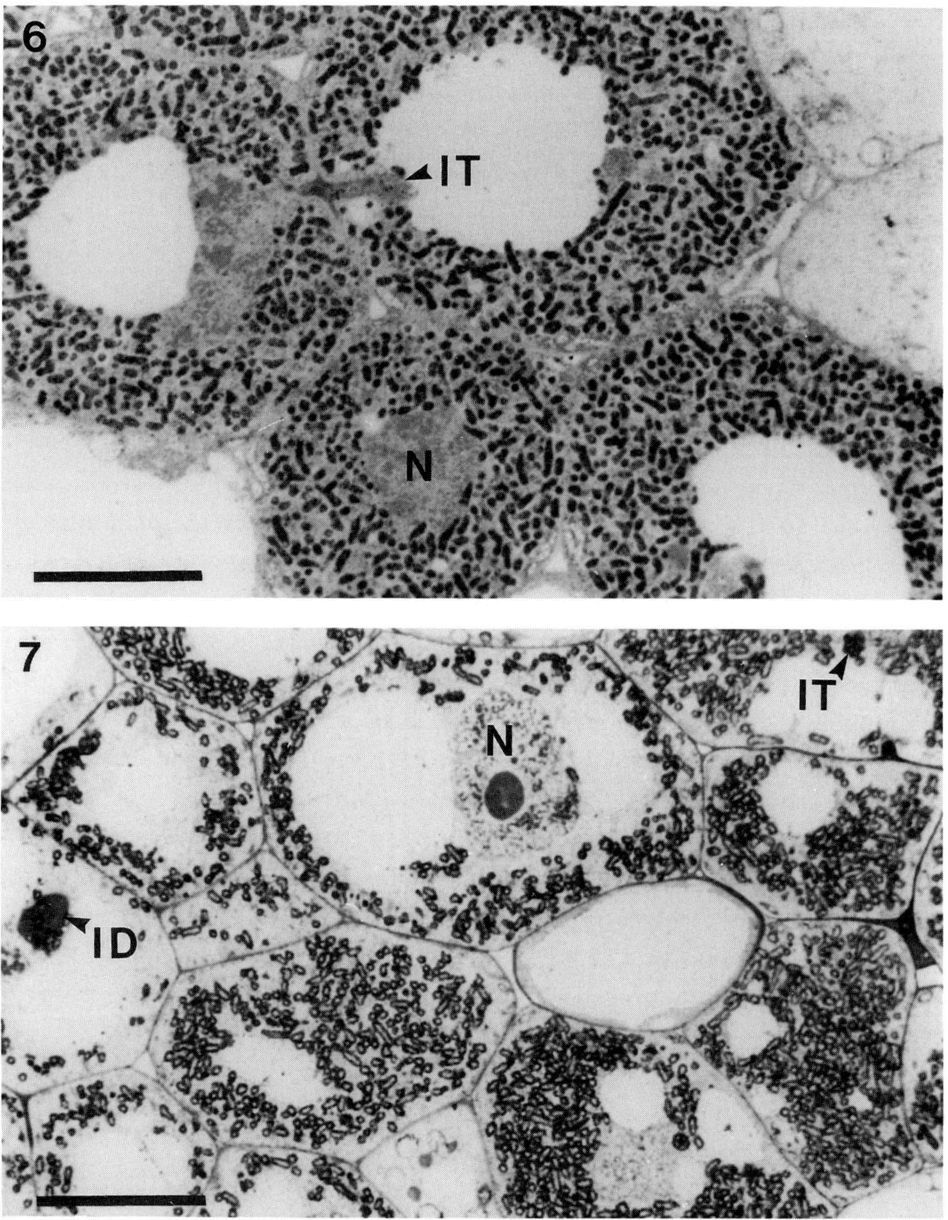

Figures 6 and 7. Silver-enhanced immunogold labelling of bacterial antigens in pea. Nodule tissue was processed as described for Figure 5. bars = 20 μm. Fig. 6: Localisation of nitrogenase in bacteroids. Note the lack of expression of nitrogenase in bacteria in the infection thread. Semi-thin sections were incubated in rabbit anti-nitrogenase, followed by goat anti-rabbit gold. Gold labelling was enhanced for light microscopy by treating labelled sections with silver developer. Fig. 7: Localisation of bacterial lipopolysaccharide. Note that bacteria in all stages of development express the antigen. Sections were first incubated in the rat McAb MAC 57, followed by goat anti-rat gold, followed by silver developer.

5. FUTURE DEVELOPMENTS

Although the development of acrylic resins for immunocytochemistry represents an increase in sensitivity, the technique may lack the sensitivity to detect rare or easily denatured antigens. Because the specimen is embedded in plastic, the antigenic sites are only exposed at the cut surface of sections. This is a problem particularly when a structure of interest, such as a cytoskeletal element or a small membrane vesicle, is smaller in diameter than the section thickness.

Reversible embedding methods, in which the embedding medium is removed after sectioning, enhance the sensitivity of immunolabelling because antigens are available for binding with the antibody throughout the entire section. In cryo-ultramicrotomy, the embedding medium is ice (Tokuyasu, 1986). Specimens are infused with a sucrose solution, quick frozen and sectioned at low temperatures. The sections are brought to room temperature for labelling, and then may be post-fixed in osmium and re-embedded in plastic for observation in the electron microscope (Keller _et al_, 1984). In reversible embedment cytochemistry, the embedment medium is the linear polymer polymethyl-methacrylate, which can be removed from sections with organic solvents (Gorbsky and Borisy, 1986). Sections are similarly labelled and then examined in the electron microscope.

Neither cryo-ultramicrotomy nor reversible embedment cytochemistry has been used in immunocytochemical studies of nitrogen-fixing symbioses. Both techniques hold considerable promise for revealing rare and recalcitrant antigens in nodule tissue.

6. CONCLUSIONS

Immunogold labelling provides a powerful tool for investigating the cellular organisation of symbioses. The technique has proved successful in examining compartmentalisation of symbiosis-related metabolism in nitrogen-fixing root nodules. Several key nodule proteins, such as leghaemoglobin, uricase and nitrogenase, have been localised at the ultrastructural level using polyclonal antibodies as probes. Monoclonal antibodies have been used to examine the composition of cell surfaces, focussing on plant glycoproteins and bacterial lipopolysaccharides. In the future, the study of cell surface differentiation will be extended with a greater range of monoclonal antibodies and the new techniques of reversible embedment that will provide greater sensitivity.

ACKNOWLEDGEMENTS

I gratefully acknowledge the receipt of monoclonal antibodies against *Rhizobium leguminosarum* lipopolysaccharide obtained from Desmond Bradley and Nicholas Brewin, and polyclonal antiserum against *Rhodospirillum rubrum* nitrogenase, furnished by Paul Ludden. Thanks are also due to Brian Wells and Graham Hills for advice and assistance with specimen preparation.

REFERENCES

Bradley, D. J., Butcher, G. W., Galfre, G., Wood, E. A. and Brewin, N. J. (1986). Physical association between the peribacteroid membrane and lipopolysaccharide from the bacterial outer membrane in Rhizobium-infected pea root nodule cells. J. Cell. Sci. 85, 47-61.

Bradley, D. J., Wood, E. A., Galfre, G., Butcher, G. W. and Brewin, N. J. Isolation of monoclonal antibodies reacting with peribacteroid membranes and other components of pea root nodules containing Rhizobium leguminosarum. Manuscript in preparation.

Brewin, N. J., Robertson, J. G., Wood, E. A., Wells, B., Larkins, A. P., Galfre, G., and Butcher G. W. (1985). Monoclonal antibodies to antigens in the peribacteroid membrane from Rhizobium-induced root nodules of pea cross react with plasma membranes and golgi bodies. EMBO J. 4, 605-611.

Brewin, N. J., Wood, E. A., Larkins, A. P., Galfre, G. and Butcher, G. W. (1986). Analysis of lipopolysaccharide from root nodule bacteroids of Rhizobium leguminosarum using monoclonal antibodies. J. Gen. Microbiol. 132, 1959-1968.

Danscher, G. and Nörgaard, J. O. R. (1983). Light microscopic visualisation of colloidal gold on resin-embedded tissue. J. Histochem. Cytochem. 31, 1394-1398.

Gorbsky, G. and Borisy, G. G. (1986). Reversible embedment cytochemistry (REC): A versatile method for the ultrastructural analysis and affinity labelling of tissue sections. J. Histochem. Cytochem. 34, 177-188.

Hanks, J. F., Schubert, K., and Tolbert, I. N. E. (1983). Isolation and characterisation of infected and uninfected cells from soybean nodules. Plant Physiol. 71, 869-873.

Holgate, C. S., Jackson, P., Cowen, P. N. and Bird, C. C. Immunogold-silver staining: new method with enhanced sensitivity. J. Histochem. Cytochem. 31, 938-944.

Keller, G. A., Tokyuasu, K. T., Dutton, A. H. and Singer, S. J. (1984). An improved procedure for immunoelectron microscopy: Ultrathin plastic embedding of immunolabelled ultrathin frozen sections. PNAS 81, 5744-5747.

Robertson, J. G., Wells, B., Bisseling, T., Farnden, K. J. F. and Johnston, A. W. B. (1984). Immunogold localisation of leghaemoglobin in cytoplasm in nitrogen-fixing root nodules of pea. Nature 311, 254-256.

Selker, J. M. L. and Newcomb, E. H. (1985). Spatial relationships between uninfected and infected cells in root nodules of soybean. Planta 165: 446-454.

Slot, J. W. and Geuze, H. J. (1985). A new method of preparing gold probes for multiple-labelling cytochemistry. Eur. J. Cell Biol. 38, 87-93.

Titus, D. E. and Becker, W. M. (1985). Investigation of the glyoxysome-peroxisome transitions in germinating cucumber cotyledons using double-label immunoelectron microscopy. J. Cell Biol. 101, 1288-1299.

Tokuyasu, K. T. (1986). Application of cryoultramicrotomy to immunocytochemistry. J. Microscopy 143, 139-149.

VandenBosch, K. A. (1986). Light and electron microscopic visualisation of uricase by immunogold labelling of sections of resin-embedded soybean nodules. J. Microscopy 143, 187-197.

VandenBosch, K. A. and Newcomb, E. H. Occurrence of leghemoglobin in uninfected cells in soybean root nodules. Manuscript in preparation.

VandenBosch, K. A. and Newcomb, E. H. (1986). Imunogold localization of nodule-specific uricase in developing soybean root nodules. Planta 167, 425-436.

CONCEPTS LEADING TO AN UNDERSTANDING OF RECOGNITION AND SIGNALLING BETWEEN HOSTS AND SYMBIONTS

By D. C. Smith
Department of Plant Sciences, University of Oxford, U.K.

Introduction

'Recognition' in the establishment of a symbiosis has been defined as the set of phenomena resulting in the expression of specificity or selectivity in associations between hosts and symbionts (Smith, 1981). The term has a somewhat different meaning in discussions of 'cell-to-cell' recognition, where it has the rather precise connotation of a process involving the transmission from one cell of information which elicits a response from another cell. The information is usually in the form of a particular molecule (the signal) whose binding by the 'target' cell constitutes 'recognition' and completely determines the specificity of the cell-to-cell interaction.

It is usually not possible to discern a single interaction which completely determines specificity in the formation of a symbiosis. There is usually a series of stages which together determine the ultimate specificity. The question therefore arises of whether recognition in symbiosis ('symbiotic' recognition) results from a complex sequence of separate 'cell-to-cell' recognition events, or whether there are other kinds of interaction between hosts and symbionts not involving precise molecular signals.

The essential features of 'cell-to-cell' recognition will therefore be considered before the more complex question of 'symbiotic' recognition.

'Cell-to-cell' recognition

This may be initiated in one of three main ways: (a) binding between stereospecific molecules on cell surfaces; (b) secretion of a signal molecule from one cell which moves to a target cell where it is bound by surface or internal receptor molecules; or (c) if the cells are already in contact, the movement of a signal molecule through connections such as gap

NATO ASI Series, Vol. H17
Cell to Cell Signals in Plant, Animal and Microbial Symbiosis. Edited by S. Scannerini et al.

junctions in animals and possibly plasmodesmata in plants. A variety of events may occur subsequently to this initial interaction. In the simplest case, cell surfaces may simply adhere to each other. In more complex responses, the initial signal may be transduced after reception, and a second, intracellular signal or 'messenger' is produced which initiates or mediates appropriate responses within the cell.

Certain types of animal cell have been particularly amenable to experimental investigation, and especially in cases where signals such as water soluble peptides or neurotransmitters are secreted. These signal molecules bind with high affinity to specific receptor proteins or glycoproteins on the plasma membranes of the target cells. The signal-receptor binding initiates second messengers, of which at least two are known, cyclic AMP and calcium ions, and there is now substantial evidence for a third, inositol phosphate. Transient changes in the intracellular concentration of second messengers initiate other reactions which result in the activation of specific proteins, so giving the cellular response to recognition.

A further mechanism by which intracellular responses may be initiated is when the bound signal-receptor complex moves laterally through the membrane into 'coated pits'. The signal-receptor complexes undergo endocytosis, and the resultant endosome usually then fuses with a lysosome, with subsequent changes to the signal or receptor molecule or both.

'Cell-to-cell' recognition involving plant cells has proved less easy to investigate experimentally, primarily because of the various technical difficulties posed by the existence of cell walls. Nevertheless, it is believed to proceed along broadly similar lines to that outlined above for animal cells. There is evidence for the existence of specific receptors, although none has yet been purified and characterised, and their precise location (i.e., whether on cell wall or plasma membrane) has yet to be established. However, many plant cell walls have been shown to contain glycoproteins, some of which are likely to be receptors. It has been further suggested that the complex structure of plant cell walls may act as a store of polysaccharide fragments ('oligosaccharins') which can be released by enzymic action (e.g., during cell wall degradation by attacking pathogens) and then act as signals to elicit specific defense responses by the host plant cells (Darvill et al., 1985). Evidence is now beginning to accumulate that calcium ions may also act as second messengers in plant cells.

'Symbiotic' recognition

The term 'symbiosis' as originally coined by De Bary included both mutualistic and parasitic associations. The essential difference between them is that in parasitic associations infected hosts are mostly at a selective disadvantage compared to uninfected, whereas in mutualistic associations the host is mostly at a selective advantage.

The difference between parasitic and mutualistic associations is very important to considerations of 'recognition'. In parasitic associations, the host is at an advantage if it can avoid being recognised. The host also benefits from mechanisms to identify parasites at an early stage of infection so that defences can be deployed. Although processes involved in the later establishment of a mature infection may be complex, the early stages often involve some cell-to-cell interactions capable of analysis at the molecular as well as the genetic level. Thus, 'gene-for-gene' hypotheses provide a reasonable explanation for the resistance/susceptibility, virulence/avirulence phenomena observed in some of the highly specific interactions between plants and biotrophic fungal pathogens. Likewise, a number of the molecular signals which are involved in eliciting host defence responses have now been isolated and characterised.

In mutualistic associations, by contrast, successful recognition is advantageous to the host. The development of the mature association is usually more complex: for example, regulatory mechanisms are developed to keep the growth of partners in balance, and many associations often have morphological or biochemical features not shown by either partner on their own. In no case has it been demonstrated that any one of the sequence of stages involved in forming the symbiosis is of overriding importance in determining the ultimate specificity.

Rather, a consistent pattern is emerging from a number of different associations that specificity on initial contact is often quite low, and then increases progressively during the subsequent establishment of the symbiosis. Specificity between host and symbiont is generally much lower than for biotrophic parasitic associations. With regard to mycorrhizas, Harley and Smith (1983) comment "... we have as yet no evidence whatever of specific genes that can affect virulence or avirulence in mycorrhizal fungi or concern susceptibility or resistance in the host .....". The specificity of mycorrhizal symbionts for their hosts is so low that Harley (1984) states "... we are driven to accept the view that a very large number of fungi are compatible with a very large number of host plants because there is some very common property that is held by these fungi, which allows them to associate with most land plants provided the latter do not possess some property antagonistic to them".

The rest of this article will be concerned only with those symbioses which are traditionally regarded as mutualistic.

Concepts of particular relevance to understanding recognition in mutualistic symbioses

(a) Specificity.

The manifestation of specificity is integral to the concept of recognition in symbiosis, so it is important to clarify what is meant by specificity and the techniques used to assess it. Specificity refers to the degree of taxonomic difference between the partners with which an organism can associate (Smith and Douglas, 1987). It is implicit that it refers only to formation of a complete and effective symbiosis, and not just to the early stages of contact.

Two approaches, the 'experimental' and the 'ecological', have been used to assess the level of specificity in symbiosis. In the 'experimental' approach, isolated hosts or symbionts are offered a range of potential partners under laboratory or green-house conditions, and the level of specificity is assessed from the range of organisms with which they form an effective symbiosis. In the 'ecological' approach, specificity is assessed from the range of partners observed to associate with a given organism in nature. Where the two approaches have been compared, as in ectomycorrhizas (Harley and Smith, 1983), a higher level of specificity is nearly always found by the 'ecological' approach. There may be several reasons for this. Fewer potential partners may be available in natural habitats, and the difficulty of identifying symbionts in the field may result in some being overlooked. Further, some associations which are viable in the laboratory do not survive under the less favourable conditions which may occur in nature.

In the 'experimental' approach, it is important to remember that associations can be divided into two groups according to how the symbiosis is perpetuated. The first group comprises those in which the host, in nature, is freshly reinfected each generation, acquiring a complement of symbionts from the environment. Offering such hosts a range of potential partners in the laboratory may result in a reasonable simulation of the natural sequence of stages by which a symbiosis is established and specificity manifested. In the second group are associations in which symbionts are transmitted directly from each host generation to the next without an intervening free-living phase. Symbiont-free hosts do not occur naturally, but some can be artificially prepared in the laboratory. The reinfection of such aposymbiotic hosts may involve stages which occur only rarely or not at all in nature, and they may discriminate between potential symbionts in a way which does not give a true picture of specificity. For example, the surface charge of *Chlorella* symbionts has a profound effect on the rate at which they enter the digestive cells of aposymbiotic green hydra by phagocytosis (McNeil *et al*., 1981). In nature, however, symbionts are transmitted directly from generation to generation, and surface charge may well be an irrelevant character in the determination of specificity.

(b) The advantage to mutualistic hosts and symbionts of low specificity.

In those associations where the host is reinfected each generation, the specificity of symbionts for their host is often low, or sometimes very low (with symbionts infecting hosts from different phyla). The advantage of low specificity for both host and mutualistic symbiont is very clear. It increases the chance of forming symbioses (and perhaps even survival) in situations where partners are sparsely distributed, and it allows the possibility of different host/symbiont combinations developing in different environments. It avoids the potential danger of a host being restricted to a single symbiont strain whose effectiveness of 'value' to the host might decline.

(c) Signals and the nature of 'symbiotic' recognition.

The establishment of a symbiosis involves a progressive sequence of interactions between organisms. Each interaction consists of a stimulus produced by one organism which elicits a response from its partner. Three types of interaction can be distinguished.

i) The sole role of the stimulus is evocation of a specific response from the partner. An example is the release from legume roots of flavonoids which activate the regulator gene _nod D_ in rhizobium. The gene product of _nod D_ induces the expression of the genes responsible for initiating the hair curling response (see the paper by Okker _et al_. in this volume). This type of interaction resembles signalling in 'cell-to-cell' recognition (although the receptor molecule has yet to be identified), and so the stimuli involved will be called _signals_ here.

ii) The stimulus is not solely produced to evoke a specific response from the partner, and it appears to be a routine consequence of normal metabolic processes; nevertheless, the other organism responds in a way which is a specific adaptation to the establishment of a symbiosis. An example would be how the close proximity of a host root stimulates the hypha of a vesicular-arbuscular mycorrhizal fungus to branch into a fan-shaped complex; there is no evidence that roots secrete specific compounds to elicit this response. Subsequently, a hyphal tip swells into a characteristic appressorium when it encounters the surface of the host cell wall. These types of interaction are clearly not 'signalling' in the strict sense used in 'cell-to-cell' recognition, and it would be unnecessarily confusing to refer to the stimuli involved as 'signals'. A better term would be recognition cues, and this will be used here.

iii) Neither the stimulus nor the response is specific. An example would be the guts of animals which provide the 'stimulus' of an anaerobic environment, and the response of the symbionts is simply to carry out anaerobic metabolism. This will be called a non-specific interaction.

Experimentally, the stimuli which are best understood are those involved in the early phases of the establishment of a symbiosis. Once the relationship between host and symbiont has become intimate and morphologically complex, the nature of the interactions becomes difficult to discern precisely.

In many types of symbiosis where each generation of hosts acquires symbionts by reinfection from an external source, specificity is low, and as argued above, this may be advantageous to both partners. The early phases of establishment of the symbiosis usually do not involve highly specific cell-to-cell signals, and the predominant type of interactions are symbiont or host responses to recognition cues. Only in legume/ rhizobium associations is there good evidence for cell-to-cell signals in the initial phases, and these take the form of the release of flavonoids and lectins from host roots. However, unusually for 'signals', a single flavonoid may elicit specific responses from more than one *Rhizobium* species (see the paper by Okker *et al.*, this volume), so they may not impose very high specificity in the selection of symbionts. With regard to lectins, evidence for their role generally in legume/rhizobium associations is confusing and there is not always evidence even for their presence (e.g., Beringer *et al.*, 1979; Etzler, 1985). While lectins are of undoubted importance in the adhesion of specific rhizobia to the root hairs of a variety of herbaceous legumes, e.g. Trifoliin A in white clover, there are many other legumes, especially numerous woody genera, in which rhizobia do not enter via root hairs (see the article by Torrey, this volume), the specificity for the type of rhizobium is low, and attachment to roots may result simply from responses to

TABLE 1. Examples of types of interaction in the establishment of mutualistic symbioses.

CONTACT OR EARLY PHASES OF ESTABLISHMENT

Type of Interaction	Type of Symbiosis	Nature of Stimulus	Response
Specific Signals	legume/rhizobium	lectins	attachment
		flavonoids	regulator gene activated
Recognition cues	legume/rhizobium	root surface	adhesion of symbiont
	V.A. Mycorrhiza	root exudate	hyphal branching
	" "	root surface	appressoria formed
	lichen	algal cell surface	hyphal extension
	Convoluta roscoffensis	algal cell wall	entry to host
Non-specific interactions	gut symbionts	anaerobic atmosphere	anaerobic metabolism
	green hydra	contact with digestive cell surface	phagocytosis

Continued ...

Table 1. continued

LATER PHASES OF ESTABLISHMENT

Type of Interaction	Type of Symbiosis	Nature of Stimulus	Response
Specific Signals	legume/rhizobium	?host + symbiont gene products	?specific cell-to-cell interactions
	mycorrhizas	?wall proteins	binding of symbiont to host wall
	lichens	?symbiont product	synthesis of secondary metabolic products by host
	green hydra	vacuolar pH	maltose release by symbiont
	" "	maltose release	avoidance of lysosome fusion
	" "	?'division factor'	symbiont cell division
	anemones, corals	?'host factor'	photosynthate release from symbionts
	N_2-fixing symbioses	?	inhibition of symbiont glutamine synthetase
Recognition cues	Ruminants	release of VFA by symbionts	input of saliva to maintain pH

recognition cues since there is no evidence for the involvement of lectins.

In most associations, therefore, symbiont and host do not come together as a result of the emission of specific host signals to attract symbionts. Rather, contact occurs because there is a sufficient natural abundance of symbionts in the immediate region of the host, and specific responses to recognition cues are sufficient for associations to be initiated. In symbioses where each generation of hosts acquires its symbionts directly from the preceding generation, signals or recognition cues to achieve contact between host and symbiont are presumably unnecessary.

It seems very likely that cell-to-cell signalling is involved in the later stages of the formation of a symbiosis. The identification of numerous symbiosis-specific genes in legumes and rhizobia is particularly good indirect evidence for this. However, in no symbiosis so far studied have signal molecules yet been isolated and characterised; evidence for their existence is almost entirely indirect.

Table 1 summarises examples of probable and possible types of interaction which occur in the establishment of a symbiosis. The Introduction to this paper posed the question of whether the difference between 'cell-to-cell' and 'symbiotic' recognition was that the latter was simply a complex sequence of many instances of the former. The experimental evidence given in many papers in this workshop shows that, apart from legume/rhizobium associations, responses to general recognition cues rather than specific signals predominate in the early phases of the formation of a symbiosis; while cell-to-cell signalling is very likely to predominate in the later stages, partner interactions in a mutualistic symbiosis are so complex that the evidence for molecular signals is almost entirely indirect. Hence, 'symbiotic' recognition involves more than a complex series of 'cell-to-cell' recognition events.

REFERENCES

BERINGER, J.E., BREWIN, N., JOHNSTON, A.W.B., SHULMAN, H.M. and HOPWOOD, D.A. (1979). The Rhizobium-legume symbiosis. Proceedings of the Royal Society of London, Series B, 204, 219-233.

DARVILL, A.G., ALBERSHEIM, P., MCNEIL, M., LAU, J.H., YORK, W.S., STEVENSON, T.T., THOMAS, J., DOARES, S., GOLLIN, D.J., CHELF, P. and DAVIS, K. (1985). Structure and function of plant cell wall polysaccharides. Journal of Cell Science Supplement, 2, 203-217.

ETZLER, M.E. (1985). Plant lectins: molecular and biological aspects. Annual Review of Plant Physiology, 36, 209-234.

HARLEY, J.L. (1984). The mycorrhizal association. In Cellular Interactions. Encyclopedia of Plant Physiology New Series, Vol. 17 (Linskens, H.F., and Heslop-Harrison, J., eds) pp. 148-186.

HARLEY, J.L. and SMITH, S.E. (1983). Mycorrhizal Symbiosis. 483 pp. Academic Press, London.

MCNEIL, P.L., HOHMAN, T.C. and MUSCATINE, L. (1981). Mechanisms of nutritive endocytosis. II. The effect of charged agents on phagocytotic recognition by digestive cells. Journal of Cell Science, 52, 243-269.

SMITH, D.C. (1981). The role of nutrient exchange in recognition between symbionts. Berichte Deutsche Botanische Gesellschaft, 94, Supplement 517-518.

SMITH, D.C. and DOUGLAS, A.E. (1987). The Biology of Symbiosis. 302 pp. Edward Arnold, London.

SUBJECT INDEX

(A pagenumber marked* refers to the whole paper)

NATO ASI Series H

Vol. 1: **Biology and Molecular Biology of Plant-Pathogen Interactions.** Edited by J.A. Bailey. 415 pages. 1986.

Vol. 2: **Glial-Neuronal Communication in Development and Regeneration.** Edited by H.H. Althaus and W. Seifert. 865 pages. 1987.

Vol. 3: **Nicotinic Acetylcholine Receptor: Structure and Function.** Edited by A. Maelicke. 489 pages. 1986.

Vol. 4: **Recognition in Microbe-Plant Symbiotic and Pathogenic Interactions.** Edited by B. Lugtenberg. 449 pages. 1986.

Vol. 5: **Mesenchymal-Epithelial Interactions in Neural Development.** Edited by J.R. Wolff, J. Sievers, and M. Berry. 428 pages. 1987.

Vol. 6: **Molecular Mechanisms of Desensitization to Signal Molecules.** Edited by T.M. Konijn, P.J.M. Van Haastert, H. Van der Starre, H. Van der Wel, and M.D. Houslay. 336 pages. 1987.

Vol. 7: **Gangliosides and Modulation of Neuronal Functions.** Edited by H. Rahmann. 647 pages. 1987.

Vol. 8: **Molecular and Cellular Aspects of Erythropoietin and Erythropoiesis.** Edited by I.N. Rich. 460 pages. 1987.

Vol. 9: **Modification of Cell to Cell Signals During Normal and Pathological Aging.** Edited by S. Govoni and F. Battaini. 297 pages. 1987.

Vol. 10: **Plant Hormone Receptors.** Edited by D. Klämbt. 319 pages. 1987.

Vol. 11: **Host-Parasite Cellular and Molecular Interactions in Protozoal Infections.** Edited by K.-P. Chang and D. Snary. 425 pages. 1987.

Vol. 12: **The Cell Surface in Signal Transduction.** Edited by E. Wagner, H. Greppin, and B. Millet. 243 pages. 1987.

Vol. 13: **Toxicology of Pesticides: Experimental, Clinical and Regulatory Perspectives.** Edited by L.G. Costa, C.L. Galli, and S.D. Murphy. 320 pages. 1987.

Vol. 14: **Genetics of Translation. New Approaches.** Edited by M.F. Tuite, M. Picard, and M. Bolotin-Fukuhara. 524 pages. 1988.

Vol. 15: **Photosensitisation. Molecular, Cellular and Medical Aspects.** Edited by G. Moreno, R.H. Pottier, and T.G. Truscott. 521 pages. 1988.

Vol. 16: **Membrane Biogenesis.** Edited by J.A.F. Op den Kamp. 477 pages. 1988.

Vol. 17: **Cell to Cell Signals in Plant, Animal and Microbial Symbiosis.** Edited by S. Scannerini, D. Smith, P. Bonfante-Fasolo, and V. Gianinazzi-Pearson. 414 pages. 1988.